Penguin Books
The Face of Battle

John Keegan was born in London in 1934, educated at King's College, Taunton, Wimbledon College and Balliol College, Oxford. Since 1960 he has taught at the Royal Military Academy, Sandhurst, where he is Senior Lecturer in War Studies. He has also written *Six Armies in Normandy* (forthcoming in Penguin) and is the co-author, with Andrew Wheatcroft, of *Who's Who in Military History*. He also wrote the text for *The Nature of War*, in which Joseph Darracott brought together some 250 paintings to illustrate Keegan's powerful statement about the experience of war and its place in history. He is currently writing *The Mask of Command*, in which he examines the changing role of generalship.

He is married, has two daughters and two sons, and lives, between Sandhurst terms, near Bruton in Somerset.

John Keegan

The Face of Battle

Penguin Books

Penguin Books Ltd, Harmondsworth, Middlesex, England
Penguin Books, 625 Madison Avenue, New York, New York 10022, U.S.A.
Penguin Books Australia Ltd, Ringwood, Victoria, Australia
Penguin Books Canada Ltd, 2801 John Street, Markham, Ontario, Canada L3R 1B4
Penguin Books (N.Z.) Ltd, 182–190 Wairau Road, Auckland 10, New Zealand

First published in Great Britain by Jonathan Cape Ltd 1976
First published in the United States of America
by The Viking Press 1976
Published in Penguin Books 1978
Reprinted 1983

Acknowledgement is made for material reprinted
by permission of Faber and Faber Limited
from *If Germany Attacks* by G. C. Wynne

Made and printed in Great Britain by
Richard Clay (The Chaucer Press) Ltd,
Bungay, Suffolk
Set in Monotype Plantin

In memory of my father and my father-in-law

Contents

Acknowledgements 11

Chapter 1 Old, Unhappy, Far-off Things 13

A Little Learning 13

The Usefulness of Military History 20

The Deficiencies of Military History 25

The 'Battle Piece' 35

'Killing No Murder?' 45

The History of Military History 53

The Narrative Tradition 61

Verdict or Truth?' 72

Chapter 2 Agincourt, 25 October 1415 78

The Campaign 78

The Battle 86

Archers versus Infantry and Cavalry 92

Cavalry versus Infantry 94

Infantry versus Infantry 97

The Killing of the Prisoners 108

The Wounded 112

The Will to Combat 114

Chapter 3 Waterloo, 18 June 1815 117

The Campaign 121

The Personal Angle of Vision 128

The Physical Circumstances of Battle 134

Categories of Combat 144

Single Combat 145

Cavalry versus Cavalry 147

Cavalry versus Artillery 151

Cavalry versus Infantry 154

Artillery versus Infantry 160

Infantry versus Infantry 162
Disintegration 195
Aftermath 197
The Wounded 200

Chapter 4 The Somme, 1 July 1916 207
The Battlefield 207
The Plan 213
The Preparations 216
The Army 219
The Tactics 229
The Bombardment 231
The Final Preliminaries 241
The Battle 246
Infantry versus Machine-Gunners 247
Infantry versus Infantry 251
The View from across No-Man's-land 259
The Wounded 268
The Will to Combat 274
Commemoration 285

Chapter 5 The Future of Battle 290
The Moving Battlefield 290
The Nature of Battle 301
The Trend of Battle 303
The Inhuman Face of War 320
The Abolition of Battle 331

Bibliography 345

Index 351

List of Illustrations

between pages 184 and 185

1 Skull of a Swedish soldier killed in 1361
2 The effect of archery on cavalry at short range, 1356
3 A 'wall of bodies' of the dead and wounded, fifteenth century
4 Men-at-arms at the mercy of archers, fifteenth century
5 A square of Highlanders receiving cavalry at Waterloo
6 *Scotland for Ever*: Waterloo
7 A German attack on a British line of 'scrapes', autumn 1914
8 Russians charging a German or Austrian trench, autumn 1914
9 A French counter-attack at Dien Bien Phu, spring 1954

List of Maps

	page
1 Agincourt, 25 October 1415	82
2 Waterloo, 18 June 1815	124-5
3 The Somme, 1 July 1916	244
4 Relative positions of the battles	289

Acknowledgements

This book has been written chiefly from printed sources, some of which are listed in the bibliography. But I have also derived much information and many ideas from colleagues, pupils and friends (it is one of the pleasures of teaching at Sandhurst that these categories overlap), in particular from the following serving or retired soldiers: Brigadier Peter Young, D.S.O., M.C., Brigadier D. W. V. P. O'Flaherty, D.S.O., Major-General A. H. Farrar-Hockley, D.S.O., M.B.E., M.C., Colonel E. M. P. Hardy, Lieutenant-Colonel Michael Barclay, Royal Scots Dragoon Guards, Lieutenant-Colonel Jeremy Reilly, D.S.O., Royal Regiment of Fusiliers, Major Charles Messenger, Royal Tank Regiment, Major Michael Dewar, Royal Green Jackets, Captain Terence Johnston, Coldstream Guards, Lieutenant Timothy Weeks, Light Infantry, and Lieutenant Hugh Willing, Royal Green Jackets; from the following members of the Sandhurst academic staff: Dr Christopher Duffy, Dr Richard Holmes, Dr Gwynne Dyer, Dr John Sweetman, Mr David Chandler, Mr Peter Vigor and Mr William McElwee; from our *homologues* at the École spéciale militaire de St-Cyr, Lieutenant-Colonel Michel Camus, Légion étrangère, and Commandant Marc Neuville, Chasseurs à pied; from Major-General Alastair Maclennan, O.B.E., of the Royal Army Medical College, Mr A. S. Till, F.R.C.S., of the United Oxford Hospitals, Dr John Cule, Dr H. Bleckwenn of Osnabruck and Dr. T. F. Everett, my father-in-law, who died before this book was finished; from Dr M. Haisman and Dr M. Allnut of the Army Personnel Research Establishment; from Mr Michael Howard, Professor Richard Cobb, Professor Geoffrey Best, Mr Harmut Pogge von Strandmann and Brigadier Shelford Bidwell. I also corresponded fruitfully with Professor Bernard Bergonzi and Dr C. T.

Allmand. Brigadier Young, Mr Howard and Mr Chandler were kind enough to give permission for extracts from their books to be used for purposes which do not do justice to their quality. I owe special debts of gratitude to Mr Barrie Pitt and Mr Derek Anyan. Mr Anthony Sheil, Mr Alan Williams and Mr David Machin have been unfailingly encouraging; I hope I have not disappointed them. It is a pleasure to thank Lieutenant-Colonel Alan Shepperd, M.B.E., the Librarian (and creator) of the Central Library, R.M.A. Sandhurst, and his friendly, helpful and efficient staff for all their help; Mr R. W. Meadows is particularly to be thanked for procuring books through the inter-library loan service. I am also grateful to Mr Kenneth White, of the Staff College Library, and to the staff of the London Library. Mrs Valerie Horsfield typed much of the manuscript and has my thanks. My wife Susanne would have typed it all, had I not insisted that her hands were already overfull with her own writing and the care of four children; and were the title and subject of this book not so inappropriate, I would have dedicated it to her, for all she has done.

JOHN KEEGAN

Royal Military Academy, Sandhurst
10 *December* 1974

1 Old, Unhappy, Far-off Things*

A Little Learning

I have not been in a battle; not near one, nor heard one from afar, nor seen the aftermath. I have questioned people who have been in battle – my father and father-in-law among them; have walked over battlefields, here in England, in Belgium, in France and in America; have often turned up small relics of the fighting – a slab of German 5.9 howitzer shell on the roadside by Polygon Wood at Ypres, a rusted anti-tank projectile in the orchard hedge at Gavrus in Normandy, left there in June 1944 by some highlander of the 2nd Argyll and Sutherlands; and have sometimes brought my more portable finds home with me (a Minié bullet from Shiloh and a shrapnel ball from Hill 60 lie among the cotton-reels in a painted papier-mâché box on my drawing-room mantelpiece). I have read about battles, of course, have talked about battles, have been lectured about battles and, in the last four or five years, have watched battles in progress, or apparently in progress, on the television screen. I have seen a good deal of other, earlier battles of this century on newsreel, some of them convincingly authentic, as well as much dramatized feature film and countless static images of battle: photographs and paintings and sculpture of a varying degree of realism. But I have never been in a battle. And I grow increasingly convinced that I have very little idea of what a battle can be like.

Neither of these statements and none of this experience is in the least remarkable. For very, very few Europeans of my generation – I was born in 1934 – have learned at first hand that knowledge of battle which marked the lives of millions of their fathers and grandfathers. Indeed, apart from the four or five

*'Will no one tell me what she sings? Perhaps the plaintive numbers flow For old, unhappy, far-off things, And Battles long ago.' Wordsworth: *The Solitary Reaper*.

thousand Frenchmen who, with their German, Spanish and Slav comrades of the Foreign Legion, survived Dien Bien Phu, and the slightly larger contingents of Britons who took part in the campaign in central Korea in 1950–51, I cannot identify any group of people, under forty, in the Old World, who have been through a battle as combatants. My use of the words 'battle' and 'combatants' will indicate that I am making some fairly careful exceptions to this generalization, most obviously in the case of all those continental Europeans who were children during the Second World War and over whose homes the tide of battle flowed, often more than once, between 1939 and 1945; but also in the case of the thousands of British and French soldiers who carried arms in Africa and South-East Asia during the era of decolonization, to whose number I ought to add the Portuguese conscripts still campaigning in Mozambique and Angola, and the British regulars policing the cities and countryside of Ulster.

The first group exclude themselves from my generalization because none of them was old enough to have had *combatant* experience of the Second World War; the second because their experience of soldiering, though often dangerous and sometimes violent – perhaps very violent if they were French and served in Algeria – was not an experience *in* and *of* battle. For there is a fundamental difference between the sort of sporadic, small-scale fighting which is the small change of soldiering and the sort we characterize as a battle. A battle must obey the dramatic unities of time, place and action. And although battles in modern wars have tended to obey the first two of those unities less and less exactly, becoming increasingly protracted and geographically extensive as the numbers and means available to commanders have grown, the action of battle – which is directed towards securing a decision by and through those means, on the battlefield and within a fairly strict time-limit – has remained a constant. In Europe's wars of decolonization, the object of 'the other side' has, of course, been to avoid facing a decision at any given time or place, rightly presuming the likelihood of its defeat in such circumstances; and 'the other side', whether consciously fighting a war of evasion and delay, as were the communist guerrillas in Malaya or the nationalist partisans in Algeria, or merely conducting a campaign of raiding and subversion because they implicitly recognized their inability to risk anything else, as did the Mau Mau in Kenya, has accordingly shunned battle. I do not think therefore that my Oxford contemporaries of the 1950s, who had spent their late teens combing the jungles of Johore or searching the forests on the slopes of Mount Kenya, will hold it against me if I suggest that, though they have been soldiers and I have not and though they have seen active service besides, yet they remain as innocent as I do of the facts of battle.

But what, it might be fairly asked at this stage, is the point of my re-emphasizing how little, if at all, unusual is my ignorance of battle? Ignorance has been bliss in Europe for nearly thirty years now, and in the United States there has been little thanks given for the lessons its young men have been forced to learn at Pleiku and Khe San. The point is, I had better admit, a personal one – not so personal that it cannot be revealed but one which, over the years, has grown to something of the dimensions of a Guilty Secret. For I have spent many of those years, fourteen of them – which is almost the whole of my working life – describing and analysing battles to officer cadets under training at Sandhurst; class after class of young men, all of whom stand a much better chance than I do of finding out whether what I have to say on the subject is or is not true. The inherent falsity of my position should be obvious. It has always been clear to me, but at Sandhurst, which carries almost to extremes the English cult of good manners, the cadets I have taught have always connived at the pretence that I and they are on a master-and-pupil footing and not, as I know and they must guess, all down together in the infant class. I for my part, anxious not to overtax their politeness, have generally avoided making any close tactical analysis of battle, entailing as that would my passing judgement on the behaviour of men under circumstances I have not had to meet, and have concentrated the weight of my teaching on such subjects as strategic theory, national defence policy, economic mobilization, military sociology and the like – subjects which, vital though they are to an understanding of modern war, nevertheless state what, for a young man training to be a professional soldier, is the central question: what is it like to be in a battle?

That this – or its subjective supplementary, 'How would *I* behave in a battle?' – is indeed the central question reveals itself when it is raised in a roomful of cadets – and probably at any gathering of young men anywhere – in a number of unmistakable ways: by a marked rise in the emotional temperature, in the pitch of voices, and in what a sociologist might call 'the rate and volume of inter-cadet exchanges'; by signs of obvious physical tenseness in the ways cadets sit or gesticulate – unless they assume, as some do, a deliberately nonchalant attitude; and by the content of what they have to say – a noisy mixture of slightly unconvincing bombast, frank admissions of uncertainty and anxiety, bold declarations of false cowardice, friendly and not-so-friendly jibes, frequent appeal to fathers' and uncles' experience of 'what a battle is really like' and heated argument over the how and why of killing human beings, ranging over the whole ethical spectrum from the view that 'the only good one is a dead one' to very civilized expressions of reluctance at the prospect of shedding human blood at all. The discussion, in short, takes on many of the characteristics of a group-therapy session, an analogy which will not, I know, commend itself to many professional soldiers but which I think none the less apt. For the sensations and emotions with which the participants are grappling, though they relate to a situation which lies in a distant and perhaps never-to-be-realized future rather than in a disturbed and immediate present, are real enough, a very powerful, if dormant, part of every human being's make-up and likely therefore, even when artificially stimulated, to affect the novice officer's composure to an abnormal and exaggerated extent. These feelings, after all, are the product of some of man's deepest fears: fear of wounds, fear of death, fear of putting into danger the lives of those for whose wellbeing one is responsible. They touch too upon some of man's most violent passions; hatred, rage and the urge to kill. Little wonder that the officer cadet, who, if he is one day to quell those fears and direct those passions, must come to terms with their presence in his make-up, should display classic signs of agitation when the subject of battle and its realities is raised. Little wonder either that my soldier colleagues regard their 'leadership' lectures, in which the psychological problems of controlling oneself and one's men in battle are explicitly reviewed, as the most taxing of their assignments in the military training programme. Few of them, I know, would think that they handle the subject satisfactorily. Most, I suspect, would agree that it is only an exceptional man who can.

Of course, the atmosphere and surroundings of Sandhurst are not conducive to a realistic treatment of war. Perhaps they never are in any military academy. But Sandhurst is a studiedly unmilitary place. Its grounds are serenely parklike, ornamentally watered and planted and landscaped, its buildings those of an English ducal mansion, fronted by nearly a square mile of impeccably mown playing-field, on which it is difficult to imagine anything more warlike being won than a hard-fought game of hockey. And the bearing and appearance of the students helps to foster the country-house illusion; as often to be seen in plain clothes as in uniform, for they are encouraged from the outset to adopt the British officer's custom of resuming his civilian identity as soon as he goes off duty, they unfailingly remind me, with their tidy hair and tweed jackets, of the undergraduate throng I joined when I went up to Oxford in 1953. It is a reminder which strikes all the more vividly those who teach in universities today. 'They look,' exclaimed an Oxford professor whom I had brought down to lecture, 'like the people I was in college with before the war.'

'Before the war'; the pun is a little too adventitious to stand very much elaboration. But 'before the war' is, after all, the spiritual state in which the pupils of a military academy exist. For however strong their motivation towards the military life, however high their combative spirit, however large the proportion who are themselves the sons, sometimes the grandsons and great-grandsons of soldiers – and the proportion at Sandhurst, as at St-Cyr, remains surprisingly large – their knowledge of war is theoretical, anticipatory and second-hand. What is more, one detects in one's own attitudes, and in those of one's colleagues, in those who know and in those who don't, in the tough-minded almost as much as in the tender-hearted, an implicit agreement to preserve their ignorance, to shield the cadets from the worst that war can bring. In part, this agreement stems from an aesthetic reflex, a civilized distaste for the discussion of what might shock or disgust; in part, too, it reflects a moral inhibition, an unwillingness to give scandal to the innocent. And it may also be a manifestation of a peculiarly English reticence. French officers, certainly, show a readiness, in reminiscing over the wars in Indo-China or Algeria, to dwell on the numbers of deaths their units have suffered or inflicted – usually inflicted – which I have seen bring physical revulsion to the faces of British veterans, and which I do not think can be wholly explained in terms of the much greater ferocity of the French than the British army's most recent campaigns.

But Sandhurst and St-Cyr would agree over a quite different justification for the de-sensitized treatment of war which in practice characterizes instruction at both academies, and at all others of which I have any knowledge. And that is that the deliberate injection of emotion into an already highly emotive subject will seriously hinder, if not indeed altogether defeat, the aim of officer-training. That aim, which Western armies have achieved with remarkably consistent success during the 200 years in which formal military education has been carried on, is to reduce the conduct of war to a set of rules and a system of procedures – and thereby to make orderly and rational what is essentially chaotic and instinctive. It is an aim analogous to that – though I would not wish to push the analogy too far – pursued by medical schools in their fostering among students of a detached attitude to pain and distress in their patients, particularly victims of accidents.

The most obvious manifestation of the procedural approach to war is in the rote-learning and repeated practice of standard drills, by which one does not only mean the manual of arms practised by warriors since time immemorial to perfect their individual skills but a very much more extended range of procedures which have as their object the assimilation of almost all of an officer's professional activities to a corporate standard and a common form. Hence he learns 'military writing' and 'voice procedure' which teach him to describe events and situations in terms of an instantly recognizable and universally comprehensible vocabulary, and to arrange what he has to say about them in a highly formalized sequence of 'observations', 'conclusions' and 'intentions'. He learns to interpret a map in exactly the same way as every other officer will interpret it (the celebrated story of Schlieffen's reply to his adjutant, who had drawn to his attention a vista of the River Pregel – 'an inconsiderable obstacle, Captain' – was only an exaggeration of a reflex response to the accidents of geography which military academies devote much effort to producing in their pupils). Personal, or personnel, relationships are book-taught too: he learns 'rights' and 'wrongs' in the treatment of prisoners, whether of his own petty defaulters or of enemy captives, by reference to simplified manuals of military and international law – and to ensure that he will get his decisions straight he watches and eventually takes part in a series of 'playlets' in which the more common military offences and submissions are simulated. Simulated for him also, of course (both in the classroom and on the ground), are the most frequently encountered combat problems, which he is asked to analyse and, on the basis of his analysis, to solve, usually only on paper, but sometimes by taking command of a group of fellow cadets or occasionally even of 'real' soldiers borrowed for the exercise. His analysis, solution and mistakes are then criticized by reference to the 'school solution' (called in the British army 'the pink', from the colour of the paper on which it is always mimeographed), which he is only then allowed to see (and not allowed to argue about).

Officer-training indeed makes use of simulation techniques to a far greater extent than that for any other profession; and the justification, which is a sound justification, for the time and effort and thought put into these not very exciting routines is that it is thus only that an army can be sure – hopeful would be more accurate – of its machinery operating smoothly under extreme stress. But besides the achievement of this functional and corporate aim, the rote-learning and repetitive form and the categorical, reductive quality of officer-training has an important and intended – if subordinate – psychological effect. Anti-militarists would call it de-personalizing and even dehumanizing. But given – even if they would not give – that battles are going to happen, it is powerfully beneficial. For by teaching the young officer to organize his intake of sensations, to reduce the events of combat to as few and as easily recognizable a set of elements as possible, to categorize under manageable headings the noise, blast, passage of missiles and confusion of human movement which will assail him on the battlefield, so that they can be described – to his men, to his superiors, to himself – as 'incoming fire', 'outgoing fire', 'airstrike', 'company-strength attack', one is helping him to avert the onset of fear or, worse, of panic and to perceive a face of battle which, if not familiar, and certainly not friendly, need not, in the event, prove wholly petrifying.

The Usefulness of Military History

History, too, can be pressed into the service of familiarizing the young officer with the unknown. One does not mean here the history of myth, of the Legion at Camerone or the Fusiliers at Albuera, though Moltke, the great nineteenth-century Chief of the German General Staff and himself an academic historian of distinction, 'held it "a duty of piety and patriotism" not to destroy certain traditional accounts' if they could be used for an inspirational end, as indeed they can; one is thinking rather of a sort of history, to the launching of which Moltke gave a weighty shove, usually known as 'Official' or 'General Staff' history. Official history can be bad and good. At its best, modern British, and even more so American official history is a model of what conscientious and at times inspired scholarship can be. But the General Staff variety of official history often took in the past, and still can take, a peculiarly desiccated and didactic form, dedicated to demonstrating, at the cost if necessary of dreadful injury to the facts, that all battles fall into one of perhaps seven or eight types: battles of encounter, battles of attrition, battles of envelopment, battles of break-through and so on. Now there is no doubt a certain brutal reality in this approach, just as there is a certain rough-and-ready applicability about the seven or eight or nine 'immutable and fundamental' Principles of War (Concentration, Offensive Action, Maintenance of the Aim, etc.) which derive from it by another route and which military academies used to, as some in the ex-colonial countries working off out-of-date training-manuals still do, teach to their students.

But it is not a reality that the university-trained historian can grant more than the shakiest foundation. He, after all, has been trained to detect what is different and particular about events, about individuals and institutions and the character of their relationships. He cannot easily accept, therefore, as the typical survey-course text of *Military History from Hannibal to Hitler* might ask him to, that the battle of Cannae, 216 B.C., and the Battle of Ramillies, A.D. 1706, still less the Battle of the Falaise Gap, 1944, are all the same sort of battle because each culminated in an encirclement of one army by the other. He may admire the painstakingly reconstructed and often beautifully drawn maps which accompany these texts, usually embellished with neat, conventional NATO symbols (infantry division symbol *equals* a Roman legion; armoured brigade symbol *equals* cavalry of the Maison du Roi) but he ought not to be persuaded that, because the course of battles fought two thousand years apart in time can be represented in the same cartographic shorthand, the victor in each case was obeying, even if unwittingly, the rules of some universal Higher Logic of War. He will, or should, want to know a great deal more about many things – arms, equipment, logistics, morale, organization, current strategic assumptions – than the General Staff text will tell him, before he will feel able to generalize about anything with the confidence that its author displays about everything.

No doubt, however, he will – as I have done frequently – adopt the General Staff approach and make use of its material. But he will do so with the mental reservation that once off the nursery slopes, he will introduce his pupils to the real thing, the hard stuff. 'Let them get hold of the distinction between strategy and tactics' (a distinction as elusive as it is artificial), he may say to himself, 'and then we'll get down to some really serious discussion of the Schlieffen Plan, look at the documents, scrutinize the railway time-tables, mobilization schedules, read some Nietzsche, talk about Social Darwinism' ... but in the meantime, 'Gentlemen, I want you to think about these two maps of the German invasions of France in 1914 and 1940 which I'm going to project on the screen. Notice the similarities between ...' He may reconcile this rough-and-readiness to himself, as do a thousand American professors who silently – or audibly – curse World Civilization XP49 but teach it all the same, with the thought that no economic historian would consider discussing the pre-market economy with a class which did not understand the law of supply and demand; no social anthropologist bother embarking on an analysis of the master-man relationship for the benefit of students who did not grasp that there had once been a world without class-structures. And he would be right to do so. We all have to begin somewhere.

There are, however, two obstacles, one minor, one major, to a military historian making with his pupils the intellectual transition from the nursery slopes to the slalom *piste* which the economic historian or social anthropologist can always look forward to achieving with his (even if he does not get them that far). The first, and lesser, is that the student-officer, and it is he we are discussing, for almost no one else systematically studies military history, is simultaneously undergoing two processes of education, each with a dissimilar object. The one, highly vocational as we have seen and best described by the French word *formation*, aims if not to close his mind to unorthodox or difficult ideas, at least to stop it down to a fairly short focal length, to exclude from his field of vision everything that is irrelevant to his professional function, and to define all that he ought to see in a highly formal manner. Hence, as he is to begin his career as a leader of a small unit of professional soldiers, it is at leadership and small-unit morale that he is asked to look; and, as he may later become a general, then let him also study generalship, strategy, logistics; no matter in either case whether the raw material of his study is culled from the Crusades or the Crimea. The difference between warfare then and now is in a sense unimportant, for it will be his task to bring his enemies to battle on his own terms and force them to fight by his rules, not theirs.

For the other process of education the student-officer undergoes is the normal, 'academic' one, which aims to offer the student not a single but a variety of angles of vision; which asks him to adopt in his study of war the standpoint not only of an officer, but also of a private soldier, a non-combatant, a neutral observer, a casualty; or of a statesman, a civil servant, an industrialist, a diplomat, a relief worker, a professional pacifist – all valid, all documented points of view. It will be obvious that any of these viewpoints, adoptable readily enough by the schoolboy or undergraduate, are reconciled much less easily by the student-officer with the stern, professional, monocular outlook he is learning to bring to bear on the phenomena of war.

However, it is by no means the case that all, or even many, regular officers find it difficult to talk or think about war from an unprofessional point of view. We are most of us capable of compartmentalizing our minds, would find the living of our lives impossible if we could not, and flee the company of those who can't or won't: zealots, monomaniacs, hypochondriacs, insurance salesmen, the love-sick, the compulsively argumentative. One of the pleasures of mixing in military society is the certainty that one will meet there no representatives of most of these categories and few of the rest. The military zealot is, in particular, a rare bird, at least among British officers, who deliberately cultivate a relaxed and undogmatic attitude to the life of *Grandeur* and *Servitude*. Indeed the frankness and lack of hypocrisy with which they, having as it were declared by their choice of career where they stand over the ethics of violence and the role of force, are able to discuss these questions makes much mess conversation a great deal more incisive, direct and ultimately illuminating than that of club bars or university common-rooms.

'Of course, killing people never bothered me,' I remember a grey-haired infantry officer saying to me, by way of explaining how he had three times won the Military Cross in the Second World War. In black and white it looks a horrifying remark; but to the ear his tone implied, as it was meant to imply, not merely that the act of killing people might legitimately be expected to upset others but that it ought also to have upset him; that, through his failure to suffer immediate shock or lasting trauma, he was forced to recognize some deficiency in his own character or, if not that, then, regrettably, in human nature itself. Both were topics he was prepared to pursue, as we did then and many times afterwards. He was, perhaps, an unusual figure, but not an uncommon one. Fiction knows him well, of course, a great deal of Romantic literature having as its theme the man-of-violence who is also the man of self-knowledge, self-control, compassion, *Weltanschauung*. He certainly exists in real life also, and as often in the army as elsewhere, as the memoirs of many professional soldiers – though few successful generals – will testify. Perhaps – it is only an impression – he is more typically a French or British than a German or American figure, the horizons of the Sahara or the North-West Frontier encouraging a breadth of outlook denied to the Hauptmann or the First Lieutenant on dreary garrison duty in Arizona or Lorraine. And although there is a German 'literary' literature of military life, it is very much more a literature of leadership, as in Bloem's *Vormarsch*, or of the exaltation of violence, as in Jünger's *Kampf als innere Erlebnis*, than of adventure, exploration, ethnography, social – sometimes even spiritual – fulfilment, the themes which characterize the novels of Ernest Psiachari or F. Yeats-Brown, or the memoirs of Lyautey, Ian Hamilton, Lord Belhaven, Meinertzhagen and a host of other major and minor servants of British and French imperialism in this century and the last who, by design or good luck, chose soldiering as a way of life and found their minds enlarged by it.

If literature of this latter sort reinforces, as I think it does, my personal view that there exists in the military mind neither a psychological barrier nor an institutional taboo against free discussion of the profession of arms, its ethics, dimensions, rewards, shortcomings, if military society is, as I have found it to be, a great deal more open than its enemies will admit or recognize, what then is this other and more important obstacle which I have suggested stands in the way of an intellectual transition from the superficial and easy to the difficult and profound in the study of war – or more particularly of battle – which lies at its heart? If the student-officer can pigeon-hole at will the highly polarized view of combat which his military training gives him, in which people are either 'enemy' (to be fought), 'friend' (to be led, obeyed or supported as rank and orders prescribe), 'casualties' (to be evacuated), 'prisoners' (to be interrogated and escorted to the rear), 'non-combatants' (to be protected where possible and ignored where not) or 'dead' (to be buried when time permits); if he can set aside this stark, two-dimensional picture of battle and prepare to look at it in the same light as a liberal-arts student might, or a professional historian, or a strategic scientist, or a member of that enormous general readership of military history which has come into being in the last twenty years, what difficulty will prevent his – and their – seeing what they want to see and being shown what they ought?

The Deficiencies of Military History

The difficulty, in a sentence, is with 'military history' itself. Military history is many things. It is, and for many writers past and present is not very much more than, the study of generals and generalship, an approach to the subject which can sometimes yield remarkable results – the American historian Jac Weller's three modern studies of Wellington in India, the Peninsula and at Waterloo, for example, convey a powerful sense of character and are informed by a deep and humane understanding of the nature of early nineteenth-century warfare at every level from the general's to the private soldier's – but which, by its choice of focus, automatically distorts perspective and too often dissolves into sycophancy or hero-worship, culminating in the odd case in a bizarre sort of identification by the author with his subject – an outcome common and understandable enough in literary or artistic biography but tasteless and even mildly alarming when the Ego is a man of blood and iron, his Alter someone of scholarly meekness and suburban physique.

Military history is also the study of weapons and weapon systems, of cavalry, of artillery, of castles and fortifications, of the musket, the longbow, the armoured knight, of the ironclad battleship, of the strategic bomber. The strategic-bombing campaign against Germany, its costs and benefits, its rights and wrongs, engages the energies of some of the most powerful minds at work in the field of military history today and has fomented one of the subject's few real intellectual antagonisms, comparable in the intensity and the scholarly rigour with which it is carried on to that sustained by seventeenth-century historians over the Rise or Decline of the Gentry; like those exercised by that long-running feud, its initiates seek constantly to widen the arena of their private conflict and to add to the list of combatants, so that all manner of passers-by – mild strategic-theorists, visiting demographers and uncommitted economic historians en route between a pre- and post-war Index of Gross National Product – find themselves challenged to stand and declare their colours over the ethics of area bombing or the practicability of bottleneck targeting. Tiresome though this faction-fighting can be, it justifies itself, quite apart from the importance of the moral issues at stake, by the high level of scholarship at which it is conducted and by the network of connections its participants, unlike so many other kinds of military historian, maintain with the wider world of historical (principally economic historical) inquiry.

Strongly economic in flavour too is a great deal of naval history, built as it must be around the study of weapon systems, of the big-gun battleship of the First World War and the aircraft carrier of the Second. And very precise, from the professional point of view very satisfying, history it can be. For modern naval warfare is, as correspondents with the Eighth Army were fond of reporting of the Desert campaign, very nearly 'pure' warfare, a war without civilians (on the whole) and one in which the common sailor cannot, as the common soldier can, by running away or sitting tight, easily confound his commander's wishes. All being in the same boat, a ship's company generally does as its captain directs, until all are sunk together; fleets, by extension, until beaten, move as their admirals order. And since naval orders must be transmitted mechanically and are logged as sent and received, navies accumulate archives whose contents are pure historical gold-dust: precisely noted changes of course, the weather reports of trained meteorologists, damage-control reports by professional engineers, accurately timed sightings of friendly and enemy units, hard nuggets of fact about visibility, casualties, sinkings, fall of shot, sea conditions, facts of a density and volume to crush the spirit and blind the imagination of all but the most inspired and dedicated scholar. For inexplicable reasons, it is American rather than British historians who have triumphed in the long-distance event that the writing of naval history is, and this although, by the majority vote of historical events, it is the doings of Royal rather than U.S. Navy which has compelled their attention. (One of them at least, Professor Arthur Marder, has achieved in his study of the British navy in the First World War standards of archival research and organization of material which defy betterment.)

Military history furthermore is the study of institutions, of regiments, general staffs, staff colleges, of armies and navies in the round, of the strategic doctrines by which they fight and of the ethos by which they are informed. At the most elevated level, this branch of the subject shades off, through the history of strategic doctrine, into the broader field of the history of ideas, and in another direction, through the study of 'civil-military' relations, into political science. 'Elevated' should of course be understood here in a very relative sense, for though academic interest in civil-military relations, particularly in those between the German army and the German state, has produced a large, satisfying and in parts distinctly exciting literature, it is elsewhere prone to clothe itself in the drab garments of sociology at its most introspective; while the history of strategic doctrine, with some notable exceptions, of which Jay Luvaas's *Military Legacy of the Civil War* is a glittering example, suffers markedly from that weakness endemic to the study of ideas, the failure to demonstrate connection between thought and action.

That weakness is not, however, peculiar to this sub-branch of military history. Action is essentially destructive of all institutional studies; just as it compromises the purity of doctrines, it damages the integrity of structures, upsets the balance of relationships, interrupts the network of communication which the institutional historian struggles to identify and, having identified, to crystallize. War, the good quartermaster's opportunity, the bad quartermaster's bane, is the institutional military historian's irritant. It forces him, whose urge is to generalize and dissect, to qualify and particularize and above all to combine analysis with narrative – the most difficult of all the historian's arts. Hence his preference, paradoxically, for the study of armed forces in *peacetime*. And excellent many works of that sort turn out to be. But, as Mr Michael Howard concluded at the end of a long, very painstaking and generally warm review, 'the trouble with this sort of book is that it loses sight of what armies are *for*.' Armies, he implied, are for fighting. Military history, we may infer, must in the last resort be about battle.

That certainly reflects Clausewitz's view. In an economic analogy, which delighted Engels and has helped to ensure this Prussian (admittedly vaguely Hegelian) general an unobtrusive niche in the Marxist *Temple du Génie*, he suggested that 'fighting is to war' (the paraphrase is Engels's) 'what cash payment is to trade, for however rarely it may be necessary for it actually to occur, everything is directed towards it, and eventually it must take place all the same and must be decisive.' Battle history, or campaign history, deserves a similar primacy over all other branches of military historiography. It is in fact the oldest historical form, its subject matter is of commanding importance, and its treatment demands the most scrupulous historical care. For it is not through what armies *are* but by what they *do* that the lives of nations and of individuals are changed. In either case, the engine of change is the same: the infliction of human suffering through violence. And the right to inflict suffering must always be purchased by, or at the risk of, combat – ultimately of combat *corps à corps*.

Combat *corps à corps* is not of course a subject which historians, any more than other sorts of writer, can be accused of ignoring.

The 'battle piece', as a historical construction, is as old as Herodotus; as a subject of myth and saga it is even more antique. It is an everyday theme of modern journalistic reportage and it presents a literary challenge which some of the world's masters have taken up. Stendhal, Thackeray and Hugo each offer us a version of the battle of Waterloo – as seen through the eyes of a shell-shocked survivor, of a distracted bystander, of a stern and unrelenting Republican deity; while Tolstoy, in his reconstruction of the battle of Borodino, which had for nineteenth-century Russians the same historical centrality as Waterloo for contemporary Western Europeans, not only brought off one of the most spectacular set-pieces in the development of the novel-form, but also opened the modern case for the prosecution against the Great Man theory of historical explanation.

Imagination and sentiment, which quite properly delimit the dimensions of the novelist's realm, are a dangerous medium, however, through which to approach the subject of battle. Indeed, in that sub-world of imaginative writing which Gillian Freeman has called the undergrowth of literature, calculated indulgence in imagination and sentiment have produced, and regrettably continue to produce, some very nasty stuff indeed, which at its Zap-Blatt-Banzai-Gott im Himmel-Bayonet in the Guts worst may justifiably be condemned by that overworked phrase, 'pornography of violence'.

Historians, traditionally and rightly, are expected to ride their feelings on a tighter rein than the man of letters can allow himself. One school of historians at least, the compilers of the *British Official History of the First World War*, have achieved the remarkable feat of writing an exhaustive account of one of the world's greatest tragedies without the display of any emotion at all. A brief, and wholly typical, extract will convey the flavour; it describes a minor trench-to-trench attack by infantry, supported by artillery, on 8 August 1916, at Guillemont, in the second month of the Battle of the Somme:

> Some confusion arose on the left brigade front, where the 166th Brigade (Brigadier-General L. F. Green Wilkinson) was replacing the 164th – a very difficult relief – and although the 1/10th King's (Liverpool Scottish), keeping close behind the barrage, approached the

German wire, it lost very heavily in two desperate but unavailing attempts to close with the enemy. Nearly all the officers were hit, including Lieutenant-Colonel J. R. Davidson, who was wounded. Next on the left, the 1/5th Loyal North Lancashire (also 155th Brigade) was late through no fault of its own; starting after the barrage had lifted, it stood no chance of success. Subsequently the 1/7th King's attacked from the position won by its own brigade (the 165th) on the previous day, but could make no headway.

Agreed that this is technical history; that it is intended as a chronological record of military incident to provide, among other things, material for Staff College lectures and authoritative source references for other historians to work from. But is this featureless prose appropriate to the description of what we may divine was something very nasty indeed that happened that morning at Guillemont fifty-eight years ago to those 3,000 Englishmen, in particular to those of the 1/10th Battalion of the King's Regiment?* That it was something very nasty is revealed by a footnote: 'The Victoria Cross was awarded to the medical officer of the 1/10th King's, Captain N. C. Chavasse, for his exceptionally gallant work in rescuing wounded under heavy fire.' For most of us know, even if nothing else about the British army, that the Victoria Cross can be won, and then very rarely, only at the risk, often at the cost, of death. If we also know that Chavasse is but one of three men ever to have won the Cross twice, his second being a posthumous award, and that his battalion was a Kitchener unit, composed of enthusiastic but half-trained volunteers; if we guess that 'could make no headway' and 'stood no chance of success' means that its neighbouring battalions returned precipitately to their trenches or did not leave them, then we can glimpse, in this episode in no-man's-land at Guillemont on 8 August 1916, a picture in miniature of

*It is revealing to contrast the mealy-mouthedness of the official historians with what Dr Anthony Storr has to say on the language of scholarship: 'The words we use to describe intellectual effort are aggressive words. We *attack* problems, or *get our teeth* into them. We *master* a subject when we have *struggled with* and *overcome* its difficulties. We *sharpen* our wits, hoping that our mind will develop a *keen edge* in order that we may better *dissect* a problem into its component parts.' *Human Aggression* (Allen Lane, The Penguin Press, 1968), p.x. Dr Storr would be better qualified than I to suggest explanations for military historians' habitual reluctance to call a spade a spade.

the First World War at, for those compelled to fight it, almost its very worst.

But if we may conclude that the official historians' decision to deal with the emotive difficulty in military historiography by denying themselves any explicit emotional outlet whatsoever was unsatisfactory, and that some exploration of the combatants' emotions, if not the indulgence of our own, is essential to the truthful writing of military history, we are still left with the problem of how it is to be done. 'Allowing the combatants to speak for themselves' is not merely a permissible but, when and where possible, an essential ingredient of battle narrative and battle analysis. The almost universal illiteracy, however, of the common soldier of any century before the nineteenth makes it a technique difficult to employ. Dr Christopher Duffy, by heroic labour among little-known Prussian and Austrian archives, has pushed use of the technique backwards into the eighteenth century; but it is not until the coming of the wars of the French Revolution that we find any extensive deposit even of officers' memoirs and not until the First World War that we hear the voice of the common man (though infant murmurs can be detected during the American Civil War). Robert Rhodes James, who is one of a handful of historians to have discussed the technical difficulties of writing military history, holds strongly to the view that battles ought to be and are best described through the words of participants; and in *Gallipoli* he gave a master's demonstration of how it may be done.

There are, however, objections to general dependence on the technique and not wholly those concerned with the paucity or absence of material from which to work. One, well known to all scholars, is the danger of reconstructing events solely or largely on the evidence of those whose reputations may gain or lose by the account they give: even if it is only a warrior's self-esteem which he feels to be at stake, he is liable to inflate his achievements – what we might call 'the Bullfrog Effect' – and *old* warriors, particularly if surrounded by Old Comrades who will endorse his yarn while waiting the chance to spin their own on a reciprocal basis, are notoriously prone to do so. Contemporary letters and even more so, genuinely private diaries (if such exist) are a much more reliable source; but they must be used in the right way. Too often they are not. At worst, they are mined for 'interest', to produce anthologies of 'eye-witness accounts' in series with titles like *Everyman at War* (*The Historian as Copy-typist* would be altogether more frank); at best, they serve as the raw material for what is not much more than anecdotal history, yielding a narrative with a great deal of pungency and a high surface shimmer but without any of that intense particularity or energetic and confident generalization which are the trademarks of the historical *maître-ouvrier*.

Anecdote should certainly not be despised, let alone rejected by the historian. But it is only one of the stones to his hand. Others – reports, accounts, statistics, map-tracings, pictures and photographs and a mass of other impersonal material – will have to be coaxed to speak, and he ought also to get away from papers and walk about his subject wherever he can find traces of it on the ground. A great pioneer military historian, Hans Delbrück in Germany in the last century, demonstrated that it was possible to prove many traditional accounts of military operations pure nonsense by mere intelligent inspection of the terrain, and an English follower of his, Lt-Colonel A. H. Burne, proposed the applicability of a principle he had tested on every major English battlefield (Inherent Military Probability) and which, used with circumspection, is a rewarding as well as intriguing concept.* I would also argue that military historians should spend as much time as they can with soldiers, not on the grounds that 'armies always remain the same at heart', a notion which any historian with a sense of professional self-preservation would dismiss out of hand, but because the quite chance observation of trivial incidents may illuminate his private understanding of all sorts of problems from the past which will otherwise almost certainly remain obscured.

Christopher Duffy, who was lucky enough to spend some weeks teaching Yugoslav militia the elements of Napoleonic drill for a film enactment of *War and Peace*, described to me the thrill of comprehension he experienced in failing to manoeuvre

*The solution of an obscurity by an estimate of what a trained soldier would have done in the circumstances.

his troops successfully across country 'in line' and of the comparative ease with which he managed it 'in column', thus proving to his own satisfaction that Napoleon preferred the latter formation to the former not because it more effectively harnessed the revolutionary fervour of his troops (the traditional 'glamorous' explanation) but because anything more complicated was simply impracticable. I myself recall a similar archaeological pang in catching a glimpse of a Guards sergeant marching backwards before his squad who were learning the slow-march on the Sandhurst drill-square; the angle of his outstretched arms and upraised stick, his perfectly practised disregard for any obstacle in his backward path, the exhortatory rictus of his expression exactly mirrored the image, sketched from life by Rowlandson, of a Guards sergeant drilling his recruits on Horse Guards parade 170 years before; and through that reflection I suddenly understood the function – choreographic, ritualistic, perhaps even aesthetic, certainly much more than tactical – which drill plays in the life of long-service armies. The insight which intimacy with soldiers at this level can bring to the military historian enormously enhances his surety of touch in feeling his way through the inanimate landscape of documents and objects with which he must work. It will, I think, rob him of patience for much that passes as military history; it will diminish his interest in much of the 'higher' study of war – of strategic theory, of generalship, of grand strategic debate, of the machine-warfare waged by air forces and navies. And that, perhaps, is a pity. But if it leads him to question – as I have found it does me – the traditional approach to writing about combat *corps à corps*, to decide that, after he has read the survivors' letters and diaries, the generals' memoirs, the staff officers' dispatches, there is yet another element which he must add to anything he writes – an element compounded of affection for the soldiers he knows, a perception of the hostilities as well as the loyalties which animate a society founded on comradeship, some appreciation of the limits of leadership and obedience, a glimpse of the far shores of courage, a recognition of the principle of self-preservation ever present in even the best soldier's nature, incredulity that flesh and blood can stand the fears with which battle will confront it and which his own deeply felt timidity will highlight – if, in short, he can learn to make up his mind about the facts of battle in the light of what all, and not merely some, of the participants felt about their predicament, then he will have taken the first and most important step in understanding battle 'as it actually was'.

For if to propagate understanding of, not merely knowledge about, the past is the historian's highest duty, making up his own mind is the essential precondition to that end. Making up one's mind about anything, let alone a large and complicated body of material, is always a difficult and often painful task but it is one which many military historians would seem to shun altogether. The anecdotal historian avoids it, since he has already decided that his only responsibility is to entertain the reader and he can therefore discard whatever material he judges will not. The anthologist historian avoids it absolutely, usually justifying this abdication of his function by the plea that he prefers to let the reader make up his mind for himself – as if someone he impropriates of only a fraction of the record is thereby put in any position to do so. The 'General Staff' historian also avoids the responsibility, for his mind is made up for him by prevailing staff doctrine about the proper conduct of war and he will accordingly select whatever facts endorse that view, while manhandling those which offer resistance. The technological, the economic, the strategic, the biographic historians will all in their turn approach the subject of battle with their attitudes somewhat pre-cast, though they are usually well trained enough to advise the reader of their bias from the outset. But even the all-round military historian tends, in my experience, however perceptive, innovative, forthright, even downright disrespectful he is in his discussion of staffwork, leadership, strategic decision and the like, to shy away from the challenge of planting the impress of his own mind on his battle descriptions. One would certainly not suggest that he does so consciously, nor that the battle pieces he writes are not the fruit of careful research and skilful organization. But the trouble precisely is that what most military historians write about battle are indeed 'battle pieces', that is to say essays in a highly traditional form, which no amount of labour to fill out with new information will materially alter so long as the historian accepts the conventions within which he is working. To suggest that most military historians do accept those conventions is not to accuse them of that beginner's error, the transmission of traditional accounts ('For want of a nail the kingdom was lost . . .'); nor is it to impugn them of unreflectingly adopting the modes of thought of this or that great historian of the past. It is rather to argue that what has been called the 'rhetoric of history' – that inventory of assumptions, and usages through which the historian makes his professional approach to the past – is not only, as it pertains to the writing of battle history, much more strong and inflexible than the rhetoric of almost all other sorts of history, but is so strong, so inflexible and above all so time-hallowed that it exerts virtual powers of dictatorship over the military historian's mind.

The 'Battle Piece'

What do I mean by the 'rhetoric of battle history'? And what are its usages and assumptions? They are demonstrated in an extreme form in a passage which, though I have already dismissed it as 'myth history', is so famous and so striking an example of the 'battle piece' that I cannot resist reproducing it. It is General Sir William Napier's account of the advance of the Fusilier Brigade (7th Royal and 23rd Royal Welch Fusiliers) at the battle of Albuera, 16 May 1811, generally regarded as the crucial moment of the battle (of which Napier was not an eye-witness, having been wounded at Fuentes d'Onoro a fortnight before):

Such a gallant line, issuing from the midst of the smoke and rapidly separating itself from the confused and broken multitude, startled the enemy's masses, then augmenting and pressing forward as to an assured victory; they wavered, hesitated and, vomiting forth a storm of fire, hastily endeavoured to enlarge their front, while a fearful discharge of grape from all their artillery whistled through the British ranks. Myers was killed, Cole, the three colonels, Ellis, Blakeney and Hawkshawe, fell wounded, and the fusilier battalions, struck by the iron tempest, reeled and staggered like sinking ships: but suddenly and sternly recovering, they closed on their terrible enemies, and then was seen with what strength and majesty the British soldier fights. In vain did Soult with voice and gesture animate the Frenchmen; in vain did the hardiest veterans, breaking from the crowded columns, sacrifice their lives to gain time for the mass to open out on such a far field; in vain did the mass itself bear up, and fiercely striving fire indiscriminately upon friends and foes, while the horsemen hovering on the flank threatened to charge the advancing line. Nothing could stop that astonishing infantry. No sudden burst of undisciplined valour, no nervous enthusiasm weakened the stability of their order, their flashing eyes were bent on the dark columns in their front, their measured tread shook the ground, their dreadful volleys swept away the head of every formation, their deafening shouts overpowered the dissonant cries that broke from all parts of the tumultuous crowd, as slowly and with a horrid carnage it was pushed by the incessant vigour of the attack to the farthest edge of the height. There the French reserve, mixing with the struggling multitude, endeavoured to restore the fight but only augmented the irremediable disorder, and the mighty mass, giving way like a loosened cliff, went headlong down the steep: the rain flowed after in streams discoloured with blood, and eighteen hundred unwounded men, the remnant of six thousand unconquerable British soldiers, stood triumphant on the fatal hill.

Now, as Romantic prose passages go, this is clearly a very remarkable achievement, rich in imagery, thunderous in rhythm and immensely powerful in emotional effect; it almost vibrates on the page, towards its climax threatens indeed to loosen the reader's hold on the book. Quite understandably it has become one of the most frequently quoted of all descriptive accounts of the British army's battles in the Peninsula and a firm favourite with compilers of military anthologies. But 'descriptive' begs, of course, an important, not to say vital, question. Just what does it tell us about the Fusiliers' advance; and is what it tell us credible?

Well, we would probably all accept that 'their measured tread shook the ground' is merely metaphor and that the difference between the British soldiers' 'deafening shouts' and the French soldiers' 'dissonant cries' is a literary sound-effect – as 'streams discoloured with blood' is probably a visual one; 'reeled and staggered like sinking ships' is a variation on a traditional simile, no more to be taken *au pied de la lettre* than is 'vomited forth a storm of fire'. But when we have made allowances for permissible over-writing, when we have stripped away the verbal superstructure of the passage, we are still left with a picture of events to which it is difficult wholly to lend credence. Am I alone in wondering whether a body of men, admittedly trained soldiers, but of whom two out of three were to suffer wounds or death as a consequence of their acts, really advanced uphill under heavy fire without once showing 'nervous enthusiasm' or indeed anything but 'disciplined valour' and 'stability' and 'order'? And how exactly, to ask another sort of question, was the 'loosened cliff' of the French mass thrust down the 'steep'; by weight of superior numbers, by hand-to-hand brawling, by push of bayonet, by the sudden onset of panic in their own ranks? These are only some of a large number of uncertainties which one would like to have one's mind set straight upon but which Napier, having successfully aroused, leaves frustratingly unresolved. It may be that the episode was as extraordinary as he makes out – by comparison at once with everyday human behaviour and by the norms of military performance. But if so, and he as a veteran was in a position to say, he owed it to the reader, one may think, to make that clear. As it is, he seems to suggest that it is by no means abnormal ('Then was seen with what strength and majesty the British soldier fights') that a leaderless brigade of infantry (the brigadier and his three colonels had been disabled) should overcome, at the cost of over half its number, a very much stronger combined force of infantry, cavalry and artillery led by one of the foremost soldiers of the age (Soult was already a marshal).

It may be thought that the evidence in the case against the 'battle piece' is being stacked by adduction of so over-written an example of the form. There are, however, besides the extravagance of his language, other elements in Napier's account of the Fusiliers' advance which deserve attention because we will find them recurring in the work of other much more sober, much more 'scientific', historians. The first is the extreme uniformity of human behaviour which he portrays: the British are all attacking and all with equal intensity ('no sudden burst of undisciplined valour . . .'); the French likewise are all resisting (though some admittedly super-energetically – 'the hardiest veterans, breaking from the crowded columns, sacrifice their lives . . .'); no individual turns tail and runs, drops down to sham dead or stands thunder-struck at the indescribable horror of it all. Second, there is the very abrupt, indeed quite discontinuous, movement of the piece; the British advance, they and the French exchange volleys, 'carnage' ensues, and then suddenly the French are over the steep. Third, there is a ruthlessly stratified characterization; the British soldiers are either 'fusiliers' or one of five named people, all senior officers; the French, except for Soult and the 'hardiest veterans' (a surprisingly short-lived bunch for old soldiers) are members either of 'crowded columns', a 'tumultuous crowd', a 'struggling multitude', a 'mighty mass', or, a most unsoldierly formation, that 'loosened cliff'. This traffic in collective images, approbatory as applied to the British ('gallant line'), pejorative in the case of the French ('dark columns'), reveals a fourth, and the most important element, in Napier's approach; a highly over-simplified depiction of human behaviour on the battlefield. Implicit rather than explicit in his prose, it is clearly discernible none the less, and amounts to an absolute division of all present into 'leaders' and 'led', who conduct themselves accordingly: for the whole point of the passage is that the French, despite the exhortations of Soult and the exemplary self-sacrifice of the 'hardiest veterans', do not prevail against the British fusiliers who, *even though* they have been deprived of their senior commanders, nevertheless fight heroically and bring the advance to a successful conclusion. Finally, though this does not exhaust the list of noteworthy elements in the passage, there is no explanation of what happened to the dead and wounded; nor surely is it facetious to seek one. Men advancing in close order across a constricted space against an enemy with whom they exchange effective fire will have to step over the bodies first of their own dead and wounded comrades, then over those of the enemy; would not that have interrupted – it is only a quibble – the

Fusiliers' 'measured tread'? And what did the wounded – combatant beings no longer but none the less, indeed, perhaps all the more, sentient for that – do with themselves while the struggle raged round them? In Napier's account, the dead and wounded apparently dematerialize as soon as struck down, exactly the contrary to what was supposed to happen in the Norse paradise – where warriors killed in combat instantly sprang up to resume the fight – but equally as puzzling.

The length and tone of this critique may be thought unfair to Napier, who was merely trying, in a limited space and for an audience unaccustomed to thinking of private soldiers as individuals worthy of mention by name, to describe what by any reckoning was one of the high points of the British effort in the wars against Napoleon – which had, for Englishmen of his own time and class, the same quality of national epic as did the struggle to overthrow Hitler for their descendants five generations later. Churlishly, it fails to pay tribute to the pioneering quality of his work. No Englishman before him had written such energetic, many-sided, informative and explicative military history; even a century after its publication, its standard could prompt a doyen of English academic historians to describe *British Battles and Sieges* as 'the finest military history in English and perhaps in any language'. Moreover, none of this taking-to-task is original. Napier, by his own admission, was psychologically a hero-worshipper and artistically a big-production man ('It is the business of the historian . . . to bring the exploits of the hero into broad daylight . . . the multitude must be told where to stop and wonder and to make them do so, the historian must have recourse to all the power of words'); while it was a perceptive contemporary critic who charged that he 'sacrificed to the general grand effect all minor and apparently trifling things.'

In short, I *am* being unfair; and, since historians of the modern school have long been taught that the sacrifice of the 'general grand effect' is a necessary preliminary to the achievement of anything professionally worth-while, I also appear to be labouring a point. But am I? Modern military historians have certainly shown themselves to be as keen as the next man in pursuit of the 'minor and apparently trifling', at least as far as the non-combatant aspects of their subject are concerned; one has only to think of a book like Quimby's *Background to Napoleonic Warfare*, which dissects the pre-Revolutionary French drill regulations with Thomist rigour, or S. P. G. Ward's *Wellington's Headquarters*, which might almost be used as a text in an enlightened school of management studies, to be satisfied on that score – and to be filled with a sense of humility at one's own scholarly shortcomings. But when one turns from drill and logistics to the battle descriptions of even the best trained modern historians, it is to find Napierism as alive as ever; less sonorous to the ear, perhaps, certainly less xenophobic, but still trading in his limited stock of assumptions and assertions about the behaviour of human beings in extreme-stress situations. Here are three passages, all the work of distinguished English historians trained in the Oxford school of Modern History.

The first, from *The British Army* 1642–1970 by Brigadier Peter Young, D.S.O., M.C., describes the charge of the British Heavy Brigade of cavalry against the Russians at Balaclava, 25 October 1854. This successful action just preceded the disastrous charge of the Light Brigade:

> As the Royals passed the vineyard they saw the Greys ahead of them, hacking their way through the main body of the Russians, while other squadrons threatened to envelop them. An ancient friendship existed between the Greys and the Royals, and a voice from the latter was heard to cry, 'By God, the Greys are cut off. Gallop. Gallop.' The regiment gave a cheer, the trumpets sounded, and with ranks imperfectly formed, fell upon the flank and rear of the wheeling Russian squadrons, catching the outer troops as they tried to face outwards and routing them utterly. The Royals pressed on into the enemy mass, but Colonel Yorke hād a grip of his men and, before more than a few had galloped off in pursuit of the enemy, halted and reformed them ... The 4th Dragoon Guards had also made themselves felt, and by this time the Russians were galloping rearwards, broken and disordered, followed by a few of the 'Heavies' and sped on their way by the Horse Artillery. In this splendid charge ten squadrons routed some 3,000 men for the loss of some eighty casualties.

The second passage by David Chandler is from his exhaustive study of *The Campaigns of Napoleon* and describes the charge of the French Reserve Cavalry against the Russians at Eylau, 8 February 1807:

In marvellous fettle, eighty squadrons of splendidly accoutred horsemen swept forward over the intervening 2,500 yards. It was one of the greatest cavalry charges in history. Leading the attack rode Dahlmann at the head of six squadrons of chasseurs, followed by Murat and the cavalry reserve, supported in due course by Bessières with the Cavalry of the Guard. The troopers of Grouchy, d'Hautpol, Klein and Milhaud swept forward in turn. First, Murat's men swept through the remnants of the Russian force retiring from Eylau, before dividing into two wings, one ploughing into the flank of the Russian cavalry force attacking St Hilaire's embattled division, the other sabering its way through the troops surrounding the square of dead men at the 14th Regiment's last stand. Even then the impetus of this fantastic charge did not slacken. Driving forward, the two cavalry wings crashed through the serried ranks of Sacken's centre, pierced them, re-formed into a single column once more in the Russian rear and then plunged back the way they had come through the disordered Russian units to cut down the gunners who had done so much harm to Augereau's men. As the stunned Russians attempted to reform their line, a relieved Napoleon ordered forward the Cavalry of the Guard to cause more disorder and thus cover the safe retirement of Murat's weary but elated squadrons ... For the loss of 1,500 men, Murat had won Napoleon a vital respite

The third, from Michael Howard's *Franco-Prussian War*, describes the attack of the infantry of the Prussian Guard against the French positions at St-Privat, 18 August 1870:

So the skirmishing lines of the Guard, with thick columns behind them, extended themselves over the bare fields below St-Privat and began to make their way up the slopes in the face of the French fire ... The result was a massacre. The field officers on their horses were the first casualties. The men on foot struggled forward against the *chassepot* fire, as if into a hailstorm, shoulders hunched, heads bowed, directed only by the shouts of their leaders and the discordant noise of their regimental bugles and drums. All formations disintegrated; the men broke up their columns into a single thick and ragged skirmishing line and inched their way forward up the bare glacis of the fields until they were within some six hundred yards of St-Privat. There they stopped. No more urging could get the survivors forward. They could only crouch in firing positions and wait for the attack of the

Saxons, which they had so disastrously anticipated, to develop on their left flank. The casualty returns were to reveal over 8,000 officers and men killed and wounded, mostly in twenty minutes; more than a quarter of the entire corps strength. If anything was needed to vindicate the French faith in the *chassepot*, it was the aristocratic corpses which so thickly strewed the fields between St-Privat and St-Marie-les-Chênes.

Stylistically, of course, these three pieces differ considerably from each other. Brigadier Young's is a jolly *genre* scene, the violence he portrays no more hurtful than the knocks exchanged in a Dutch 'Low Life' painting of a beerhouse brawl; David Chandler's is Second Empire Salon School, a large canvas, highly coloured and animated by a great deal of apparent movement but conveying no real sense of action; Michael Howard's is Neo-Classical, severe in mood, sombre in tone, his subjects frozen in the attitudes of tragedy to which fate, deaf to appeals of compassion, has consigned them.

They differ too in the demands they make on the reader's credulity. Brigadier Young is content to be very vague about what actually passed between the Heavy Brigade and their Russian adversaries, perhaps because he has been in too many battles himself to think that this or any other can be explained in simple terms. Nevertheless, the factors he isolates as significant – the 'ancient friendship', the voice from the ranks and the chance which caught the Russian squadrons wheeling as they were struck by the charge – do not of themselves supply a sufficient explanation of how so small a force came to rout so large a one at such little cost.

David Chandler tells us a good deal more; the exact number of squadrons committed to the charge, the distance they covered, how many lines of resistance they broke and more besides. He is also quite specific how the French manoeuvred during this episode: after an initial sweep forward they divided into two wings, each of which fought a separate running battle before jointly breaking through a densely packed Russian formation, after which they re-formed into a single column, turned about, passed once more through the Russians, attacked with their swords a fourth enemy body and only then withdrew from action. It sounds unbelievably complicated; indeed, it reads like something from a military Kama Sutra, exciting, intriguing, but likely to have proved a good deal more difficult in practice than it reads on the printed page. And to fortify one's doubts about whether all went as smoothly as the narrative depicts it to have done are the questions which the presence of the Russians raises. What, in the path of a manoeuvre which would have been regarded as a *tour de force* if executed on a peacetime parade-ground, were all those thousands of Russians doing with themselves? The narrative implies that they stood their ground, neither falling beneath nor running clear of the French onslaught. But fallen or run the Russians must have, for otherwise the French could not have passed from in front of their formations to the other side. In falling, however, must they not have brought down numbers of French horses and riders, either by acting as stumbling-blocks or by causing collisions as horses swerved to avoid them? Both things certainly happen on the far side of a big jump in a steeplechase (for horses, even when frightened or excited, never like to tread on a living object or bump into one). And would running really have done much to clear the course? A man cannot out-distance a horse – unless, of course, he is given a considerable head start. But if one supposes a head start long enough to clear the French path of obstacles, then sentences like 'two cavalry wings crashed through the serried ranks of Sacken's centre' lose much of their meaning. It is all very baffling. And to say that is not to imply disbelief that the episode happened, nor that it happened much as described. It is only to say that one does not see how.

Michael Howard's description of the advance of the Prussian Guard leaves one with no such list of unanswered 'hows'. He sets out, as he makes clear, to give a straightforward description of a straightforward massacre and he does so in prose which is one of the many gifts serving to elevate his work above that of all other contemporary military historians. He leaves us nevertheless with a mighty 'why'. Why did the Guard not turn and flee before that terrible fire? He, having himself been decorated for bravery in leading Guardsmen of his own against the enemy – possibly, it is not completely fanciful, against great-grandsons of men killed at St-Privat – may feel no need to ask himself that question and in consequence does not seek to answer it for the reader. The question, which a less successful evocation of mood might not have posed, stands none the less. And with it a number of supplementaries. Did the whole lot, every last *Grenadier* and *Fusilier*, stick where they crouched on the open hillside? Were the bonds of discipline and group loyalty so strong that no one made a bolt for the rear, or burrowed for cover between the corpses of his comrades? We know from many other accounts that large bodies of men can display a sheep-like docility under heavy fire, often for hours at a time – the infantry of Ostermann-Tolstoi's corps are reported to have stood for two hours under point-blank artillery fire at Borodino 'during which the only movement was the stirring in the lines caused by falling bodies' – but the temporary extinction of the survival instinct that behaviour of this sort implies is beyond the ordinary reader's comprehension. Unless it is faced square by the author, and some attempt made to discuss it, his reader, fairly or unfairly, is going to feel that something more than the 'minor and trifling' has been 'sacrificed to the general grand effect'.

That something has been sacrificed in these passages their three authors would probably all concede, for sacrifice is a necessary exercise for the historian, who would befuddle himself and his audience if he tried to write down everything he could find out about an episode from the past. But they would probably also seek to justify it on particular grounds: Brigadier Young, that limitations of space precluded his attempting anything more than an atmospheric sketch of the Heavy Brigade's charge; David Chandler, that he was writing a military life of Napoleon and that it was the thought-processes of the master, not the acts of his men that he had contracted to describe; Michael Howard, that he was writing a political and strategic history of the war of 1870 and hence that it was its influence on the political and military future of Europe, rather than on the lives of combatants, that he sought to portray. All, if this accurately anticipates the sense of their rejoinders, would in short be arguing that the events and characters of a battle are subordinate in importance to its outcome; that, for the development of the British army, for the fulfilment of Napoleon's strategy, for the settlement of French and Prussian rivalry over European primacy, it was the results of Balaclava, of Eylau, of Gravelotte-St-Privat which counted, not the experience of those who took part, which becomes, therefore, of marginal relevance. Arguing at that historical level and in those terms, it would indeed be difficult to frame a reasonable opposition case.

'Killing No Murder?'

An opposition case can nevertheless be framed by asking why, if a historian is interested only in the outcome of a battle, he should trouble to provide any sort of narrative at all? The answer, at one level, would be that battles are deliberate, not chance, happenings; commanders plan battles and must pit their wits against each other to make their plans succeed. Exactly how they manoeuvre their men around the constricted arena of a battlefield, in the race against time which the limits of daylight, of human resilience, and of available material will measure them out, is therefore of obvious importance to an understanding of the success of one commander and the failure of the other – or of both, if the battle ends, as it so often does, in stalemate. But at another level that answer will not do. For the 'outcome' approach to military history, like the time-honoured but outmoded 'causes and results' approach to general history, prejudges the terms in which the narrative can be cast. That is so because the concepts 'win' and 'lose' through which a commander and his chronicler approach a battle are by no means the same as those through which his men will view their own involvement in it. Their view, like that of all human beings confronted with the threat or reality of extreme personal danger, will be a much simpler one: it will centre on the issue of personal survival, to which the commander's 'win/lose' system of values may be, indeed often proves, irrelevant or directly hostile.

But the soldier's view will also be much more complicated

than the commander's. The latter fights his battle in a comparatively stable environment – that of his headquarters, peopled by staff officers who will, because for efficiency's sake they must, retain a rational calm; and he visualizes the events of and parties to the battle, again because for efficiency's sake he must, in fairly abstract terms: of 'attack' and 'counter-attack', of the 'Heavy Brigade', of the 'Guard Corps' – large, intellectually manageable blocks of human beings going here or there and doing, or failing to do, as he directs. The soldier is vouchsafed no such well-ordered and clear-cut vision. Battle, for him, takes place in a wildly unstable physical and emotional environment; he may spend much of his time in combat as a mildly apprehensive spectator, granted, by some freak of events, a comparatively danger-free grandstand view of others fighting; then he may suddenly be able to see nothing but the clods on which he has flung himself for safety, there to crouch – he cannot anticipate – for minutes or for hours; he may feel in turn boredom, exultation, panic, anger, sorrow, bewilderment, even that sublime emotion we call courage. And his perception of community with his fellow-soldiers will fluctuate in equal measure. Something like the Guard Corps, an important reality for the German commander-in-chief at St-Privat, whether he could see it or not, would probably have ceased to have much meaning for the ordinary Guardsman once it had deployed beyond the boundaries of his vision; but he may still have felt some sense of belonging, possibly to his battalion, probably to his company, until confronted by some dramatic personal threat; then it must only have been the circle of his most immediate comrades which would have retained for him any extra-personal identity and only their survival, so much bound up with his own, for which he would have striven.

In circumstances of extreme personal danger, in short, the wishes of the commander, which the individual soldier apprehends only in the most abbreviated sense – 'Forward!' or 'Form square!' or 'Fire at will!' – (though conforming to a 'win/lose' programme of events at his superior's level) will influence his behaviour to only a marginal extent; and the commander's 'win/lose' conceptions will have no relevance to his personal predicament. 'Battle', for the ordinary soldier, is a very small-scale situation which will throw up its own leaders and will be fought by its own rules – alas, often by its own ethics.

I am not, of course, claiming personal experience as verification of these statements; as I began by saying, I have not been in a battle. I have, however, picked up haphazardly in the course of a great deal of reading about battle a large reference-stock of incidents which seem to me to bear out the points I have been making above. Those quoted below have been chosen because each concerns the conduct of the one army I know well, the British, whose norms of behaviour and code of training I can therefore use to measure the 'rightness', 'wrongness' and military utility of the incidents described.

The first passage – from the remarkably frank *Australian Official History of the Great War* – describes an episode in the middle stages of the Third Battle of Ypres (Passchendaele). The fighting by that time had resolved itself into a struggle for possession of a belt of German pillboxes, which commanded the surrounding desolation almost completely. The witness is an Australian officer, Lt W. P. Joynt, who was later to win the Victoria Cross. On 20 September 1917, he came upon

> a wide circle of troops of his brigade surrounding a two-storey pillbox, and firing at a loophole in the upper storey from which shots were coming. One man, coolly standing close below and firing up at it, fell back killed but the Germans in the lower chamber soon after surrendered. The circle of Australians at once assumed easy attitudes, and the prisoners were coming out when shots were fired, killing an Australian. The shot came from the upper storey, whose inmates knew nothing of the surrender of the men below; but the surrounding troops were much too heated to realize this. To them the deed appeared to be the vilest treachery, and they forthwith bayoneted the prisoners. One [Australian], about to bayonet a German, found that his own bayonet was not on his rifle. While the wretched man implored him for mercy, he grimly fixed it and then bayoneted the man.

'The Germans in this case', the official historian platitudinously continues, 'were entirely innocent, but such incidents are inevitable in the heat of battle, and any blame for them lies with those who make wars, not with those who fight them.'

The second incident is narrated by Professor Guy Chapman, at the time in question a young officer in a Kitchener battalion which had just taken part in one of the attacks which formed part of the Battle of the Somme in 1916.

Blake's face was slack and haggard, but not from weariness. He greeted me moodily, and then sat silent, abstracted in some distant perplexity. 'What's the matter, Terence?' I asked.

'Oh, I don't know. Nothing – at least. Look here, we took a lot of prisoners in those trenches yesterday morning. Just as we got into their line, an officer came out of a dugout. He'd got one hand above his head, and a pair of field glasses in the other. He held his glasses out to S ..., you know, the ex-sailor with the Messina earthquake medal – and said, "Here you are, Sergeant, I surrender." S ... said, "Thank you, sir," and took the glasses with his left hand. At the same moment, he tucked the butt of his rifle under his arm, and shot the officer straight through the head. What the hell ought I to do?' ... 'I don't see that you can do anything,' I answered slowly. 'What can you do? Besides, I don't see that S... is really to blame. He must have been half mad with excitement when he got into the trench. I don't suppose he even thought what he was doing. If you start a man killing, you can't turn him off again like an engine. After all, he is a good man. He was probably half off his head.'

'It wasn't only him: another did exactly the same thing.'

'Anyhow, it's too late to do anything now. I suppose you ought to have shot both on the spot. The best thing is to forget it.'

The third extract is from the *History of the Irish Guards in the Second World War*. The battalion was fighting in a mountainous region of Italy in 1943. A company officer is relating his experience:

We ran straight into a large body of Germans and, after a few bursts of Bren and Tommy gun fire, about forty ran out with their hands up. Elated by this, we proceeded to winkle them out at a great pace. Wheeling round the next corner, Lance-Sergeant Weir led his section in a charge against another group of Germans. These Germans were ready for them and met them with long bursts of fire ... Weir was shot through the shoulder, but the bullet only stopped him for a moment, while he recovered his balance. He led his men full tilt into the Germans and they killed those who delayed their surrender with the *traditional comment*, 'Too late, chum.' [Italics supplied.]

Now what does all this add up to? In each case, what is described is 'improper violence' – unqualifiably improper in the case of the Australians, circumstantially excusable in the case of the 'ex-sailor with the Messina earthquake medal', just barely licit by a pretty rough-and-ready code of justice in the case of the Irish Guardsmen. These, at any rate, are the verdicts which a dispassionate reader might reasonably be expected to enter. By the army's official code, however, all would be categorized as offences and dealt with accordingly. Indeed, Guy Chapman's piece of dialogue – a reconstruction, but by an author whose reputation guarantees its veracity – might be lifted straight into one of those training playlets, described earlier, in which are dramatized for subsequent discussion by officer cadets issues of 'right' and 'wrong' behaviour. And it is almost possible to predict word for word the conclusion that the 'Directing Staff's Solution' would come to: 'Incidents of this sort will not occur if soldiers are properly briefed and kept under strict control by their officers. If a soldier does unlawfully kill a prisoner, he should at once be placed under close arrest and evacuated for psychiatric examination; if found fit to plead, he will be dealt with under Section ...' For to the army, quite as much as to the courts, hard cases make bad law. It wants dependable and conformist junior leaders, men who will neither fight private battles by local rules – as did the Australians at Ypres; nor seek, like Guy Chapman and his comrade, to 'understand' the behaviour of soldiers who break the Geneva convention; nor, when a beaten enemy proves momentarily uncooperative, revert, like the Irish Guardsmen, to the grim traditions of a mercenary past.

It would be pleasing to think that the British army takes the view it does for reasons of humanity; and my judgement would in passing be that it does cultivate an admirably humane attitude to the use of violence, notably by its propagation of the doctrine of 'minimum necessary force' which, though it applies most strictly to its role as an arm of the civil power in domestic disorder, also colours its attitude to battlefield action. But it would be more realistic to recognize that the army seeks to instil in its leaders the attitudes it does because experience has taught it that its mechanisms of command and control can only be kept functioning under stress if officers will scrupulously obey the rules of procedure. Those rules allot fixed values to all individuals and groups on the battlefield – 'friend', 'enemy', 'prisoner', 'casualty' – and impose strict limits upon which can be offered violence, and in what circumstances. Hence the impropriety – by military as well as humanitarian standards – of these three lethal encounters.

Their propriety or impropriety does not, however, concern the military historian, at least at a professional level. He is a judge of the significance of events, not of their morality or even strictly of their utility. But significance for what, in these cases? That poses an awkward question. In 'win/lose' terms, the three incidents are absolutely meaningless. As we saw, the Germans done to death either had given their surrender or were on the point of doing so; and in that sense each of these particular episodes in the larger set of events we call 'the battle' and been 'won'. How, therefore, should the historian treat them? Undoubtedly, he would find it most convenient, from every point of view, to ignore them. Like a man who finds himself left with a collection of screws and cogs after he has made a watch 'work' again, he can tell himself that they are clearly not essential components and slide them into his miscellaneous tray. The fact that his concluding résumé of results – so many 'killed' in the battle, so many 'prisoners' – will conceal an overlap (some of his 'killed' having been momentarily 'prisoners') need not be mentioned or can be glossed over by some reference to the 'uncivilized behaviour of small groups of soldiers'.

The introduction of the concept of 'small groups', however, deals a body blow to the assumptions underlying the 'win/lose' approach. For if one once admits that the behaviour of a group of soldiers on any part of the battlefield ought to be understood in terms of their corporate mood, or of the conditions there prevailing at the time, indeed in terms of anything but their willingness to do as duty, discipline and orders demand, then the whole idea of the outcome of a battle being determined by one commander's defter manipulation of his masses against his opponent's crumbles. Students of generalship will object that this is an overstatement; and so of course it is. But because the decisions and acts of a commander apparently contribute more to the outcome of a battle than the decisions and acts of any single group of his subordinates, it does not follow that what he does is more important than what *all* his subordinates do, nor that his behaviour is a more valid subject of study than theirs. On the contrary: their relative importance is an unresolved question, and since we appear to know a great deal more about generalship than we do about how and why ordinary soldiers fight, a diversion of historical effort from the rear to the front of the battlefield would seem considerably overdue. All the more does it seem desirable in the light of what little reliable information we do have about what goes on at the place soldiers call 'the sharp end'. Most of it we owe to the American army historical service which, during the Second World War, undertook the first systematic study of human behaviour in combat, a study which yielded remarkable results.

Foremost among them was the revelation that ordinary soldiers do not think of themselves, in life-and-death situations, as subordinate members of whatever formal military organization it is to which authority has assigned them, but as equals within a very tiny group – perhaps no more than six or seven men. They are not exact equals, of course, because at least one of them will hold junior military rank and he – through perhaps another, naturally stronger character – will be looked to for leadership. But it will not be because of his or anyone else's leadership that the group members will begin to fight and continue to fight. It will be, on the one hand, for personal survival, which individuals will recognize to be bound up with group survival, and, on the other, for fear of incurring by cowardly conduct the group's contempt. The American army, and subsequently the British, has taken the findings of the U.S. army's historical teams very much to heart, trying as far as possible to adjust the internal organization of their fighting units to a pattern which will take advantage of what they now know of 'small group dynamics'. Each, as a result, has tended to find itself speaking with two voices about the problem of human behaviour in battle; with a newly found private voice which admits that everything ultimately rests with the ordinary soldier's 'motivation to combat'; and with a traditional public voice, the one heard in military academy leadership lectures, which continues to emphasize the primary role of discipline and command. There is no real inconsistency in this duality of attitude. It merely marks an acceptance by the armies concerned that combat is as complicated and multiform as any other sort of human activity, and given the stakes at issue more so than most.

But, it is not unreasonable to inquire, if soldiers themselves have come to recognize that what they would like to happen on a battlefield is by no means the same thing as what does happen, why do so many military historians continue to write as if generalship and the big battalions were their only proper study? This question would be a great deal easier to answer if military historiography – using the word in its alternative meaning of 'the history of military history writing' – were a properly developed subject as, given the centrality which historians have accorded to war since the earliest times, one might expect it to be. Alas, one's expectations would be false. Although general historians have long recognized that what a historian will see as significant in his chosen subject, and how he will write about it, is almost always heavily influenced by the view that other historians have already taken, and although even the most casual dipper into history books is aware that there are schools of historians – he will certainly have heard of Marxist historians, probably of Freudian historians, perhaps of Whig historians – not even the beginnings of an attempt have been made by military historians to plot the intellectual landmarks and boundary stones of their own field of operations. This makes any sketch of military historiography – which I must attempt here – a matter of guesswork.

The History of Military History

It might seem a safe guess that the figure who bestrides the military historian's landscape is the great nineteenth-century Prussian, Hans Delbrück, a pupil of pupils of Ranke, the first Rankeian to concern himself with military history and therefore the pioneer of the modern 'scientific' and 'universal' approach to the subject. And immensely influential he undoubtedly was – with other Germans. In the highly militarized Second Reich, however, anything to do with war was so intertwined with national policy and national myth that no study of it could reasonably hope to achieve either the autonomy of an academic discipline or the aesthetic freedom of genuine literature. Military history was too loaded a subject, loaded with questions of national unity, of national survival, of dynastic prestige, for any German to feel ultimate detachment about it; and without a measure of intellectual detachment, of course, any historian is bound to become either an obscurantist or a publicist. Delbrück became the latter, and achieved enormous standing thereby, ending the life he had begun as tutor to the Kaiser's grandson as strategic schoolmaster of the German nation. It was not, at the finish, a job anyone would have envied him, for having spent four years teaching his countrymen in monthly articles how Germany ought to win, he found himself in 1918 landed with the responsibility of explaining to the Reichstag why she had lost. Inevitably, if unfairly – for he had almost always talked sense – his reputation was ruined. He remains none the less a significant figure, if not as a historian, then as the honorary colonel of that monstrous modern regiment, the academic strategists. Herman Kahn is but Hans Delbrück writ large.

The great nineteenth-century school of French historians fails equally to yield us an example of a seminal mind. In that often defeated country, too, a genuinely objective approach to military history always risked incurring the slur of carrying comfort to the enemy, and its development was further hindered by the endemic national neurosis of Napoleon-worship. One or two names – Palat and Colin – stand out, but both were soldiers, their intellectual credentials accordingly widely suspect in that divided society, and their genuine talents without influence or recognition outside professional and 'patriotic' circles.

It is really only in the English-speaking countries, whose land campaigns, with the exception of those of the American Civil War, have all been waged outside the national territory, that military history has been able to acquire the status of a humane study with a wide, general readership among informed minds. The reasons for that are obvious; our defeats have never threatened our national survival, our wars in consequence have never deeply divided our countries (Vietnam may – but probably will not – prove a lasting exception) and we have never therefore demanded scapegoats (like Bazaine, the 'traitor' of 1870) or Titans (like Hindenburg). In that vein, it is significant that the only cult general in the English-speaking world – Robert E. Lee – was the paladin of its only component community ever to suffer military catastrophe, the Confederacy. For the privileged majority of our world, land warfare during the last hundred and fifty years – the period which coincides with the emergence of modern historical scholarship – has been in the last resort a spectator activity. Hence our demand for, and pleasure in, well-written and intelligent commentary. Hence too our limited conception of military-historical controversy, which does not extend much further than the discussion of whether Montgomery let Rommel slip through his fingers by negligence after Alamein or whether Patton ought or ought not to have slapped the face of a shell-shocked soldier. It does not comprehend questions about whether or not, by better military judgement, we might still govern ourselves from our national capital – as it does for the Germans; whether or not we might have avoided four years of foreign occupation – as it does for the French; whether or not we might have saved the lives of twenty millions of our fellow-countrymen – as it does for the Russians. Had we to face questions like that, were not military history for us a success story, our military historiography would doubtless bear all the marks of circumscription, over-technicality, bombast, personal vilification, narrow xenophobia and inelegant style which, separately or in combination, disfigure – to our eyes – the work of French, German and Russian writers.

But there is another reason which explains why continental scholarship, as represented by Delbrück, has failed to influence the way in which military history as a humane study has developed in the English-speaking world. It is not that Delbrück remains untranslated; nor is it that his idiosyncratic critical method – what he called *Sachkritik* – renders the rest of his work suspect to Anglo-Saxon minds; nor is it even that his 'philosophy of history' carries too thorough-going a Prussian flavour for liberal Western taste. Chronological considerations apart, the latter might well indeed have been the factor which embargoed the export of his ideas and methods, for though, unlike Treitschke, he does not exalt warfare and exult in violence, he accepts the normality of both with a readiness which few American or British scholars could find it in themselves to do. One of the main objects of his work, after all, was to demonstrate that every political system is, if not actually determined by, then in a symbiotic relationship with its own form of military organization; and, to citizens of countries which had always ridden their armies on a very tight political rein (without necessarily perceiving that it was the all-, or almost all-, surrounding sea which allowed them to do so), that proposition alone might have been enough to brand him as a bad as well as a dangerous thinker. But it is, in the last resort, none of these things which serves to disprove Delbrück's formative influence on British and American military historiography. The chronological factor is decisive, for, fifty years before Delbrück began to publish, England had already produced her own philosopher of war.

He was, characteristically, an amateur historian, an Etonian barrister, ultimately to become Chief Justice of Ceylon, who was merely interrupting his legal career to hold a chair of history at London University at the time he published the book for which he is chiefly remembered. It is not, to be frank, a book whose title trips readily off the tongue of many modern historians, nor one which they commonly put on their reading-lists, but there can be few who have not, at one time or other, had it in their hands or nodded at it on the shelf. For Sir Edward Creasy's *Fifteen Decisive Battles of the World* was one of the great Victorian best-sellers, rivalling Darwin's *Origin of Species* in the frequency with which it was republished – thirty-eight times in the forty-three years between 1851 and 1894 – and Samuel Smiles's *Self-Help* in the approbation it won from parents and schoolteachers. Its success is easily explained. It resolved a great Victorian dilemma, of the same sort which Darwin and Smiles resolved, and by the same method. Darwin, whatever cats he released in spiritual dovecotes, at least persuaded many Victorians that the tide of competition which had sprung up to sweep through their society was the manifestation of a natural order of things. Smiles, through his doctrine of self-help, further showed that competition, for all its harsh impact on the lives of individuals, might have a morally good and socially useful result by its stimulation of effort and thrift even among the very poor. Creasy, whose book had appeared eight years before either of theirs (both, by chance, were published in 1859) was quite as attuned as any Victorian to the difficulty of reconciling Chrisfian compassion and a belief in Progress with the inhumanity of a getting-and-spending world. But unlike Smiles, who saw the issue defined principally in terms of class-struggle, Creasy confronted, and sought to outface, the issue of conflict in a yet more extreme form, that of war itself.

'It is,' he wrote in his preface, 'an honourable characteristic of the Spirit of this Age that projects of violence and warfare are regarded among civilized states with gradually increasing aversion' – a faultlessly Victorian sentiment, not least in its delicate allusion to the fact that such things undoubtedly go on and that large numbers of people are deeply if secretly interested by them. And he hastens to explain that he is not pandering to debased instincts in bringing such projects before the public. It would be evidence, he says, of 'strange weakness or depravity of mind for a writer ... of the present day to choose battles for his favourite topic merely because they were battles ... and so many hundreds or thousands of human beings stabbed, hewed or shot each other to death during them'; nor does the display of human courage or of the intellectual talents of command associated with battle furnish an excuse, for such 'qualities ... are to be found in the basest as well as the noblest of mankind.' No; if we are to study battles – and the logic of his argument implies that it can be only *some* battles we may read about – it is because 'independent of the moral worth of the combatants' some battles 'have helped to make us what we are ... [For] the interests of many states are often involved in the collisions of but a few ... and ... the effect of these collisions is not limited to a single age, but may give an impulse which will sway the fortunes of mankind.'

He does not claim the originality of this idea for himself, which he ascribes to his contemporary Henry Hallam, but he appears to have been the first systematically to develop it. He does so in a way one can only describe as the 'Whig interpretation of history writ in blood', the gist of his argument being that everything admirable to the Victorian world – Greek wisdom, Roman virtue, Saxon bravery. Norman centralism, Christian faith in a specifically Protestant form, English liberty and French democracy – had each been saved from extinction by some brilliant military exploit; and that such threats to that world as remained, notably Russian autocracy, might equally have been extinguished had fortune favoured the good on this battlefield or that.

It is not, by any reckoning, a particularly sophisticated philosophy of history. Creasy, indeed, was too talented a writer for it to have been likely that he took it very seriously himself, his main energy in the writing of the book clearly having been devoted to making it a jolly good read. And a jolly good read it remains. But, whatever his philosophical intentions or literary achievements, he had launched, through the book's eye-catching title and runaway commercial success, an immensely powerful idea into the English historical vocabulary. And it is one which has never lost its impetus. Almost as soon as Creasy was dead, Malleson, the historian of the Indian Mutiny, had published his *Decisive Battles of India* (1883); four years later, the American Thomas Knox published *Decisive Battles since Waterloo* (the last of Creasy's 'Fifteen'), which had as 'its ... purpose the idea of presenting an outline survey of the history of the Nineteenth Century, considered from the point of view of its chief military events' – a slightly less emphatic restatement of Creasy's approach, but a restatement all the same, not least in its assumption that the nineteenth century was a high moment in the history of man. The end of the First World War, which had yielded a new crop of Decisive Battles, impelled Colonel Whitton to produce the *Decisive Battles of Modern Times* (1922) which was followed in 1929 by Liddell Hart's *Decisive Wars of History* and 1939 by his great rival J. F. C. Fuller's first attempt at a major reworking of Creasy's idea, *Decisive Battles of the World* (*From Salamis to Madrid* in its pre-war two-volume version, *From Salamis to Leyte Gulf* in three volumes after 1945). The post-war period has seen the American official historians won over to Creasy's method, *Command Decisions* (1960) being an epitome of their multi-volume history of the Second World War, and its best-selling item; while, by an ironic stroke, the Germans themselves have succumbed to him, two of their best-known post-mortems on their defeat being *The Fatal Decisions* (1956) and *The Decisive Battles of the Second World War* (1965) while at some comparatively recent moment *Entscheidungschlacht* has replaced *Hauptschlacht* in the German military vocabulary to convey the idea of a crucial engagement.

There have been a number of popular variations on, even whole series devoted to the theme. But it is the attention which serious military historians, like Liddell Hart and Fuller, have given to the idea which reveals its importance. And that surely lies, whatever disclaimers Creasy made to the contrary, in the moral freedom of action it conferred on historians reared upon and working within the Western tradition. For whether or not an individual historian accepts the Christian ethic which supplies that tradition with its dynamic, the Christian revulsion from war hedges about any humane intellectual approach to the subject with formidable difficulties. War, in Christian theology, is a sinful activity, unless carried on within a framework of rules which few commanders are in practice able to obey; in particular those which demand that he shall have a just aim and a reasonable expectation of victory. Any objective study quickly reveals, however, that most wars are begun for reasons which have nothing to do with justice, have results quite different from those proclaimed as their objects, if indeed they have any clear-cut result at all, and visit during their course a great deal of casual suffering on the innocent. Western historians, whether monastic chroniclers or Gibbonian sceptics, had always therefore tended to depict war as a calamity, a scourge, or a foolishness, unless it could be represented as a crusade (always a Just War in Christian terms) or be used to exemplify the life and exploits of great men. Great national triumphs, like Waterloo, always found their epic-writers; but serious historians, though compelled to write about war, were generally unanimous in deprecating the necessity. The intellectual movements of the nineteenth century heightened scholarly uncertainties about the ethics and role of warfare. On the one hand, the school of Ranke advanced a view of history which looked for much deeper and more complicated explanations of historical change than surface events like military victory or defeat could supply. On the other, the economic school, which Marx was about to capture outright, argued that it was in the dynamic relationship between capital and labour that the explanation of human conduct lay, and to this armies and their doings were an irrelevance. Parallel to these ideas, and not inconsistent with either, lay that of Progress itself, one of the most potent that the nineteenth century was to produce, so powerful that, though terribly wounded, it is still with us today. A belief in Progress was indeed already promising to supplant a belief in God. And the phenomenon of war offers, if anything, greater offence to the former than the latter. For Christians have always accepted that Man, whether individually or en masse, can and will behave badly, cruelly, and violently. The vision of the future which the idea of Progress holds out, however, demands much greater optimism about human nature. How, in these intellectual and moral circumstances, were scholars to justify to themselves or their readers any discussion of war which did not condemn it outright as an aberration on the face of human history?

Creasy supplied the formula. War had a purpose; it had made the nineteenth century. Moreover the study of war is also a study of human free will:

I am aware [he wrote] that ... the reproach of Fatalism is justly incurred by those, who, like the writers of a certain school in a neighbouring country, recognize in history nothing more than a series of necessary phenomena, which follow inevitably one upon the other. But when, in this work, I speak of probabilities, I speak of human probabilities only. And the occurrence of war in the past in no way determined the recurrence of war in the future: In closing our observations in this the last of the Decisive Battles of the World [Waterloo], it is pleasing to contrast the year which it signalized with the year that is now passing over our heads. We have not (and long may we be without) the stern excitement of martial strife and we see no captive standards of our European neighbours brought in triumph to our shrines. But we behold an infinitely prouder spectacle. We see the banners of every civilized nation waving over the arena of our competition with each other, in the arts that minister to our race's support and happiness, and not to its suffering and destruction. 'Peace hath her victories/No less renowned than War;' and no battle-field ever witnessed a victory more noble than that which England, under her Sovereign Lady and her Royal Prince, is now teaching the peoples of the earth to achieve over selfish prejudices and international feuds, in the great cause of the general promotion of the industry and welfare of mankind.

The delicate hypocrisy of Creasy's formula provided every historian who wished to write about battles with the excuse he needed. Battles are important. They decide things. They improve things. Exactly what, and how, are questions that the individual historian is left free by Creasy's *nihil obstat*, his grant of moral approval, to judge for himself. It is a dispensation which whole squads of modern military historians have seized on to justify an endless, repetitive examination of battles which by no stretch of the imagination can be said to have done anything but make the world worse; to justify their ascription to strategically piffling, pointless bloodbaths of the cachet 'decisive' on the grounds that they must have decided something, even if what exactly that might have been escapes elucidation; to wallow in battles for battles' sake; and to evade any really inquisitive discussion of what battles might be like by recourse to the easy argument that one must stick to the point, which is decision, results, winning or losing. Against the power and simplicity of that argument, any other – poor old Delbrück's advocacy, for example, of the notion that battle is warp to the weft of a whole social fabric – makes slow headway in the competition for a public hearing. A minority have heard him: Michael Howard, whose interweaving of diplomatic with military events is always a tour de force; in a different vein, Alistair Horne; and, though the sheer bulk of their enterprise overshadows its guiding theme, the American official historians. But for the majority it is the Decisive Battle idea which persists – both for readers and writer. Hence the form which almost all modern writing about battle takes.

The Narrative Tradition

The aim which the majority pursues does not serve to explain, however, the peculiar narrative style in which most battle writing, the typical 'battle piece', with its reduction of soldiers to pawns, its discontinuous rhythm, its conventional imagery, its selective incident and its high focus on leadership, is cast. To explain that, one must look beyond Creasy – since it is already so highly developed in the work of Napier – to another source. Modern historiography, like modern warfare, began with the Renaissance. And it is obvious that the writings of the Ancients, which served as models for the writing of all modern history from the Renaissance onwards, must have done so for military history too. The question is, which classical writers? A great deal of controversy has flowed round the issue of exactly how influential classical writers were on Renaissance military affairs. Vegetius, a late Roman author, is known to have been widely read. But F. L. Taylor, historian of *The Art of Warfare in Italy 1494–1529*, came to the conclusion, after reviewing what authors the Condottieri might have studied, that 'the influence of classical history and literature was mainly academic. We view the warfare of the Renaissance through the academic medium of contemporary historians and teachers and are consequently apt to form an exaggerated opinion of the effect of theoretical writings on military operations.' Michael Mallet, a modern expert and the author of a brilliant, many-sided study of mercenary warfare, concurs.

> The fifteenth-century captain learnt the art of war as an apprentice to an established condottiere, not from books. He may have been gratified to learn from one of the humanists in his entourage that his tactics resembled those of Caesar in Gaul, but it is unlikely that he consciously intended it to be so. It was not a study of the Roman republican army which produced a revived interest in infantry but the practical necessities of fifteenth-century warfare.

For our purpose, however, what the soldiers did or did not read is irrelevant. For, if soldiers did not learn to fight their battles from reading books, neither is it likely that military historians learnt to write their books from watching battles. Battles are extremely confusing; and confronted with the need to make sense of something he does not understand, even the cleverest, indeed pre-eminently the cleverest man, realizing his need for a language and metaphor he does not possess, will turn to look at what someone else has already made of a similar set of events as a guide for his own pen. To whom might he have turned? Caesar has just been mentioned. And although Caesar's *Commentaries* had only been recently rediscovered, they had achieved a wide popularity in fifteenth-century Italy and were being translated into other European languages by the beginning of the sixteenth (French, 1488; German, 1507; English, 1530). A bibliographer would no doubt be able to show by what routes his ideas and methods percolated thereafter into European historiography; though, to my knowledge, it has not been done for military historiography. It is, however, usually claimed that two of the most important military reformers of the late sixteenth and early seventeenth centuries, Maurice of Nassau and Gustavus Adolphus, were consciously influenced in the making of their armies by what they had learnt about the Roman Legion from Caesar's *Gallic War*, about which, thanks to his writing, 'we know more ... than any other military operation of the

Ancient World'. And it is obvious, in a much more general way, that from the seventeenth century onwards, it is Roman military practices – drill, discipline, uniformity of dress – and Roman military ideas – of intellectual leadership, automatic valour, unquestioning obedience, self-abnegation, loyalty to unit – which are dominant in the European soldier's world. By the end of the eighteenth century, the Neo-Classical revival had made fashionable an outward assumption of Roman symbols, to express an attitude which was already internalized: the Frenchmen whom the Fusiliers drove off the hill at Albuera marched, after all, behind Eagles which were facsimiles of those carried by Caesar's Legions; the Greys and Royals who charged at Balaclava wore on their uniforms miniature representations of those same Eagles which their ancestors in the regiments had captured at Waterloo – their proudest achievements; the Prussian Guardsmen died at St-Privat in head-dress which mimicked the Legionary helmet. And by this date we know enough of what the leading soldiers studied to be able to demonstrate convincingly that the Roman military authors, Caesar foremost among them, had helped to furnish their minds. The schoolboy Napoleon noted Caesar among his list of books read; Schlieffen, cynosure of Prussian Great General Staff officers, nurtured an obsession with the Roman defeat at Cannae that helped to precipitate the First World War.

There is, however, no need to proceed deductively to the conclusion that because the Romans, and particularly Caesar, were an important influence on post-Renaissance armies, then it was probably Caesar who most influenced the way in which military history was written from the Renaissance onwards. We can reach the same point by a single inductive leap, for the distinctive features of the 'battle piece' will all be found in any of Caesar's narratives of his own victories that one cares to turn up. Take, for example, his description of the defeat of the Nervii, on the River Sambre in modern Belgium, in B.C. 57:

Caesar proceeded, after encouraging the Tenth Legion, to the right wing, where he saw that his men were hard pressed. The soldiers were crowded too closely together to be able to fight easily, because the standards of the Twelfth Legion had been massed in one place.

All the centurions of the fourth cohort had been killed, together with its standard-bearer, and its standards had been lost. In the other cohorts almost all the centurions were dead or wounded, and the chief centurion, Sextius Baculus, a very brave man, was so exhausted by the wounds, many and severe, that he had suffered that he could hardly stand up. Caesar also noticed that the rest of the soldiers in this Legion were giving up the fight and that some were leaving the battle to join those in the rear ranks who were already making off. The enemy, though advancing uphill, were maintaining the pressure on their front and at the same time pushing hard on both flanks. Caesar recognized that a crisis was at hand. He had no reserves left to commit, so, snatching a shield from one of the soldiers in the rear (he himself having come without one) he put himself in the front rank. Calling to the centurions by name, and shouting encouragements to the rest he ordered them to advance the standards and deploy into extended order, so that they could use their swords more easily. His appearance brought hope to the soldiers and restored their courage. Under his eye, each man strove his utmost and the enemy's onset was checked.

Here it all is – *disjunctive movement*: 1. the Legion is hard pressed, some of the soldiers are slinking away; 2. Caesar arrives and has the standards advanced; 3. the enemy's attack loses its impetus; *uniformity of behaviour*: the enemy are all attacking, the legionaries are either resisting feebly or drifting off until Caesar's arrival makes them all fight with fervour; *simplified characterization*: only two people are mentioned by name, of whom only one is accorded an important role – the author; *simplified motivation*: the led have lost the will to fight until the leader restores it to them by some simple orders and words of encouragement.

We now know that Caesar composed his *Commentaries* for a carefully calculated political end. And intelligent readers, whether so aware or not, have probably always guessed that he overdid the descriptions of his own exploits. Yet, surprisingly and exceptionally, military historians have never seriously questioned the realism of his battle-scenes, viewed as reportage, have indeed generally used his depiction of how his legionaries fought as a truth to which they had to adapt whatever facts they could glean of the battles of their own times. Some may be excused for doing so. The humanists of the Renaissance, groping their way towards critical standards in historiography and unacquainted with legionary-style armies, could all too easily have taken at face value Caesar's account of the legionary's pliability and automatism, giving it fresh currency in their own narration of battle, told as they thought they ought to go. But later historians, working to established standards and living in countries garrisoned by disciplined, salaried armed forces, should have known better. All the more is this the case because Antiquity yielded an alternative tradition in military historiography, a great deal richer, more subtle, more psychological, above all more frank in its treatment of how men behave in battle, which, though slower than the Roman tradition to make its way into the stream of modern European scholarship, should, once present, have prompted them to a reappraisal of how they might conduct their business. That tradition is the Greek, fathered by Herodotus at the beginning of the fifth century B.C. and already elevated by Thucydides, at the end of that century, to a scientific and artistic level which European historians would not have regained until two hundred years ago.

Here is part of Thucydides' account of the battle of Mantinea, 418 B.C., between the Lacedaemonians (Spartans) and their allies, and the Argives and theirs:

The Lacedaemonian army looked the largest; though as to putting down the numbers of either host, I could not do so with accuracy ... and men are so apt to brag about the forces of their country that the estimate of their opponents was not trusted. The following calculation makes it possible however to estimate the numbers [some scholarly calculations follow]. The armies now being on the eve of engaging, each contingent received some words of encouragement from its own commander. The Mantineans were reminded that they were going to fight for their country and to avoid returning to slavery ... the Argives ... to punish an enemy and a neighbour for a thousand wrongs ... The Lacedaemonians, meanwhile, exhorted each brave comrade to remember what he had learnt before, well aware that ... long training ... was of more virtue than any brief verbal exhortation.

After this they joined battle, the Argives and their allies advancing with haste and fury, and Lacedaemonians slowly and to the music of many flute-players – a standing institution in their army which has nothing to do with religion, but is meant to make them advance slowly, stepping in time, without breaking their order, as large armies are apt to do in the moment of engaging.

Just before the battle joined, King Agis [of the Lacedaemonians] resolved upon the following manoeuvre. All armies are alike in this: on going into action, they get forced out rather on their right wing, and one and the other overlap with this their adversary's left; because fear makes each man do his best to shelter his unarmed side with the shield of the man next to him on the right, thinking that the closer the shields are locked together the better will he be protected. The man primarily responsible for this is the first upon the right wing who is always striving to withdraw from the enemy his unarmed side; and the same apprehension makes the rest follow him. [The two armies each outflanked the other's left, so Agis, having more men, ordered some of his to move leftwards. They disobeyed, however, the two responsible leaders, Hipponoidas and Aristocles being 'afterwards banished from Sparta, as having been guilty of cowardice', and, while Agis was dealing with this insubordination, the Argives made a sudden attack.] Now it was that the Lacedaemonians, utterly worsted in respect of skill, showed themselves as superior in point of courage. As soon as they came to close quarters with the enemy [they succeeded in beating them].

In fact the account of the action which follows is rather complicated, being full of the names of the two sides' minor allies, and contains furthermore that deadly, non-explicative phrase, 'instantly routed them', that came so easily to Caesar's pen. But in almost every other respect, how very much superior to Caesar's is Thucydides' style of battle narrative. Where Caesar's soldiers are automatons, Thucydides' are human beings; where their actions depend on his presence or absence, Thucydides' are motivated by self-concern (like the 'man on the right wing') or by stuffiness (like Hipponoidas and Aristocles); where Caesar can only introduce the position of the standards as an external influence on their behaviour, Thucydides mentions the appeals of music (and, by implication, of religion), patriotism, xenophobia, professional pride; where Caesar's subordinate figures are cardboard – if Sextius Baculus was not 'a very valiant man', which is all he tells us about him, what was he doing as senior centurion? – Thucydides' are individuals, with wills of their own, who suffer for mis-employing them (banishment from Sparta); and where the intervention of the leader in Caesar's battle sets things to rights, in Thucydides', King Agis's change of plan actually makes things worse for his side. Moreover, the general feeling of the two pieces is quite different: Caesar tells us nothing about his army, except that it obeyed his orders; the most interesting thing about it, the narrative implies, is that he was its leader. Thucydides' army, on the other hand, is one of a species of institutions interesting in themselves, with well-known but by no means uniform patterns of behaviour ('large armies are apt ...' – meaning that small armies may not be; 'all armies are alike in this ...' – meaning not necessarily so in other ways) and these patterns of behaviour are the product of human conduct and character at every level. In short, while Caesar is writing particular history, Thucydides is writing general history, by every test a more useful, a more difficult and a more illuminating form of the art.

The objection to this depreciation of Caesar's historical skills – that he was a no-nonsense soldier whose simplifications of issues and motives was the fruit of a successful ruthlessness with concrete military difficulties – cannot be sustained. For Thucydides was also a practising soldier, whose history of the Peloponnesian War was based upon his own experience, eye-witness or collection of first-hand reports. A better objection is that Caesar was describing the operations of armies quite different from those of the Greek city-states; while the latter were part-time militias of free men, his were long-service, regular, mercenary formations, recruited by voluntary enlistment but ruled by the whip and the sword; and if they appeared automatous in their battlefield behaviour, it was because they were trained to be so. This is a good and strong objection. But not a clinching one. Maurice of Nassau and Gustavus Adolphus may have believed that, given money, time and effort, they could recreate armies in the image Caesar had revealed to them. Modern classical scholars, increasingly inclined to fret at the lack of real understanding of the inner life of the Legions which the Ancients have left them, suspect that they were far more complex, fickle and individual in their behaviour than Caesar lets on. If this is so, then Maurice and Gustavus were chasing a chimera. Certainly no military institution of which we have detailed, objective knowledge has ever been given the monumental, marmoreal, almost monolithic uniformity of character which classical writers conventionally ascribe to the Legions. That being so, we may safely guess that such an ascription was, indeed, a convention, of the same sort which always made priests holy, temples sacred and old men wise.

The difference between Roman and Greek historiography, in the words of Professor Michael Grant, is that the former 'began with politics and the state', while the latter 'sprang from geography and human behaviour'. It was appropriate, therefore, that the Greek historians should have begun to make their influence felt on European historiography at the precise moment when an interest in 'geography and human behaviour', an interest whose intellectual and artistic manifestation we call Romanticism, was replacing a dry-as-dust legalistic concern with 'politics and the state' as the motive force of historical inquiry. Appropriate, and probably consequential upon; for the foremost practitioner of the new history, Leopold von Ranke, insisted on regarding Thucydides as the greatest of all historians, living or dead. Ranke's new history or 'general history' did not, of course, descend from the Greeks. It was a conception independently arrived at, and under continual development throughout his long life (1795–1886). But because of his championship of the Greeks, something of their spirit – practical, realistic, speculative, witty, humane – in each of those qualities an important corrective to the plodding laboriousness of the German school from which he emerged, made its way through his into the work of lesser, often unacademic historians, some of whom were no doubt quite ignorant of the debt they owed him.

One of these may have been Ardant du Picq, who made in the middle of the nineteenth century a strikingly novel approach to the study of battle via the study of human behaviour. Du Picq was an infantry officer of the French army, a veteran of the Crimea and Algeria, who was to be killed outside Metz in August 1870. In strict fact, his military career was not all that distinguished, and a great deal of what he had to say about armies and battle now looks, or has been made to look by a century of warfare, almost deliberately perverse; he believed, for example, in the necessity of cultivating an officer-aristocracy. Nevertheless, his military-historical method was unique; wishing to get at the 'truth' about battle, he circulated among his brother officers a questionnaire soliciting their precise answers to a long list of very detailed inquiries about what had happened to them and their soldiers when in close contact with the enemy. The questionnaire was not a success, most who received it finding its tone impertinent or its completion tedious. But the questions were intelligent and original and, when applied by du Picq (whose rebuff by his brother officers had not extinguished his curiosity) to documentary material, elicited fascinating answers. It was upon the work of the Ancients, particularly Polybius, a follower of Thucydides, that he concentrated, for they, he felt, were franker than moderns about why and how disasters happened in war: why men ran away and what happened when they did. The conclusions to which he came were not wholly original, since Marshal de Saxe for one and Guibert for another, had anticipated him by a century in denying that men ran away as a result of 'shock'. But he much elaborated that denial and had a great deal to say about death on the battlefield and the 'will to combat'.

Du Picq did not believe in 'shock' – the collision of masses of armed men – for two reasons, one good, one less good. The good one was his demonstration, from documentary evidence, that large masses of soldiers do not smash into each other, either because one gives way at the critical moment, or because the attackers during the advance to combat lose their fainthearts and arrive at the point of contact very much inferior in numbers to the mass they are attacking. In either case, the side which turns and runs does so not because it has been physically shaken but because its nerve has given. The less good reason depends upon a more complicated argument: disciplined bodies of 'civilized' soldiers, he said, always beat undisciplined bodies of barbarians. Yet barbarians, man for man, are fiercer fighters than civilized soldiers. Therefore, no contact has taken place between a civilized force and the barbarian force which it beats, for if they had actually crossed swords the civilized force would have been beaten.

This argument is less good because he fails to show that barbarians are, mass for mass, better than civilized soldiers in hand-to-hand fighting. But he goes on from his general denial of the reality of shock to demonstrate, convincingly, a more illuminating truth about the nature of battle: that soldiers die in largest numbers when they run, because it is when they turn their backs to the enemy that they are least able to defend themselves. It is their rational acceptance of the dangers of running that makes civilized soldiers so formidable, he says, that and the discipline which has them in its bonds. And by discipline he does not mean the operation of an abstract principle but the example and sanctions exercised by the officers of an organized force. Men fight, he says in short, from fear: fear of the consequences first of not fighting (i.e. punishment), then of not fighting well (i.e. slaughter).

Du Picq's ideas were, after his death, and in an exaggerated and misinterpreted form, adopted by the French army. But they struck their most lasting response as ideas in America, where his proclamation of the dominance of fear over events on the battlefield was welcomed as much for its refreshing frankness as for its apparent truth. Fear is something everyone can understand, and fear was what thousands of American soldiers had patently felt on the battlefields of the Civil War, sometimes at moments inconvenient to their commanders. That war had already produced, by the outbreak of the First World War, a remarkable crop of soldier's literature, in which battle had been depicted very much from the private's rather than the general's angle of vision, and many of the authors had not disguised how frightened they had been. When, on America's entry into the Second World War, the United States Army decided to record in detail its war effort – something it had not done for the First – it assembled a group of historians, some soldiers, some not, who decided from the outset that in retailing the history of combat – as opposed to grand strategy or logistics – their approach should be du Picq's. Since they were Americans of their period –

patriotic, populist, self-confident, immensely optimistic – they took it as axiomatic that it was the spirit of armies which determined their success or failure, and that the spirit of America guaranteed the success of its army. The guiding theme of their history would therefore be an examination of how the American soldier overcame his fears to do his duty.

The conclusions to which the American Historical Teams came, as a result of many thousands of interviews with individuals and groups fresh from combat, are now widely known. They form the basis of the magnificent American campaign histories and have been publicized in pungent, capsule form by the leading historian of the European Theatre, General S. L. A. Marshall. Marshall is, in a sense, an American du Picq, in that, although owing to him his *idée de base* – that the battlefield is a place of terror – he has come to a radically different view of how the soldier's fears of it should be overcome. Both he and du Picq believe that an army is a genuine social organism, governed by its own social laws, and that formal discipline, imposed from above, is of limited utility in getting men to fight. But du Picq, though he uses a phrase which no doubt caught Marshall's fancy – that soldiers must develop a 'mutual acquaintanceship which establishes pride' – sees the suppression of fear chiefly as the officer's task. Marshall, in a manner distinctively American, believes it a function which falls upon everyone in the firing line. 'Whenever one surveys the forces of the battlefield,' he wrote in his masterpiece, *Men Against Fire*, 'it is to see that fear is general among men, but to observe further that men are commonly loath that their fear will be expressed in specific acts which their comrades will recognize as cowardice. The majority are unwilling to take extraordinary risks and do not aspire to a hero's role, but they are equally unwilling that they should be considered the least worthy among those present.' It is therefore, in Marshall's view, vital that an army should foster the closest acquaintance among its soldiers, that it should seek to create groups of friends, centred if possible on someone identified as a 'natural' fighter, since it is their 'mutual acquaintanceship' which will ensure no one flinches or shirks. 'When a soldier is ... known to the men who are around him, he ... has reason to fear losing the one thing he is likely to value more highly than life – his reputation as a man among other men.'

Verdict or Truth?

There is more to Marshall's historical method than an acceptance of the prevalence of fear in the hearts of soldiers on the battlefield. His work with infantrymen fresh from combat, both against the Japanese in the Pacific islands and Germans in Normandy, revealed to him a startling discovery: that, even in 'highly motivated' units, and even when hard pressed, no more than about a quarter of all 'fighting' soldiers will use their weapons against the enemy. 'The army cannot unmake [Western man],' he wrote in *Men Against Fire*.

> It must reckon with the fact that he comes from a civilization in which aggression, connected with the taking of life, is prohibited and unacceptable. The teaching and ideals of that civilization are against killing, against taking advantage. The fear of aggression has been expressed to him so strongly and absorbed by him so deeply and pervadingly – practically with his mother's milk – that it is part of the normal man's emotional make-up. This is his greatest handicap when he enters combat. It stays his trigger-finger even though he is hardly conscious that it is a restraint upon him.

It is the underlying effect of these two basic assumptions of Marshall's – that all men are afraid on the battlefield, yet that most, despite their fear, remain products of their culture and its value-system – which lends to his battle-narratives their original and unmistakable flavour. It is a flavour we can begin to call distinctively American, for his influence on military historians in his own country, particularly those who learnt their trade in the Army Historical Teams, is becoming marked. It is also appropriately American, for a focus of interest upon the common soldier, rather than upon the commander, upon the acts of the majority, rather than the decisions of a few, accords both with the spirit of American life and with the traditions of American historical scholarship.

But there are limits nevertheless to the usefulness and general applicability of the Marshall method. For his ultimate purpose in writing was not merely to describe and analyse – excellent though his description and analysis are – but to persuade the American army that it was fighting its wars the wrong way. It was his conviction that success in battle depended upon structuring an army correctly; and in arguing his case for a new structure of small groups or 'fire teams' centred on a 'natural fighter', he was undoubtedly guilty of over-emphasis and special pleading. His arguments were consonantly effective, so that he has had the unusual experience, for a historian, of seeing his message not merely accepted in his own lifetime but translated into practice. But, almost for that reason, they are arguments of which the academic historian, trained not to simplify but to portray the complexity of human affairs, ought to beware. A dose of Marshall is a useful corrective but it is not a cure-all for the ills of military history.

Nor would it be a cure-all to forswear the 'win/lose' approach of the Decisive Battlers, or the narrative focus on the doings of generals – 'strategocentric' narrative to give it a name if one is needed – bequeathed by Caesar. One clearly has to come to a judgement in writing about battles, as about anything else, and it would be perverse to ignore, or even to minimize, the influence on events of the directing class. Rather, it is over the question of *how* one should come to a judgement and in what light one should cast the central characters that the crucial touch has to be found. There cannot be any hard and fast rules. But there can be suggestions and useful analogies. The most useful, to my mind, is that of the difference between the English and French judicial systems. In England (and America), the task of the court in criminal cases, which it devolves upon a jury, is to arrive at a verdict of 'guilty' or 'not guilty' on the evidence presented by prosecuting and defending counsel in turns. Trials are conflicts and verdicts are decisions; the two sides 'win' or 'lose'. In France, and other countries which observe Roman Law, the task of the court in a criminal case is to arrive at the truth, as far as it can be perceived by human eyes, and the business of establishing the outlines of the truth falls not on a jury, which is strictly asked to enter a judgement, but upon a *juge d'instruction*. This officer of the court, unknown to English law, is accorded very wide powers of interrogation – of the suspect, his family, his associates – and of investigation – of the circumstances and scene of the crime – at which the suspect is often required to participate in a reconstruction. Only when the *juge* is satisfied that a crime has indeed occurred and that the suspect is responsible will he allow the case to go forward for prosecution.

The character of these two different legal approaches is usually defined as 'accusatorial' (English) and 'inquisitorial' (French) respectively. And it may well be that the dramatic accusatorial element in the English approach has had its effect – Creasy, after all, was a barrister – on the form in which English and, until recently, American military historiography has been cast. For most British military historians, as we have seen, implicitly put someone or something – a general or an army – in the dock, charge him or it with a crime – defeat if a friend, victory if an enemy – and marshal the evidence to show his or its responsibility. Indeed, given the accusatorial approach, there is not much else a historian can do. The inquisitorial approach, on the other hand, confers – or would confer, one is constrained to say, so infrequently is it adopted – very much greater freedom of action.

It would allow the historian, for example, to discuss battles not necessarily as conflicts for a decision, but as value-free events – for it is as events that they appear to many participants and to most non-combatant spectators – and if one began from their unpartisan stance one might well hit on a clearer view of what real significance it was that a battle held. The inquisitorial approach would also free the historian to discuss, for example, in what sense a given battle, so called, had taken place. The Battle of the Marne, it has been pointed out, was not something of which the Germans were aware at the time of fighting, and Telford Taylor has gone a long way in demonstrating that the Battle of Britain, which Churchill had suspected was 'about to begin' in June 1940, never, as far as Hitler was concerned, seriously got under way. To pursue the legal theme, the inquisitorial approach might also lead a historian to undertake a true piece of detective work, tracing messages from source to recipient, relating the times of their arrival and departure to the passage of events on the battlefield and so arriving at some balanced judgement of how influential a commander was in the determination of a battle's outcome. This is something at which the best naval historians are very accomplished and which has also been done, for example, as a study of the battle of the Ardennes, in a regular war-game at the American Command and Staff College. But it is attempted by few military historians and then sketchily, despite Tolstoy's provocative denial, in *War and Peace*, that generals influence the outcome of battles in any way at all.

The inquisitorial approach offers still larger freedoms than these, even though it also imposes wearisome burdens. It offers the freedom to consider, for instance, the long-term effects that a major battle, like any other sudden and violent occurrence, may exert on national and cultural attitudes. Just as the Lisbon earthquake is said to have given a timely stimulus to religious observance in eighteenth-century Europe, it is often adumbrated that the battle of Stalingrad has been the most important single lesson in the education of a democratic Germany. How true is that? The repression of the Paris Commune in 1871 undoubtedly left scars on the psyche of working-class Paris which ache to this day. But what exactly did the battle of France in 1940 do to the psyche of the French nation? The very size of the question ought not to deter the historian from attempting an answer. A few have already had a bite at it. Alistair Horne had tried to demonstrate that the experience of Verdun in 1916 led the French, by way of the building of the Maginot line, to the construction of the fortress of Dien Bien Phu, and its fall to the collapse of their colonial empire. But a real examination requires more room than he had left himself at the end of his book on Verdun, and a very special sort of historical expedition: not so much a plunge into the archives as a voyage through a nation's literature, from Sartre's *La Mort dans l'Ame* via the script of *Les Jeux Interdits* to the soundtrack of *Le Chagrin et la Pitié.*

The treatment of battle in fiction is a subject almost untouched by literary critics, but one which the military historian, with his specialized ability to check for veracity and probability, might very well think of tackling. He might also think of relating battles more closely to the social context of their own times. How violent, for example, was the society, and more specifically the class, from which Wellington's Peninsula scum were recruited? How sacrificial in general was the mood of Europe in 1914, when commanders by the score – it would be simple to compile a list – lost sons and sons-in-law in the battles they were directing, yet continued without flinching at their posts? How precisely, rather than in broad terms, did losses – the human result through which battles make their effect – intrude upon the feelings of a locality which suffered them in sudden excess – the Nord and the Pas de Calais, say, after the Battle of Morhange in August 1914, or Belfast after the Battle of the Somme – and how abiding was the demographic damage? More positively, how enduring were the bonds which a particular battle forged among the men who survived it and how important for their lives in after years? To have been 'out' in 1916 was a necessary passport into political life in independent Ireland and, slightly less obviously, to have been in the 2nd Free French Armoured Division in the Battle of Normandy was to draw a ticket in the eventual triumph of Gaullism. But one is also aware that there exists a subtle, unspoken regard for each other, a readiness to protect, if not to further each other's interests among men who have 'been through the same show' about which it would be very illuminating for the historian to know.

These are only some of the directions in which the study of battle, it seems to me, might be enlarged. I am tempted by many of them, but realize that their dimensions and my limitations put most of them beyond my reach. What I mean to attempt here is something altogether smaller, though still, I think, important: to tackle again the concept of the 'battle piece' and to suggest ways in which it might be wrenched out of the stereotype into which it has been set for so long by custom and unreflective imitation. I do not intend to write about generals or generalship, except to discuss how a commander's physical presence on the field may have influenced his subordinates' will to combat. I do not intend to say anything of logistics or strategy and very little of tactics in the formal sense. And I do not intend to offer a two-sided picture of events, since what happened to one side in any battle I describe will be enough to convey the features I think are salient. On the other hand, I do intend to discuss wounds and their treatment, the mechanics of being taken prisoner, the nature of leadership at the most junior level, the role of compulsion in getting men to stand their ground, the incidence of accidents as a cause of death in war and, above all, the dimensions of the danger which different varieties of weapons offer to the soldier on the battlefield. Crudely, but I think meaningfully, one may distinguish three sorts of battlefield weapons: the hand weapon – sword or lance; the single-missile weapon – musket or rifle; the multi-missile weapon – machine-gun or projector of toxic-gas particles. I have chosen three battles to describe in detail – Agincourt, Waterloo, the Somme – my basis of choice being availability of evidence, and my purpose to demonstrate, as exactly as possible, what the warfare, respectively, of hand, single-missile and multiple-missile weapons was (and is) like, and to suggest how and why the men who have had (and do have) to face these weapons control their fears, staunch their wounds, go to their deaths. It is a personal attempt to catch a glimpse of the face of battle.

2 Agincourt, 25 October 1415

Agincourt is one of the most instantly and vividly visualized of all epic passages in English history, and one of the most satisfactory to contemplate. It is a victory of the weak over the strong, of the common soldier over the mounted knight, of resolution over bombast, of the desperate, cornered and far from home, over the proprietorial and cocksure. Visually it is a pre-Raphaelite, perhaps better a Medici Gallery print battle – a composition of strong verticals and horizontals and a conflict of rich dark reds and Lincoln greens against fishscale greys and arctic blues. It is a school outing to the Old Vic, Shakespeare is fun, *son-et-lumière*, blank verse, Laurence Olivier in armour battle; it is an episode to quicken the interest of any schoolboy ever bored by a history lesson, a set-piece demonstration of English moral superiority and a cherished ingredient of a fading national myth. It is also a story of slaughter-yard behaviour and of outright atrocity.

The Campaign

The events of the Agincourt campaign are, for the military historian, gratifyingly straightforward to relate. For, as medieval battles go, it is surprisingly well-documented: the chronology can be fixed with considerable accuracy, the exact location of the culminating battle has never been in dispute, its topography has altered little over five hundred years, and there is less than the usual wild uncertainty over the numbers engaged on either side.

In the late summer of 1415 Henry V, twenty-seven years old and two years King of England, embarked on an invasion of

France. He came to renew by force the claims of his house to the lands it had both won and lost during the previous century in the course of what we now call the Hundred Years War. England had not, of course, lost all her French possessions. She retained Calais and its hinterland and Bordeaux, together with a large enclave behind it and along the coast to the south: it is an area now represented by all or parts of the departments of the Landes, Basses-Pyrénées, Gironde, Dordogne, Charente and Charente-Maritime. But the possessions to which she had been given title in 1360 at the Treaty of Bretigny, which concluded Edward III's campaign of conquest, were very much wider, embracing in Poitou and Aquitaine north and east of Bordeaux almost a third of the territory of France. It was these lands which Henry V was bent on repossessing, though he was also prepared to revive, it would appear, English claims to the Duchy of Normandy, of which King John had been disinherited in 1204.

What military strategy he had in mind for the campaign can only be reconstructed by conjecture. The contracts struck with the leaders of the major contingents accompanying him alluded to operations both in northern and southern France but it seems unlikely that he intended to strike deep into the heart of France at the immediate outset. Long-distance offensives of that sort had worn the heart and strength out of several English armies during the last thirty years of the previous century and allowed the French, who had deliberately and persistently refused battle during the period now called the Duguesclin war (after the Constable of France whose Fabian policy this was), to reduce piecemeal the extensive network of walled towns and castles through which England held her French dominions. Henry's plan seems to have been the exact contrary to that of John of Gaunt and the Black Prince. He would embark on mobile operations only after he secured a firm base, and he would seek to establish that base at the end of the shortest possible sea-route. This decision limited his choice of disembarkment place to the coasts of Normandy, Picardy, Artois or Flanders. Much the same set of considerations would cause the British and American planners of the D-Day landings to plump in their case for Normandy. Henry chose the Bay of the Seine and the port of Harfleur.

The army embarked in the second week of August at Portsmouth and set sail on 11 August. It had been gathering since April, while Henry conducted deliberately inconclusive negotiations with Charles VI, and now numbered about 10,000 in all, 8,000 archers and 2,000 men-at-arms, exclusive of camp followers. A good deal of the space in the ships, of which there were about 1,500, was given over to impedimenta and a great deal to the expedition's horses: at least one for each man-at-arms, and others for the baggage train and wagon teams. The crossing took a little over two days and on the morning of 14 August the army began to disembark, unopposed by the French, on a beach three miles west of Harfleur. Three days were taken to pitch camp and on 18 August the investment of the town began. It was not strongly garrisoned but its man-made and natural defences were strong, the Seine, the River Lézarde and a belt of marshes protecting it on the south, north and east. An attempt at mining under the moat on the western front was checked by French counter-mines so the small siege train, which contained at least three guns, undertook a bombardment of that section of the walls. It lasted for nearly a month, until the collapse of an important gate-defence, the repulse of a succession of sorties and the failure of a French relieving army to appear, convinced the garrison that they must surrender. After parleys, the town opened its gates to Henry on Sunday, 22 September.

He now had his base, but was left with neither time nor force enough to develop much of a campaign that year; at least a third of his army was dead or disabled, chiefly through disease, and the autumnal rains were due. Earlier in September he had set to paper his intention of marching down the Seine to Paris and thence to Bordeaux as soon as Harfleur fell; that had clearly become unfeasible, but honour demanded that he should not leave France without making a traverse, however much more circumspect, of the lands he claimed. At a long Council of War, held on 5 October, he convinced his followers that they could both appear to seek battle with the French armies which were known to be gathering and yet safely out-distance them by a march to the haven of Calais. On 8 October he led the army out.

His direct route was about 120 miles and lay across a succession of rivers, of which only the Somme formed a major obstacle. He began following the coast as far as the Béthune, which he crossed on 11 October, revictualling his army at Arques. The following day he crossed the Bresle, near Eu, having made eighty miles in five days, and on 13 October swung inland to cross the Somme above its estuary. On approaching, however, he got his first news of the enemy and it was grave; the nearest crossing was blocked and defended by a force of 6,000. After discussion, he rejected a retreat and turned south-east to follow the line of the river until he found an unguarded ford. For the next five days, while his army grew hungrier, the French kept pace with him on the northern bank until on the sixth, by a forced march across the plain of the Santerre (scene of the great British tank battle on 8 August 1918), he got ahead of them and found a pair of unguarded though damaged causeways at Bethencourt and Voyennes. Some hasty sappering made them fit for traffic and that evening, 19 October, the army slept on the far bank. Henry declared 20 October a day of rest, which his men badly needed, having marched over 200 miles in twelve days, but the arrival of French heralds with a challenge to fight was a reminder that they could not linger. On 21 October they marched eighteen miles, crossing the tracks of a major French army and, during the three following days, another fifty-three. They were now within two, at most three, marches of safety. All were aware, however, that the French had caught up and were keeping pace on their right flank. And late in the day of 24 October scouts came back with word that the enemy had crossed their path and were deploying for battle ahead of them. Henry ordered his men to deploy also but, as darkness was near, the French eventually stood down and withdrew a little to the north where they camped astride the road to Calais.

The English army found what shelter it could for the night in and around the village of Maisoncelles, ate its skimpy rations, confessed its sins, heard Mass and armed for battle. At first light knights and archers marched out and took up their positions

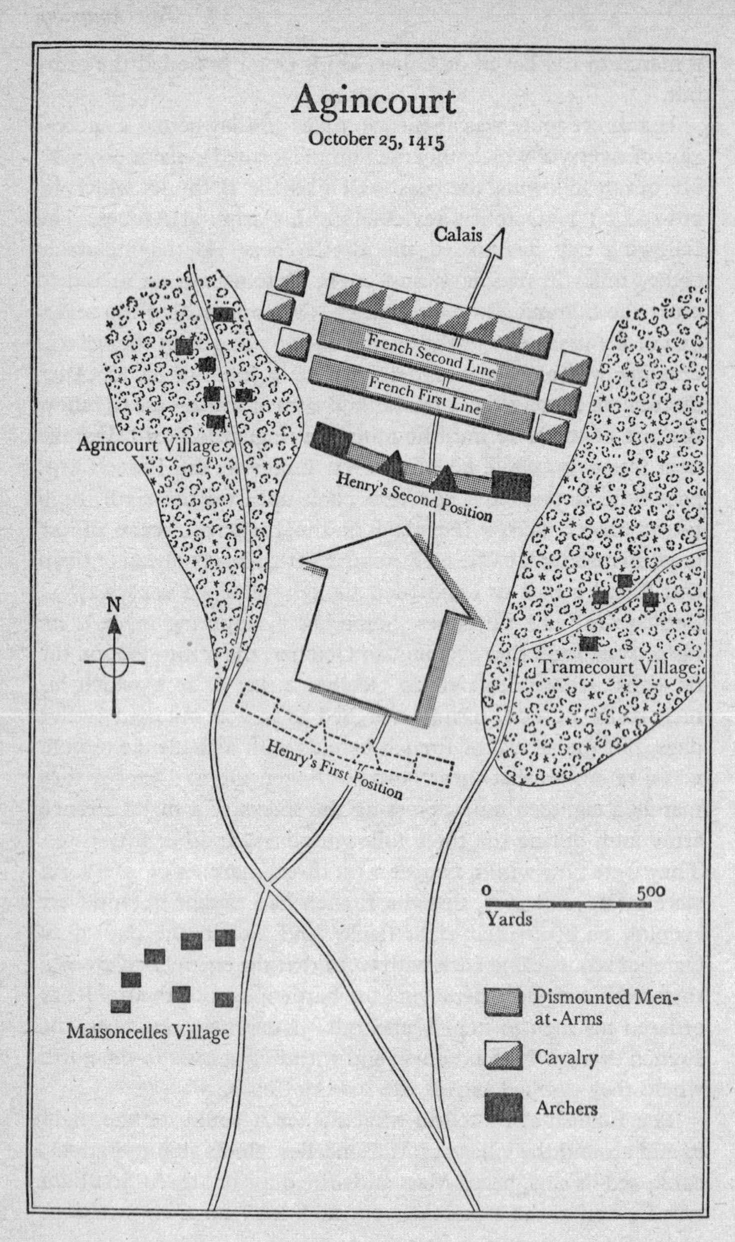
Agincourt
October 25, 1415
Calais
French Second Line
French First Line
Agincourt Village
Henry's Second Position
N
Tramecourt Village
Henry's First Position
0
500
Yards
Maisoncelles Village
Dismounted Men-at-Arms
Cavalry
Archers

between two woods. The French army, composed almost exclusively of mounted and dismounted men-at-arms, had deployed to meet them and was in similar positions about 1,000 yards distant. For four hours both armies held their ground. Henry apparently hoped that the French would attack him; they, who knew that sooner or later he would have to move – either to the attack, which suited their book, or to retreat, which suited them even better – stood or sat idle, eating their breakfasts and calling about cheerfully to each other. Eventually Henry decided to up sticks (literally: his archers had been carrying pointed stakes to defend their lines for the last week) and advance on the French line. Arrived within 300 yards – extreme bowshot – of the army, the English archers replanted their stakes and loosed off their first flights of arrows. The French, provoked by these arrow strikes, as Henry intended, into attacking, launched charges by the mounted men-at-arms from the wings of the main body. Before they had crossed the intervening space they were followed by the dismounted men-at-arms who, like them, were wearing full armour. The cavalry failed to break the English line, suffered losses from the fire of the archers, and turned about. Heading back for their own lines, many riders and loose horses crashed into the advancing line of dismounted men-at-arms. They, though shaken, continued to crowd forward and to mass their attack against the English men-at-arms, who were drawn up in three groups with archers between them and on the right and left flank. Apparently disdaining battle with the archers, although they were suffering losses from their fire, the French quickened their steps over the last few yards and crashed into the middle of the English line. For a moment it gave way. But the French were so tightly bunched that they could not use their weapons to widen the breach they had made. The English men-at-arms recovered their balance, struck back and were now joined by numbers of the archers, who, dropping their bows, ran against the French with axes, mallets and swords, or with weapons abandoned by the French they picked up from the ground. There followed a short but very bloody episode of hand-to-hand combat, in which freedom of action lay almost wholly with the English.

Many of the French armoured infantrymen lost their footing and were killed as they lay sprawling; others who remained upright could not defend themselves and were killed by thrusts between their armour-joints or stunned by hammer-blows. The French second line which came up, got embroiled in this fighting without being able to turn the advantage to their side, despite the addition they brought to the very great superiority of numbers the French already enjoyed. Eventually, those Frenchmen who could disentangle themselves from the mêlée made their way back to where the rest of their army, composed of a third line of mounted men-at-arms, stood watching. The English who faced them did so in several places, over heaps of dead, dying or disabled French men-at-arms, heaps said by one chronicler to be taller than a man's height. Others were rounding up disarmed or lightly wounded Frenchmen and leading them to the rear, where they were collected under guard.

While this went on, a French nobleman, the Duke of Brabant, who had arrived late for the battle from a christening party, led forward an improvised charge; but it was broken up without denting the English line, which was still drawn up. Henry had prudently kept it under arms because the French third line – of mounted men – had not dispersed and he must presumably have feared that it would ride down on them if the whole English army gave itself up to taking and looting prisoners. At some time in the afternoon, there were detected signs that the French were nerving themselves to charge anyhow; and more or less simultaneously, a body of armed peasants, led by three mounted knights, suddenly appeared at the baggage park, inflicted some loss of life and stole some objects of value, including one of the King's crowns, before being driven off.

Either that incident or the continued menace of the French third line now prompted Henry to order that all the prisoners instantly be killed. The order was not at once obeyed, and for comprehensible reasons. Even discounting any moral or physical repugnance on the part of their captors, or a misunderstanding of the reasons behind the order – that the prisoners might attack the English from the rear with weapons retrieved from the ground if the French cavalry were suddenly to attack their front – the poorer English soldiers, and perhaps not only the poorer, would have been very reluctant to pass up the prospects of ransom which killing the prisoners would entail. Henry was nevertheless adamant; he detailed an esquire and 200 archers to set about the execution, and stopped them only when it became clear that the French third line was packing up and withdrawing from the field. Meantime very many of the French had been killed; some of the English apparently even incinerated wounded prisoners in cottages where they had been taken for shelter.

The noblest and richest of the prisoners were, nevertheless, spared and dined that evening with the King at Maisoncelles, his base of the previous evening, to which he now returned. En route he summoned the heralds of the two armies who had watched the battle together from a vantage point, and settled with the principal French herald a name for the battle: Agincourt, after the nearest fortified place. Next morning, after collecting the army, marshalling the prisoners and distributing the wounded and the loads of booty among the transport, he marched the army off across the battlefield towards Calais. Numbers of the French wounded had made their way or been helped from the field during the night; those still living, unless thought ransomable, were now killed. On 29 October, the English, with two thousand prisoners, reached Calais. The King left for England at once, to be escorted into London by an enormous party of rejoicing citizens.

These are the bare outlines of the battle, as recorded by seven or eight chroniclers, who do not materially disagree over the sequence, character or significance of events. Of course, even though three of them were present at the scene, none was an eye-witness of everything, or even of very much, that happened. An army on the morrow of a battle, particularly an army as small as that of Agincourt, must, nevertheless, be a fairly efficient clearing-house of information, and it seems probable that a broadly accurate view of what had happened – though not necessarily why and how it had happened – would quickly crystallize in the mind of any diligent interrogator, while a popularly agreed version, not dissimilar from it, would soon circulate within, and outside, the ranks. It would seem reasonable therefore to believe that the narrative of Agincourt handed down to us is a good one; it would in any case be profitless to look for a better.

The Battle

What we almost completely lack, though, is the sort of picture and understanding of the practicalities of the fighting and of the mood, outlook and skills of the fighters which were themselves part of the eye-witness chroniclers' vision. We simply cannot visualize, as they were able to do, what the Agincourt arrow-cloud can have looked, or sounded like; what the armoured men-at-arms sought to do to each other at the moment of the first clash; at what speed and in what density the French cavalry charged down; how the mêlée – the densely packed mass of men in hand-to-hand combat – can have appeared to a detached onlooker, say to men in the French third line; what level the noise of the battle can have reached and how the leaders made themselves heard – if they did so – above it. These questions lead on to less tangible inquiries: how did leadership operate once the fighting had been joined – by exhortation or by example? Or did concerted action depend upon previously rehearsed tactics and corporate feeling alone? Or was there, in fact, no leadership, merely every man – or every brave man – for himself? Less tangible still, what did 'bravery' mean in the context of a medieval fight? How did men mentally order the risks which they faced, as we know it is human to do? Were the foot more likely to be frightened of the horses, or of the men on them? Were the armoured men-at-arms more or less frightened of the arrows than of meeting their similarly clad opponents at a weapon's length? Did it seem safer to go on fighting once hard pressed than to surrender? Was running away more hazardous than staying within the press of the fighting?

The answers to some of these questions must be highly conjectural, interesting though the conjectures may be. But to others, we can certainly offer answers which fall within a fairly narrow bracket of probability, because the parameters of the questions are technical. Where speed of movement, density of formations, effect of weapons, for example, are concerned, we can test our suppositions against the known defensive qualities of armour plate, penetrative power of arrows, dimensions and capacities of the human body, carrying power and speed of the horse. And from reasonable probabilities about these military mechanics, we may be able to leap towards an understanding of the dynamics of the battle itself and the spirit of the armies which fought it.

Let us, to begin with, and however artificially, break the battle down into a sequence of separate events. It opened, as we know, with the armies forming up in the light of early morning: whether that meant just after first light, or at the rather later hour of dawn itself – about 6.40 a.m. – is a point of detail over which we cannot expect the chroniclers to meet Staff College standards of precision. Nor do they. They are even more imprecise about numbers, particularly as they concern the French. For though there is agreement, supported by other evidence, that Henry's army had dwindled to about 5 or 6,000 archers and 1,000 men-at-arms, the French are variously counted between 10,000 and 200,000. Colonel Burne convincingly reconciles the differences to produce a figure of 25,000, a very large proportion of which represented armoured men-at-arms. Of these, about 1,000 brought their horses to the battlefield; the rest were to fight on foot.

The two armies initially formed up at a distance of some thousand yards from each other; at either end of a long, open and almost flat expanse of ploughland, bordered on each side by woodland. The width of the field, which had recently been sown with winter wheat, was about 1,200 yards at the French end. The woods converged slightly on the English and, at the point where the armies were eventually to meet, stood about 900 to 1,000 yards apart. (These measurements suppose – as seems reasonable, field boundaries remaining remarkably stable over centuries – that the outlines of the woods have not much changed.)

The English men-at-arms, most of whom were on foot, took station in three blocks, under the command of the Duke of York, to the right, the King, in the centre, and Lord Camoys, on the left. The archers were disposed between them and also on the flanks; the whole line was about four or five deep. The archer flanks may have been thrown a little forward, and the archers of the two inner groups may have adopted a wedge-like formation. This would have made it appear as if the men-at-arms were deployed a little to their rear. Opposite them, the French were drawn up in three lines, of which the third was mounted, as were two groups, each about 500 strong, on the flanks. The two forward lines, with a filling of crossbowmen between and some ineffectual cannon on the flanks were each, perhaps, 8,000 strong, and so ranked some eight deep. On both sides, the leaders of the various contingents – nobles, bannerets and knights – displayed armorial banners, under which they and their men would fight, and among the French there was a great deal of tiresome struggling, during the period of deployment, to get these banners into the leading rank.

Deployed, the armies were ready for the battle, which, as we have seen, resolved itself into twelve main episodes: a period of waiting; and English advance; an English arrow strike; a French cavalry charge; a French infantry advance; a mêlée between the French and English men-at-arms; an intervention in the mêlée by the English archers; the flight of the French survivors from the scene of the mêlée; a second period of waiting, during which the French third line threatened, and a small party delivered, another charge; a French raid on the baggage park; a massacre of the French prisoners; finally, mutual departure from the battlefield. What was each of these episodes like, and what impetus did it give to the course of events?

The period of waiting – three or four hours long, and so lasting probably from about seven to eleven o'clock – must have been very trying. Two chroniclers mention that the soldiers in the front ranks sat down and ate and drank and that there was a good deal of shouting, chaffing and noisy reconciliation of old quarrels among the French. But that was after they had settled, by pushing and shoving, who was to stand in the forward rank; not a real argument, one may surmise, but a process which put the grander and the braver in front of the more humble and timid. There is no mention of the English imitating them, but given their very real predicament, and their much thinner line of battle, they can have felt little need to dispute the place of honour among themselves. It is also improbable that they did much eating or drinking, for the army had been short of food for nine days and the archers are said to have been subsisting on nuts and berries on the last marches. Waiting, certainly for the English, must then have been a cold, miserable and squalid business. It had been raining, the ground was recently ploughed, air temperature was probably in the forties or low fifties Fahrenheit and many in the army were suffering from diarrhoea. Since none would presumably have been allowed to leave the ranks while the army was deployed for action, sufferers would have had to relieve themselves where they stood. For any afflicted man-at-arms wearing mail leggings laced to his plate armour, even that may not have been possible.

The King's order to advance, which he gave after the veterans had endorsed his guess that the French would not be drawn, may therefore have been generally and genuinely welcome. Movement at least meant an opportunity to generate body heat, of which the metal-clad men-at-arms would have dissipated an unnatural amount during the morning. Not, however, when the moment came, that they would have moved forward very fast. An advance in line, particularly by men unequally equipped and burdened, has to be taken slowly if order is to be preserved. The manoeuvre, moreover, was a change of position, not a charge, and the King and his subordinate leaders would presumably have recognized the additional danger of losing cohesion in the face of the enemy who, if alert, would seize on the eventuality as an opportune moment to launch an attack. Several chroniclers indeed mention that on the King's orders a knight, Sir Thomas Erpingham, inspected the archers before they marched off in order to 'check their dressing', as a modern drill sergeant would put it, and to ensure that they had their bows strung. The much smaller groups of men-at-arms would have moved as did the banners of their lords, which in turn would have followed the King's.

The army had about 700 yards of rain-soaked ploughland to cover. At a slow walk (no medieval army marched in step, and no modern army would have done so over such ground – the 'cadenced pace' followed from the hardening and smoothing of the surface of roads), with halts to correct dressing, it would have reached its new position in ten minutes or so, though one may guess that the pace slackened a good deal as they drew nearer the French army and the leaders made mental reckoning of the range. 'Extreme bowshot', which is the distance at which Henry presumably planned to take ground, is traditionally calculated at 300 yards. That is a tremendous carry for a bow, however, and 250 yards would be a more realistic judgement of the distance at which he finally halted his line from the French. If, however, his archer flanks were thrown a little forward, his centre would have been farther away; and if, as one chronicler suggests, he had infiltrated parties of bowmen into the woods, the gap between the two armies might have been greater still. Something between 250 and 300 yards is a reasonable bracket therefore.

There must now have ensued another pause, even though a short one. For the archers, who had each been carrying a stout double-pointed wooden stake since the tenth day of the march, had now to hammer these into the ground, at an angle calculated to catch a warhorse in the chest. Once hammered, moreover, the points had to be hastily resharpened. Henry had ordered these stakes to be cut as a precaution against the army being surprised by cavalry on the line of march. But it was a sensible improvisation to have them planted on the pitched battlefield, even if not a wholly original one. The Scots at Bannockburn, the English themselves at Crécy and the Flemings at Courtrai had narrowed their fronts by digging patterns of holes which would break the leg of a charging horse; the principle was the same as that which underlay the planting of the Agincourt archers' fence. Though it is not, indeed, possible to guess whether a fence was what the archers constructed. If they hammered their stakes to form a single row, it supposes them standing for some time on the wrong side of it with their backs to the enemy. Is it not more probable that each drove his in where he stood, so forming a kind of thicket, too dangerous for horses to penetrate but roomy enough for the defenders to move about within? That would explain the chronicler Monstrelet's otherwise puzzling statement that 'each archer placed before him a stake.' It would also make sense of the rough mathematics we can apply to the problem. Colonel Burne, whose appreciation has not been challenged, estimates the width of the English position at 950 yards. Given that there were 1,000 men-at-arms in the line of battle, ranked shoulder to shoulder four deep, they would have occupied, at a yard of front per man, 250 yards. If the 5,000 archers, on the remaining 700 yards, planted their stakes side by side, they would have formed a fence at five-inch intervals. That obstacle would have been impenetrable to the French – but also to the English archers;* and *their* freedom of movement was, as we shall see, latterly an essential element in the winning of the battle. If we want to picture the formation the archers adopted, therefore, it would be most realistic to think of them standing a yard apart, in six or seven rows, with a yard between them, also disposed chequerboard fashion so that the men could see and shoot more easily over the heads of those in front: the whole forming a loose belt twenty or thirty feet deep, with the stakes standing obliquely among them.

What we do not know – and it leaves a serious gap in our understanding of the mechanics of the battle – is how the archers were commanded. The men-at-arms stood beneath the banners of their leaders, who had anyhow mustered them and brought them to the war, and the larger retinues, those of noblemen like the Earl of Suffolk, also contained knighted men-at-arms, who must have acted as subordinate leaders. There is thus no difficulty in visualizing how command was exercised within these fairly small and compact groups – providing one makes allowances for what a modern world would regard as the unsoldierly habit in the man-at-arms of seeking to engage in 'single combat'

*Indeed, they could not have got back *behind* it after they had driven their stakes in.

and of otherwise drawing attention to his individual prowess and skill-at-arms. But if the 'officer class', even though the expression has a very doubtful meaning in the medieval military context, was wholly committed to the leadership of a single component of the army, who led the rest? For it is not naïve, indeed quite the contrary, to suppose some sort of control over and discipline within the archers' ranks. Had the groupings into twenties, under a double-pay 'vintenar' and of the twenties into hundreds, under a mounted and armoured 'centenar', which we know prevailed in the reign of Edward I, at the beginning of the fourteenth century, persisted into the fifteenth? That would be probable. But we cannot tell to whom the 'centenars' were immediately answerable, nor how the chain of command led to the King. We can only feel sure that it did.

Archers versus Infantry and Cavalry

The archers were now in position to open fire (an inappropriate expression, belonging to the gunpowder age, which was barely beginning). Each man disposed his arrows as convenient. He would have had a sheaf, perhaps two, of twenty-four arrows and probably struck them point down into the ground by his feet. The men in the front two ranks would have a clear view of the enemy, those behind only sporadic glimpses: there must therefore have been some sort of ranging order passed by word of mouth. For the archers' task at this opening moment of the battle was to provoke the French into attacking, and it was therefore essential that their arrows should 'group' as closely as possible on the target. To translate their purpose into modern artillery language, they had to achieve a very narrow 100° zone (i.e. that belt of territory into which *all* missiles fell) and a Time on Target effect (i.e. all their missiles had to arrive simultaneously).

To speculate about their feelings at this moment is otiose. They were experienced soldiers in a desperate spot; and their fire, moreover, was to be 'indirect', in that their arrows would not depart straight into the enemy's faces but at a fairly steeply angled trajectory. They need have had no sense of initiating an act of killing, therefore; it was probably their technical and professional sense which was most actively engaged in an activity which was still preliminary to any 'real' fighting that might come.

They must have received at least two orders: the first to draw their bows, the second to loose their strings. How the orders were synchronized between different groups of archers is an unanswerable question, but when the shout went up or the banner down, four clouds of arrows would have streaked out of the English line to reach a height of 100 feet before turning in flight to plunge at a steeper angle on and among the French men-at-arms opposite. These arrows cannot, however, given their terminal velocity and angle of impact, have done a great deal of harm, at least to the men-at-arms. For armour, by the early fifteenth century, was composed almost completely of steel sheet, in place of the iron mail which had been worn on the body until fifty years before but now only covered the awkward points of movement around the shoulder and groin. It was deliberately designed, moreover, to offer a glancing surface, and the contemporary helmet, a wide-brimmed 'bascinet', was particularly adapted to deflect blows away from the head and the shoulders. We can suppose that the armour served its purpose effectively in this, the opening moment of Agincourt. But one should not dismiss the moral effect of the arrow strike. The singing of the arrows would not have moved ahead of their flight, but the sound of their impact must have been extraordinarily cacophonous, a weird clanking and banging on the bowed heads and backs of the French men-at-arms. If any of the horses in the flanking squadrons were hit, they were likely to have been hurt, however, even at this extreme range, for they were armoured only on their faces and chests, and the chisel-pointed head of the clothyard arrow would have penetrated the padded cloth hangings which covered the rest of their bodies. Animal cries of pain and fear would have risen above the metallic clatter.

We can also imagine oaths and shouted threats from the French. For the arrow strike achieved its object. How quickly, the chroniclers do not tell us; but as a trained archer could loose a shaft every ten seconds we can guess that it took at most a few minutes to trigger the French attack. The French, as we know, were certain of victory. What they had been waiting for was a tactical pretext; either that of the Englishmen showing them their backs, or on the contrary, cocking a snook. One or two volleys would have been insult enough. On the arrival of the first arrows the two large squadrons of horse on either flank mounted – or had they mounted when the English line advanced? – walked their horses clear of the line and broke into a charge.

A charge at what? The two chroniclers who are specific about this point make it clear that the two groups of cavalry, each five or six hundred strong, of which that on the left hand was led by Clignet de Brébant and Guillaume de Saveuse, made the English archer flanks their target. Their aim, doubtless, was to clear these, the largest blocks of the enemy which immediately threatened them, off the field, leaving the numerically much inferior centre of English men-at-arms, with the smaller groups of their attendant archers, to be overwhelmed by the French infantry. It was nevertheless a strange and dangerous decision, unless, that is, we work on the supposition that the archers had planted their stakes among their own ranks, so concealing that array of obstacles from the French. We may then visualize the French bearing down on the archers in ignorance of the hedgehog their ranks concealed; and of the English giving ground just before the moment of impact, to reveal it.

For 'the moment of impact' otherwise begs an important, indeed a vital, question. It is not difficult to picture the beginning of the charge; the horsemen booting their mounts to form line, probably two or three rows deep, so that, riding knee to knee, they would have presented a front of two or three hundred lances, more or less equalling in width the line of the archers opposite, say 300 yards. We can imagine them setting off, sitting (really standing) 'long' in their high-backed, padded saddles, legs straight and thrust forward, toes down in the heavy stirrups, lance under right arm, left free to manage the reins (wearing plate armour obviated the need to carry a shield); and we can see them in motion, riding at a pace which took them across all but the last fifty of the two or three hundred yards they had to cover in forty seconds or so and then spurring their horses to ride down on the archers at the best speed they could manage – twelve or fifteen miles an hour.*

So far so good. The distance between horses and archers narrows. The archers, who have delivered three or four volleys at the bowed heads and shoulders of their attackers, get off one more flight. More horses – some have already gone down or broken back with screams of pain – stumble and fall, tripping their neighbours, but the mass drive on and ... and what? It is at this moment that we have to make a judgement about the difference between what happens in a battle and what happens in a violent accident. A horse, in the normal course of events, will not gallop at an obstacle it cannot jump or see a way through, and it cannot jump or see a way through a solid line of men. Even less will it go at the sort of obviously dangerous obstacles which the archers' stakes presented. Equally, a man will not stand in the path of a running horse: he will run himself, or seek shelter, and only if exceptionally strong-nerved and knowing in its ways, stand his ground. Nevertheless, accidents happen. Men, miscalculating or slow-footed, and horses, confused or maddened, do collide, with results almost exclusively unpleasant for the man. We cannot therefore say, however unnatural and exceptional we recognize collisions between man and horse to be, that nothing of that nature occurred between the archers and the French cavalry at Agincourt. For the archers were trained to 'receive cavalry', the horses trained to charge home, while it was the principal function of the *riders* to insist

*The horses were probably a big hunter type, not the carthorse of popular belief, and the weight they had to carry some 250 lbs (man 150 lbs, armour 60 lbs, saddle and trappings 40 lbs).

on the horses doing that against which their nature rebelled. Moreover, two of the eye-witness chroniclers, St Remy and the Priest of the Cottonian MS, are adamant that some of the French cavalry did get in among the archers.

The two opposed 'weapon principles' which military theorists recognized had, in short, both failed: the 'missile' principle, personified by the archers, had failed to stop or drive off the cavalry; they, embodying the 'shock' principle, had failed to crush the infantry – or, more particularly, to make them run away, for the 'shock' which cavalry seek to inflict is really moral, not physical in character. It was the stakes which must have effected the compromise. The French, coming on fast, and in great numbers over a short distance, had escaped the deaths and falls which should have toppled their charge over on itself; the English, emboldened by the physical security the hedgehog of stakes lent their formation, had given ground only a little before the onset; the horses had then found themselves on top of the stakes too late to refuse the obstacle; and a short, violent and noisy collision had resulted.

Some of the men-at-arms' horses 'ran out' round the flanks of the archers and into the woods. Those in the rear ranks turned their horses, or were turned by them, and rode back. But three at least, including Guillaume de Saveuse, had their horses impaled on the stakes, thumped to the ground and were killed where they lay, either by mallet blows or by stabs between their armour-joints. The charge, momentarily terrifying for the English, from many of whom French men-at-arms, twice their height from the ground, and moving at ten or fifteen miles an hour on steel-shod and grotesquely caparisoned war-horses, had stopped only a few feet distant, had been a disaster for the enemy. And, as they rode off, the archers, with all the violent anger that comes with release from sudden danger, bent their bows and sent fresh flights of arrows after them, bringing down more horses and maddening others into uncontrolled flight.

Infantry versus Infantry

But the results of the rout went beyond the demoralization of the survivors. For, as their horses galloped back, they met the first division of dismounted men-at-arms marching out to attack the English centre. Perhaps 8,000 strong, and filling the space between the woods eight to ten deep, they could not easily or quickly open their ranks to let the fugitives through. Of what happened in consequence we can get a clear idea, curiously, from a cinema newsreel of the Grosvenor Square demonstration against the Vietnam war in 1968. There, a frightened police horse, fleeing the demonstrators, charged a line of constables on foot. Those directly in its path, barging sideways and backwards to open a gap and seizing their neighbours, set up a curious and violent ripple which ran along the ranks on each side, reaching policemen some good distance away who, tightly packed, clutched at each other for support, and stumbled clumsily backwards and then forwards to keep their balance. The sensations of that ripple are known to anyone who has been a member of a dense, mobile and boisterous crowd and it was certainly what was felt, to a sudden and exaggerated degree, by the French men-at-arms in the face of that involuntary cavalry charge. As in that which had just failed against the archers, many of the horses would have shied off at the moment of impact. But those that barged in, an occurrence to which the chroniclers testify, broke up the rhythm of the advance and knocked some men to the ground, an unpleasant experience when the soil is wet and trampled and one is wearing sixty or seventy pounds of sheet metal on the body.

This interruption in an advance which should have brought the French first division to within weapons' length of the English in three or four minutes at most gave Henry's men-at-arms ample time to brace themselves for the encounter. It also gave the archers, both those in the large groups on the wings and the two smaller groups in the central wedges, the chance to prolong their volleying of arrows into the French ranks. The range was progressively shortened by the advance, and the arrows, coming in on a flat trajectory in sheets of 5,000 at ten-second intervals, must have begun to cause casualties among the French foot. For though they bowed their heads and hunched their shoulders, presenting a continuous front of deflecting surface (bascinet top, breastplate, 'taces' – the overlapping bands across the stomach and genitals – and leg-pieces) to the storm, some of the arrows must have found the weak spots in the visor and at the shoulders and, as the range dropped right down, might even have penetrated armour itself. The 'bodkin-point' was designed to do so, and its terminal velocity, sufficient to drive it through an inch of oak from a short distance, could also, at the right angle of impact, make a hole in sheet steel.

The archers failed nevertheless to halt the French advance. But they succeeded in channelling it – or helping to channel it – on to a narrower front of attack. For the French foot, unlike the cavalry, apparently did not make the archers' positions their objective. As their great mass came on, their front ranks 'either from fear of the arrows ... or that they might more speedily penetrate our ranks to the banners (of the King, the Duke of York and Lord Camoys) ... divided themselves into three ... charging our lines in the three places where the banners were.' We may also presume that the return of their own cavalry on the flanks would have helped to compress the infantry mass towards the centre, a tendency perhaps reinforced (we really cannot judge) by the alleged unwillingness of men-at-arms to cross weapons with archers, their social inferiors, when the chance to win glory, and prisoners, in combat with other mem-at-arms presented itself. Whatever the play of forces at work on the movement of the French first division, several narrators testify to the outcome. The leading ranks bunched into three assaulting columns and drove into what Colonel Burne, in a topographical analogy, calls the three 're-entrants' of the English line, where the men-at-arms were massed a little in rear of the archers' staked-out enclosures.

Their charge won an initial success, for before it the English men-at-arms fell back 'a spear's length'. What distance the chronicler means by that traditional phrase we cannot judge, and all the less because the French had cut down their lances in anticipation of fighting on foot. It probably implies 'just enough to take the impetus out of the onset of the French', for we must imagine them, although puffed by the effort of a jostling tramp across 300 yards of wet ploughland, accelerating over the last few feet into a run calculated to drive the points of their spears hard on to the enemy's chests and stomachs. The object would have been to knock over as many of them as possible, and so to open gaps in the ranks and isolate individuals who could then be killed or forced back on to the weapons of their own comrades; 'sowing disorder' is a short-hand description of the aim. To avoid its achievement, the English, had they been more numerous, might have started forward to meet the French before they developed impulsion; since they were so outnumbered, it was individually prudent and tactically sound for the men most exposed to trot backwards before the French spearpoints, thus 'wrong-footing' their opponents (a spearman times his thrust to coincide with the forward step of his left foot) and setting up those surges and undulations along the face of the French mass which momentarily rob a crowd's onrush of its full impact. The English, at the same time, would have been thrusting their spears at the French and, as movement died out of the two hosts, we can visualize them divided, at a distance of ten or fifteen feet, by a horizontal fence of waving and stabbing spear shafts, the noise of their clattering like that of a bully-off at hockey magnified several hundred times.

In this fashion the clash of the men-at-arms might have petered out, as it did on so many medieval battlefields, without a great deal more hurt to either side – though the French would have continued to suffer casualties from the fire of the archers, as long as they remained within range and the English had arrows to shoot at them (the evidence implies they must now have been running short). We can guess that three factors deterred the antagonists from drawing off from each other. One was the English fear of quitting their solid position between the woods and behind the archers' stakes for the greater dangers of the open field; the second was the French certainty of victory; the third was their enormous press of numbers. For if we accept that they had now divided into three *ad hoc* columns and that the head of each matched in width that of the English opposite – say eighty yards – with intervals between of about the same distance, we are compelled to visualize, taking a bird's-eye viewpoint, a roughly trident-shaped formation, the Frenchmen in the prongs ranking twenty deep and numbering some 5,000 in all, those in the base a shapeless and unordered mass amounting to, perhaps, another 3,000 – and all of them, except for the seven or eight hundred in the leading ranks, unable to see or hear what was happening, yet certain that the English were done for, and anxious to take a hand in finishing them off.

No one, moreover, had overall authority in this press, nor a chain of command through which to impose it. The consequence was inevitable: the development of an unrelenting pressure from the rear on the backs of those in the line of battle, driving them steadily into the weapon-strokes of the English, or at least denying them that margin of room for individual manoeuvre which is essential if men are to defend themselves – or attack – effectively. This was disastrous, for it is vital to recognize, if we are to understand Agincourt, that all infantry actions, even those fought in the closest of close order, are not, in the last resort, combats of mass against mass, but the sum of many combats of individuals – one against one, one against two, three against five. This must be so, for the very simple reason that the weapons which individuals wield are of very limited range and effect, as they remain even since missile weapons have become the universal equipment of the infantryman. At Agincourt, where the man-at-arms bore lance, sword, dagger, mace or battleaxe,* his ability to kill or wound was restricted to the circle centred on his own body, within which his reach allowed him to club, slash or stab. Prevented by the throng at their backs from dodging, side-stepping or retreating from the blows and thrusts directed at them by their English opponents, the individual French men-at-arms must shortly have begun to lose their man-to-man fights, collecting blows on the head or limbs which, even through armour, were sufficiently bruising or stunning to make them drop their weapons or lose their balance

*A category which includes glaive, bill, and similar weapons.

or footing. Within minutes, perhaps seconds, of hand-to-hand fighting being joined, some of them would have fallen, their bodies lying at the feet of their comrades, further impeding the movement of individuals and thus offering an obstacle to the advance of the whole column.

This was the crucial factor in the development of the battle. Had most of the French first line kept their feet, the crowd pressure of their vastly superior numbers, transmitted through their levelled lances, would shortly have forced the English back. Once men began to go down, however – and perhaps also because the French had shortened their lances, while the English had apparently not – those in the next rank would have found that they could get within reach of the English only by stepping over or on to the bodies of the fallen. Supposing continuing pressure from the rear, moreover, they would have had no choice but to do so; yet in so doing, would have rendered themselves even more vulnerable to a tumble than those already felled, a human body making either an unstable fighting platform or a very effective stumbling block to the heels of a man trying to defend himself from a savage attack to his front. In short, once the French column had become stationary, its front impeded by fallen bodies and its ranks animated by heavy pressure from the rear, the 'tumbling effect' along its forward edge would have become cumulative.

Cumulative, but sudden and of short duration: for pressure of numbers and desperation must eventually have caused the French to spill out from their columns and lumber down upon the archers who, it appears, were now beginning to run short of arrows. They could almost certainly not have withstood a charge by armoured men-at-arms, would have broken and, running, have left their own men-at-arms to be surrounded and hacked down. That did not happen. The chroniclers are specific that, on the contrary, it was the archers who moved to the attack. Seeing the French falling at the heads of the columns, while those on the flanks still flinched away from the final flights of arrows, the archers seized the chance that confusion and irresolution offered. Drawing swords, swinging heavier weapons – axes, bills or the mallets they used to hammer in their stakes – they left their staked-out positions and ran down to assault the men in armour.

This is a very difficult episode to visualize convincingly. They cannot have attacked the heads of the French columns, for it was there that the English men-at-arms stood, leaving no room for reinforcements to join in. On the flanks, however, the French cannot yet have suffered many casualties, would have had fairly unencumbered ground to fight on and ought to have had no difficulty in dealing with any unarmoured man foolish enough to come within reach of their weapons. The observation offered by two chroniclers that they were too tightly packed to raise their arms, though very probably true of those in the heart of the crowd, cannot apply to those on its fringes. If the archers did inflict injury on the men-at-arms, and there is unanimous evidence that they did, it must have been in some other way than by direct assault on the close-ordered ranks of the columns.

The most likely explanation is that small groups of archers began by attacking individual men-at-arms, infantry isolated by the scattering of the French first line in the 'reverse charge' of their own cavalry or riders unhorsed in the charge itself. The charges had occurred on either flank; so that in front of the main bodies of archers and at a distance of between fifty and 200 yards from them, must have been seen, in the two or three minutes after the cavalry had ridden back, numbers of Frenchmen, prone, supine, half-risen or shakily upright, who were plainly in no state to offer concerted resistance and scarcely able to defend themselves individually. Those who were down would indeed have had difficulty getting up again from slithery ground under the weight of sixty or seventy pounds of armour; and the same hindrances would have slowed those who regained or had kept their feet in getting back to the protection of the closed columns. Certainly they could not have outdistanced the archers if, as we may surmise, and St Remy, a combatant, implies, some of the latter now took the risk of running forward from their stakes to set about them.*

*'Soon afterwards, the English archers perceiving this disorder of the advance guard . . . and *hastening to the place where the fugitives came from*, killed and disabled the French.' (Author's italics.) Nicolas, *The History of the Battle of Agincourt*, p. 268.

'Setting about them' probably meant two or three against one, so that while an archer swung or lunged at a man-at-arms' front, another dodged his sword-arm to land him a mallet-blow on the back of the head or an axe-stroke behind the knee. Either would have toppled him and, once sprawling, he would have been helpless; a thrust into his face, if he were wearing a bascinet, into the slits of his visor, if he were wearing a closed helmet, or through the mail of his armpit or groin, would have killed him outright or left him to bleed to death. Each act of execution need have taken only a few seconds; time enough for a flurry of thrusts clumsily parried, a fall, two or three figures to kneel over another on the ground, a few butcher's blows, a cry *in extremis*. 'Two thousand pounds of education drops to a ten rupee ...' (Kipling, *Arithmetic on the Frontier*). Little scenes of this sort must have been happening all over the two narrow tracts between the woods and the fringes of the French main body within the first minutes of the main battle being joined. The only way for stranded Frenchmen to avoid such a death at the hands of the archers was to ask for quarter, which at this early stage they may not have been willing to grant, despite prospects of ransom. A surrendered enemy, to be put *hors de combat*, had to be escorted off the field, a waste of time and manpower the English could not afford when still at such an apparent disadvantage.

But the check in the front line and the butchery on the flanks appear fairly quickly to have swung the advantage in their favour. The 'return charge' of the French cavalry had, according to St Remy, caused some of the French to retreat in panic, and it is possible that panic now broke out again along the flanks and at the front.* If that were so – and it is difficult otherwise to make sense of subsequent events – we must imagine a new tide of movement within the French mass: continued forward pressure from those at the back who could not see, a rearward drift along the flanks of the columns by those who had seen all too clearly what work the archers were at, and a reverse pressure

*The sight of archers killing men-at-arms might either have provoked a counter-attack from the Frenchmen on the flanks *or* persuaded them individually that Agincourt had become no sort of battle to get killed in. There was no reputation to be won in fighting archers.

by men-at-arms in the front seeking, if not escape, at least room to fight without fear of falling, or being pushed, over the bodies of those who had already gone down. These movements would have altered the shape of the French mass, widening the gaps between its flanks and the woods, and so offering the archers room to make an 'enveloping' attack. Emboldened by the easy killings achieved by some of their number, we must now imagine the rest, perhaps at the King's command, perhaps by spontaneous decision, massing outside their stakes and then running down in formation to attack the French flanks.

'Flank', of course, is only the military word for 'side' (in French, from which we take it, the distinction does not exist) and the advantage attackers enjoy in a flank attack is precisely that of hitting at men half turned away from them. But presumably the state in which the archers found the French flanks was even more to their advantage than that. On the edge of the crowd, men-at-arms were walking or running to the rear. As they went, accelerating no doubt at the sight of the English charging down on them they exposed men deeper within the crowd who would not until then have had sight of the archers, who were not indeed expecting yet to use their arms and whose attention was wholly directed towards the banging and shouting from their front, where they anticipated doing their fighting. Assaulted suddenly at their right or left shoulders, they can have had little chance to face front and point their weapons before some of them, like those already killed by the English men-at-arms, were struck down at the feet of their neighbours.

If the archers were now able to reproduce along the flanks of the French mass the same 'tumbling effect' which had encumbered its front, its destruction must have been imminent. For most death in battle takes place within well-defined and fairly narrow 'killing zones', of which the 'no-man's-land' of trench warfare is the best known and most comprehensible example. The depth of the killing zone is determined by the effective range of the most prevalent weapon, which, in infantry battles, is always comparatively short, and, in hand-to-hand fighting, very short – only a few feet. That being so, the *longer* the winning side can make the killing zone, the more casualties can it inflict

If the English were now able to extend the killing zone from along the face to down the sides of the French mass (an 'enveloping' attack), they threatened to kill very large numbers of Frenchmen indeed.

Given the horror of their situation, the sense of which must now have been transmitted to the whole mass, the French ought at this point to have broken and run. That they did not was the consequence, once again, of their own superiority of numbers. For heretofore it had only been the first division of their army which had been engaged. The second and the third had stood passive, but as the first began to give way, its collapse heralded by the return of fugitives from the flanks, the second walked forward across the wet and trampled ground to lend it support. This was exactly *not* the help needed at that moment. Had the cavalry, in third line, been brought forward to make a second charge against the archers, now that they were outside the protection of their stakes and without their bows, they might well have achieved a rescue. But they were left where they were, for reasons impossible to reconstruct.* Instead, the second division of infantrymen arrived and, thrusting against the backs of their tired and desperate compatriots, held them firmly in place to suffer further butchery.

From what the chroniclers say, we can suppose most of those in the French first line now to be either dead, wounded, prisoner or ready to surrender, if they could not escape. Many had made their surrender (the Priest of the Cottonian MS cattily reports that 'some, even of the more noble ... that day surrendered themselves more than ten times'); some had not had it accepted: the Duke of Alençon, finding himself cut off and surrounded in a dash to attack the Duke of Gloucester, shouted his submission over the heads of his attackers to the King, who was coming to his brother's rescue, but was killed before Henry could extricate him. Nevertheless, very large numbers of Frenchmen had, on promise of ransom, been taken captive, presumably from the moment when the English sensed that the battle was

*But probably having to do (a) with the lack of effective overall command in the French army, (b) with the difficulty of seeing from the third line (c. 500 yards from the 'killing zone') what was happening at the front.

going their way. Their removal from the field, the deaths of others, and the moral and by now no doubt incipient physical collapse of those left had opened up sufficient space for the English to abandon their close order and penetrate their enemy's ranks.

This advance brought them eventually – we are talking of an elapsed time of perhaps only half an hour since the first blows were exchanged – into contact with the second line. They must themselves have been tiring by this time. For the excitement, fear and physical exertion of fighting hand-to-hand with heavy weapons in plate armour quickly drained the body of its energy, despite the surge of energy released under stress by glandular activity. Even so, they were not repulsed by the onset of the second line. Indeed, its intervention seems to have made no appreciable impact on the fighting. There is a modern military cliché, 'Never reinforce failure', which means broadly that to thrust reinforcements in among soldiers who have failed in an attack, feel themselves beaten and are trying to run away is merely to waste the newcomers' energies in a struggle against the thrust of the crowd and to risk infecting them with its despair. And it was indeed in congestion and desperation that the second line appear to have met the English. The chroniclers do not specify exactly what passed between them, presumably because it was so similar to what had gone on before during the defeat of the first line. Though we may guess that a large number of the second line, as soon as they became aware of the disaster, turned their backs and ran off the way they had come; some were dragged out by their pages or servants.

What facts the chroniclers do provide about this, the culmination of the hand-to-hand phase, are difficult to reconcile. The English appear to have had considerable freedom of movement, for they were taking hundreds prisoner and the King and his entourage are reported to have cut their way into the second line (it may have been then that he took the blow which dented the helmet which is still be be seen above his tomb in Westminster Abbey). And yet in at least three places, suggested by the priest's narrative to have been where the enemy columns initially charged the English men-at-arms, the bodies of the

French lay piled 'higher than a man'. Indeed the English are said to have climbed these heaps 'and butchered the adversaries below with swords, axes and other weapons'.

This 'building of the wall of dead' is perhaps the best known incident of the battle. If it had occurred, however, we cannot accept that the King and his armoured followers were able to range freely about the field in the latter stages, since the heaps would have confined them within their own positions. Brief reflection will, moreover, demonstrate that the 'heap higher than a man' is a chronicler's exaggeration. Human bodies, even when pushed about by bulldozers, do not, as one can observe if able to keep one's eyes open during film of the mass-burials at Belsen, pile into walls, but lie in shapeless sprawling hummocks. When stiffened by rigor mortis, they can be laid in stacks, as one can see in film of the burial parties of a French regiment carting its dead from the field after an attack in the Second Battle of Champagne (September 1915). But men falling to weapon-strokes in the front line, or tripping over those already down, will lie at most two or three deep. For the heaps to rise higher, they must be climbed by the next victims: and the 'six-foot heaps' of Agincourt could have been topped-out only if men on either side had been ready and able to duel together while balancing on the corpses of twenty or thirty others. The notion is ludicrous rather than grisly.

The dead undoubtedly lay thick at Agincourt, and quite probably, at the three places where fighting had been heaviest, in piles. But what probably happened at those spots, as we have seen, is that men-at-arms and archers achieved an envelopment of the heads of the French columns, hemmed in and perhaps completely surrounded groups of the enemy, toppled them over on top of each other with lance thrusts and killed them on the ground. The mounds thus raised were big and hideous enough to justify some priestly rhetoric – but not to deny the English entry into the French positions.

The Killing of the Prisoners

Indeed, soon after midday, the Englishmen were 'in possession of the field' – by which soldiers would understand that they were able to move freely over the ground earlier occupied by the French, of whom only dead, wounded and fugitives were now to be seen. Fugitives too slow-footed to reach hiding in the woods, or sanctuary among the cavalry of the still uncommitted third division, were chased and tackled by bounty-hunters; others, greedy for ransom, were sorting through the recumbent bodies and pulling 'down the heaps ... to separate the living from the dead, proposing to keep the living as slaves to be ransomed'. At the back of the battlefield, the most valuable prisoners were massed together under guard. They were still wearing their armour but had surrendered their right gauntlets to their captors, as a token of submission (and subsequent re-identification), and taken off their helmets, without which they could not fight.

Henry could not allow each captor individually to sequester his prisoners because of the need to keep the army together as long as the French third division threatened a charge. So while small parties, acting both on their own behalf and that of others still in the ranks, reaped the rewards of the fight, the main bodies of men-at-arms and archers stood their ground – now about two or three hundred yards forward of the line on which they had received the French charge. Henry's caution was justified. Soon after midday, the Duke of Brabant, arriving late, half-equipped, and with a tiny following, charged into these ranks. He was overpowered and led to the rear. But this gallant intervention inspired at least two French noblemen in the third division, the Counts of Masle and Fauquemberghes, to marshal some 600 of their followers for a concerted charge. They could clearly be seen massing, two or three hundred yards from the English line, and their intentions were obvious. At about the same time, moreover, shouting from the rear informed the

English of a raid by the enemy on the baggage park, which had been left almost unguarded.

It was these events which precipitated Henry's notorious order to kill the prisoners. As it turned out, the charge was not delivered and the raid was later revealed to have been a mere rampage by the local peasantry, under the Lord of Agincourt. The signs were enough, however, to convince Henry that his victory, in which he can scarcely have yet believed, was about to be snatched from him. For if the French third division attacked the English where they stood, the archers without arrows or stakes, the men-at-arms weary after a morning of hacking and banging in full armour, all of them hungry, cold and depressed by the reaction from the intense fears and elations of combat, they might easily have been swept from the field. They could certainly not have withstood the simultaneous assault on their rear, to which, with so many inadequately guarded French prisoners standing about behind them on ground littered with discarded weapons, they were likely also to have been subjected. In these circumstances, his order is comprehensible.

Comprehensible in harsh tactical logic; in ethical, human and practical terms much more difficult to understand. Henry, a Christian king, was also an experienced soldier and versed in the elaborate code of international law governing relations between a prisoner and his captor. Its most important provision was that which guaranteed the prisoner his life – the only return, after all, for which he would enter into anything so costly and humiliating as a ransom bargain. And while his treachery broke that immunity, the mere suspicion, even if well-founded, that he was about to commit treason could not justify his killing. At a more fundamental level, moreover, the prisoner's life was guaranteed by the Christian commandment against murder, however much more loosely that commandment was interpreted in the fifteenth century. If Henry could give the order and, as he did, subsequently escape the reproval of his peers, of the Church and of the chroniclers, we must presume it was because the battlefield itself was still regarded as a sort of moral no-man's-land and the hour of battle as a legal *dies non.*

His subordinates nevertheless refused to obey. Was this because they felt a more tender conscience? The notion is usually dismissed by medieval specialists, who insist that, at best, the captors objected to the King's interference in what was a personal relationship, the prisoners being not the King's or the army's but the vassals of those who had accepted their surrender; that, at worst, they refused to forgo the prospect of so much ransom money (there being almost no way for a man of the times to make a quick fortune except on the battlefield). But it is significant that the King eventually got his order obeyed only by detailing two hundred *archers*, under the command of an esquire, to carry out the task. This may suggest that, among the captors, the men-at-arms at any rate felt something more than a financially motivated reluctance. There is, after all, an important difference between fighting with lethal weapons, even if it ends in killing, and mere butchery, and we may expect it to have been all the stronger when the act of fighting was as glorified as it was in the Middle Ages. To meet a similarly equipped opponent was the occasion for which the armoured soldier trained perhaps every day of his life from the onset of manhood. To meet and beat him was a triumph, the highest form which self-expression could take in the medieval nobleman's way of life. The events of the late morning at Agincourt, when men had leapt and grunted and hacked at each other's bodies, behaving in a way which seems grotesque and horrifying to us, was for them, therefore, a sort of apotheosis, giving point to their existence, and perhaps assuring them of commemoration after death (since most chroniclers were principally concerned to celebrate individual feats of arms). But there was certainly no honour to be won in killing one's social equal after he had surrendered and been disarmed. On the contrary, there was a considerable risk of incurring dishonour, which may alone have been strong enough to deter the men-at-arms from obeying Henry's order.

Archers stood outside the chivalric system; nor is there much to the idea that they personified the yeoman virtues. The bowmen of Henry's army were not only tough professional soldiers. There is also evidence that many had enlisted in the first place to avoid punishment for civil acts of violence, including murder. The chroniclers also make clear that, in the heat of combat, and during the more leisurely taking of prisoners after the rout of the French second division, there had been a good deal of killing, principally by the archers, of those too poor or too badly hurt to be worth keeping captive. The question of how more or less reluctant they were to carry out the King's command need not therefore delay us.

But the mechanics of the execution do demand a pause. Between one and two thousand prisoners accompanied Henry to England after the battle, of whom most must have been captured before he issued his order to kill. The chroniclers record that the killers spared the most valuable prisoners and were called off as soon as Henry assured himself that the French third division was not going to attack after all. We may take it therefor that the 200 archers whom he detailed were heavily outnumbered by their victims, probably by about ten to one. The reason for wanting them killed, however, was that they were liable to re-arm themselves from the jetsam of battle if it were renewed. Why did they not do so when they saw themselves threatened with death, for the announcement of the King's order 'by trumpet' and the refusal of their captors to carry it out can have left them in no doubt of the fate he planned for them? And how were the archers able to offer them a match? It may have been that they were roughly pinioned (some contemporary pictures of battle show prisoners being led away with their hands bound); but in that case they offered no proper – or a very much reduced – menace to the army's rear, which in turn diminishes the justification for Henry's order. And even if they were tied, their actual killing is an operation difficult to depict for oneself. The act of surrender is notably accompanied by the onset of lassitude and self-reproach. Is it realistic to imagine, however, these proud and warlike men passively awaiting the arrival of a gang of their social inferiors to do them to death – standing like cattle in groups of ten for a single archer to break their skulls with an axe?

It does seem very improbable, and all the more because what we know of twentieth-century mass-killing suggests that it is very difficult for small numbers of executioners, even when armed with machine-guns, to kill people much more defenceless than armoured knights quickly and in large numbers. What seems altogether more likely, therefore, is that Henry's order, rather than bring about the prisoners' massacre, was intended by its threat to terrorize them into abject inactivity. We may imagine something much less clinical than a *Sonderkommando* at work: the captors loudly announcing their refusal to obey the proclamation and perhaps assuring their prisoners that they would see them come to no harm; argument and even scuffling between them and members of the execution squad; and then a noisy and bloody cattle-drive to the rear, the archers harrying round the flanks of the crowd of armoured Frenchmen as they stumbled away from the scene of fighting and its dangerous debris to a spot nearer the baggage park, whence they could offer no serious threat at all. Some would have been killed in the process, and quite deliberately, but we need not reckon their number in thousands, perhaps not even in hundreds.

The killing, moreover, had a definite term, for Henry ordered it to end when he saw the French third division abandon their attack formation and begin to leave the battlefield. The time was about three o'clock in the afternoon, leaving some two hours more of daylight. The English began at once to spread out over the field looking for prisoners and spoil in places not yet visited. The King made a circuit and, on turning back for his quarters at Maisoncelles, summoned to him the French and English heralds.

The Wounded

The heralds had watched the battle in a group together and, though the French army had left, the French heralds had not yet followed them. For the heralds belonged not to the armies but to the international corporation of experts who regulated civilized warfare. Henry was anxious to hear their verdict on the day's fighting and to fix a name for the battle, so that its outcome and the army's exploits could be readily identified when chroniclers came to record it. Montjoie, the principal French herald, confirmed that the English were the victors and provided Henry with the name of the nearest castle – Agincourt – to serve as eponym.

That decision ended the battle as a military and historical episode. The English drove their prisoners and carried their own wounded back to Maisoncelles for the night, where the twenty surgeons of the army set to work. English casualties had been few: the Duke of York, who was pulled from under a heap of corpses, dead either from suffocation or a heart-attack, and the Earl of Suffolk were the only notable fatalities. The wounded numbered only some hundreds. What were their prospects? In the main, probably quite good. The English had not undergone an arrow attack, so most of the wounds would have been lacerations rather than penetrations, clean even if deep cuts which, if bound up and left, would heal quickly. There would also have been some fractures; depressed fractures of the skull could not be treated – the secret of trepanning awaited rediscovery – but breaks of the arm and lower leg could have been successfully set and splinted. The French wounded enjoyed a much graver prognosis. Many would have suffered penetrating wounds, either from arrows or from thrusts through the weak spots of their armour. Those which had pierced the intestines, emptying its contents into the abdomen, were fatal: peritonitis was inevitable. Penetrations of the chest cavity, which had probably carried in fragments of dirty clothing, were almost as certain to lead to sepsis. Many of the French would have suffered depressed fractures of the skull, and there would have been broken backs caused by falls from horses in armour at speed. Almost all of these injuries we may regard as fatal, the contemporary surgeons being unable to treat them. Many of the French, of course, had not been collected from the battlefield and, if they did not bleed to death, would have succumbed to the combined effects of exposure and shock during the night, when temperatures might have descended into the middle-30s Fahrenheit. It was, therefore, not arbitrary brutality when, in crossing the battlefield next morning, the English killed those whom they found alive.

They were almost certain to have died, in any case, when their bodies would have gone to join those which the local peasants, under the supervision of the Bishop of Arras, dug into pits on the site. They are said to have buried about 6,000 altogether.

The Will to Combat

What sustained men in a combat like Agincourt, when the penalty of defeat, or of one's own lack of skill or nimbleness was so final and unpleasant? Some factors, either general to battle – as will appear – or more or less particular to this one are relatively easy to isolate. Of the general factors, drink is the most obvious to mention. The English, who were on short rations, presumably had less to drink than the French, but there was drinking in the ranks on both sides during the period of waiting and it is quite probable that many soldiers in both armies went into the mêlée less than sober, if not indeed fighting drunk. For the English, the presence of the King would also have provided what present-day soldiers call a 'moral factor' of great importance. The personal bond between leader and follower lies at the root of all explanations of what does and does not happen in battle: and that bond is always strongest in martial societies, of which fifteenth-century England is one type and the warrior states of India, which the British harnessed so successfully to their imperial purpose, are another. The nature of the bond is more complex, and certainly more materialistic than modern ethologists would like to have us believe. But its importance must not be underestimated. And though the late-medieval soldier's immediate loyalty lay towards his captain, the presence on the field of his own and his captain's anointed king, visible to all and ostentatiously risking his life in the heart of the mêlée, must have greatly strengthened his resolve.

Serving to strengthen it further was the endorsement of religion. The morality of killing is not something with which the professional soldier is usually thought to trouble himself, but the Christian knight, whether we mean by that the ideal type as seen by the chroniclers or some at least of the historical figures of whom we have knowledge, was nevertheless exercised by it. What constituted unlawful killing in time of war was well-defined, and carried penalties under civil, military and religious law. Lawful killing, on the other hand, was an act which religious precept specifically endorsed, within the circumscription of the just war; and however dimly or marginally religious doctrine impinged on the consciousness of the simple soldier or more unthinking knight, the religious preparations which all in the English army underwent before Agincourt must be counted among the most important factors affecting its mood. Henry himself heard Mass three times in succession before the battle, and took Communion, as presumably did most of his followers; there was a small army of priests in the expedition. The soldiers ritually entreated blessing before entering the ranks, going down on their knees, making the sign of the cross and taking earth into their mouths as a symbolic gesture of the death and burial they were thereby accepting.

Drink and prayer must be seen, however, as last-minute and short-term reinforcements of the medieval soldier's (though, as we shall see, not only his) will to combat. Far more important, and, given the disparity of their stations, more important still for the common soldier than the man-at-arms, was the prospect of enrichment. Medieval warfare, like all warfare, was about many things, but medieval battle, at the personal level, was about only three: victory first, of course, because the personal consequences of defeat could be so disagreeable; personal distinction in single combat – something of which the man-at-arms would think a great deal more than the bowman; but, ultimately and most important, ransom and loot. Agincourt was untypical of medieval battle in yielding and then snatching back from the victors the bonanza of wealth that it did; but it is the gold-strike and gold-fever character of medieval battle which we should keep foremost in mind when seeking to understand it.

We should balance it, at the same time, against two other factors. The first of these is the pressure of compulsion. The role which physical coercion or force of unavoidable circumstance plays in bringing men into, and often through, the ordeal of battle is one which almost all military historians consistently underplay, or ignore. Yet we can clearly see that the force of unavoidable circumstances was among the most powerful of the drives to combat at work on the field of Agincourt. The English had sought by every means to avoid battle throughout their long march from Harfleur and, though accepting it on 25 October as a necessary alternative to capitulation and perhaps life-long captivity, were finally driven to attack by the pains of hunger and cold. The French had also hoped to avoid bringing their confrontation with the English to a fight; and we may convincingly surmise that many of those who went down under the swords or mallet-blows of the English had been drawn into the battle with all the free-will of a man who finds himself going the wrong way on a moving-staircase.

The second factor confounds the former just examined. It concerns the commonplace character of violence in medieval life. What went on at Agincourt appals and horrifies the modern imagination which, vicariously accustomed though it is to the idea of violence, rarely encounters it in actuality and is outraged when it does. The sense of outrage was no doubt as keenly felt by the individual victim of violence five hundred years ago. But the victim of assault, in a world where the rights of lordship were imposed and the quarrels of neighbours settled by sword or knife as a matter of course, was likely to have been a good deal less surprised by it when it occurred. As the language of English law, which we owe to the Middle Ages, reveals, through its references to 'putting in fear', 'making an affray', and 'keeping the Queen's peace', the medieval world was one in which the distinction between private, civil and foreign war, though recognized, could only be irregularly enforced. Thus battle, though an extreme on the spectrum of experience, was not something unimaginable, something wholly beyond the peace-loving individual's ken. It offered the soldier risk in a particularly concentrated form; but it was a treatment to which his upbringing and experience would already have partially inured him.

3 Waterloo, 18 June 1815

The Duke of Wellington strongly disapproved of all attempts to turn the battle of Waterloo either into literature or history. His own account of it in his official dispatch was almost dismissive and he advised a correspondent who had requested his help in writing a narrative to 'leave the battle of Waterloo as it is'. The Duke's attitude rested in part on his disdain for sensationalism, in part on a well-founded doubt about the feasibility of establishing a chain of cause and effect to explain its outcome. 'The history of the battle,' he explained, 'is not unlike the history of a ball! Some individuals may recollect all the little events of which the great result is the battle lost or won; but no individual can recollect the order in which, or the exact moment at which, they occurred, which makes all the difference as to their value or importance.'

The Duke's wishes were disregarded, as they were bound to be, from the start. Waterloo, it seemed to contemporaries, had reversed the tide of European history and almost anyone who had taken part in the battle and could still hold a pen found a word-hungry readership. Official thanksgiving determined, moreover, that the style writers adopted should be heroic and declamatory from the outset. For the first time in the history of the British army, each veteran of the battle was awarded a commemorative medal. He was to be known as a Waterloo Man and his single day of service on 18 June 1815, was to count two years for pension. This alone was sufficient to convince even the most unimaginative private soldier that he had survived an extraordinary event; officers were in no doubt of it from the moment the battle closed and many of their letters, written on the spot to relieve the anxieties of relatives for their safety, strike a note of triumph which was taken up by almost every professional who made Waterloo his subject. Even the French, by some strange translatory process, managed to make an epic out of the defeat. And the two most distinguished literary figures whose imaginations were captured by Waterloo – Byron and Victor Hugo – turned their feelings into poetry.

Remarkably the results – *Childe Harold* and *Les Châtiments* – are still thrillingly readable. But the cumulative effect of treating the battle as a drama seen and felt as such by the participants in the heat of combat, has been to cover the human experience and military facts with a thick sedimentary deposit of romance. Even Siborne, whose methodological approach to the battle was impeccable, felt compelled to conclude his ponderous history in a sunburst of adjectives:

> Such was the termination of this ever memorable Battle – a Battle remarkable for the spectacle it afforded, on the one hand of a bravery the most noble and undaunted; of a passive endurance, the most calm, dignified, and sublime; of a loyalty and patriotism, the most stern and inflexible: and on the other, of a gallantry in assault the most daring and reckless; of a devotion to their Chief, the most zealous and unbounded; and, lastly, of a physical overthrow and moral annihilation unexampled in the history of modern warfare. Such was the consummation of a victory, the most brilliant in its development, the most decisive in its operation, and the most comprehensive in its result, of any that had occurred since the bringing to the termination so long and so ardently desired by the suffering and indignant nations of Europe.

Generations of writers, for whom the overthrow of Napoleon had little and eventually none of the moral and political significance it did for Siborne and his contemporaries, nevertheless followed him in seeing the victory as a deliverance from tyranny. The visual imagination of writer and reader was meanwhile fed by an outpouring of brightly coloured canvases from the studios of an army of successful salon painters – Dighton, Philippoteaux, Raffet, Bellangé, Caton Woodville – paintings which by their combination of photographic observation of detail with defiance of physical laws anticipate the work of the Surrealists. Much of the prose imagery in the constantly retold story of Waterloo – flashing sabres, dissolving squares and torrents of horseflesh – has its counterpart, often, one suspects, its origin, in the vision of artists who saw the battlefield, if at all, only as tourists.

Recently a number of writers have shown impatience with the ritually dramatic approach. David Howarth, in *Waterloo: Day of Battle*, was remarkably successful in narrating the battle largely through the reminiscences of combatants and eyewitnesses. Jac Weller, a specialist with a highly original eye, attempted and very largely achieved an ambitious project: to recount the battle 'from no more information than Wellington had at any one time'. He is an expert on weapons and his book also provides a valuable yardstick of the degree of damage armies could inflict on each other and the sort of harm individuals could suffer on the battlefield. But neither book is wholly satisfactory. Howarth's individuals remain individuals, leaving the reader uncertain as to how representative each was of his rank, how typical his experience of the day. Weller makes a marvellously imaginative attempt at generalization; but, having called his book *Wellington at Waterloo*, ultimately and inevitably imprisons himself within the confines of the biographical approach.

What the 'human experience' and 'military facts' of Waterloo demand of the historian is some combination of Howarth's and Weller's methods, a cooperative effort between the former's heart, the latter's head. The wonder is that it has not yet been attempted. For the military archaeology of 1815 – systems of drill, types, ranges and effects of weapons, mechanics of command, rates of movement across country – lies ready to be rediscovered, as Weller has demonstrated, only just beneath the topsoil of the past. And rich deposits of personal reminiscence have been open to prospectors for over a century. Indeed these deposits exceed in value those relating to any other battle outside the twentieth century. For besides the memoirs written at leisure by a host of British, French, German, Dutch and Belgian veterans, and the letters dashed off spontaneously by survivors the day or week following, and besides also the official documentation of the campaign, there exists, for the British army, an archive which was to have no counterpart until General S. L. A. Marshall inaugurated 'after-action reporting' in the American army during the Second World War. The quality of this archive is magnificent, its origin decidedly eccentric. Captain Siborne, its collector, conceived in the 1830s the idea of constructing an enormous model of Waterloo 'at the crisis of the battle' and was granted permission by the Commander-in-Chief to circularize surviving officers for evidence. The result might have been a hodge-podge. He was, however, a methodical man and required his correspondents to answer a precise list of questions. Naturally, not all kept within the rules he laid down, while others did so too strictly, but the overall result was to provide a sort of Argus-eye view of Waterloo from the British side. The collection has provided the principal source for all subsequent accounts and treatments of the battle. Siborne himself wrote an enormous and unreadable history based upon it; most regimental histories' Waterloo chapters are little more than thinly rewritten versions of their own officers' letters to Siborne; and jackdaw-flocks of anthologists, attracted to Waterloo by the easy pickings his work provides, have filled scores of their pages with 'finds' lifted unacknowledged from his.

The material undoubtedly exists then for a human history which would also be a military history of Waterloo. All the more is this true because so many of the sources remain curiously untouched. For Waterloo has, above other battles, most consistently been dealt with in 'win/lose' terms. It was, as we have seen, the culminating fifteenth of Creasy's *Decisive Battles* and a majority of writers have followed or anticipated him in seeing their principal task as one of explanation: an explanation of how Wellington won and Napoleon lost (or was cheated of victory or would have won if X had happened; 'after an exhaustive reading of Waterloo literature, I flinch,' writes Weller, 'when I come upon a sentence which begins with either "if" or "had".'). Moreover, the participants, to a remarkable degree, subordinated the story of their own doings to the larger story of the downfall of Napoleon. And Siborne, whose work is so rich in information about human behaviour on the battlefield, collected it incidentally while pursuing a much narrower – if absolutely fundamental – purpose: the fixing of exact locations and precise times as links in a chain of cause and effect. But it is precisely because so much of the human information comes to us incidentally that we can value it so highly; and because it is, in comparison with what we know of Agincourt, so plentiful that we can hope to reconstruct with some authenticity what the battle was like for those who took part in it.

The Campaign

To begin at the beginning of the campaign – which only briefly antedates the battle itself. Napoleon, defeated by the combined armies of Britain, Austria, Prussia and Russia in 1814 and exiled to Elba, returned to France on 1 March 1815. It very quickly became clear to Louis XVIII, whom the Allies had restored to the throne, that his armies' loyalty did not belong to him and on 20 March he left Paris for Ghent in Belgium. Napoleon entered Paris the same day. He hoped that the Allies would acquiesce in his resumption of power but a week earlier they had agreed between themselves to go to war, and to this agreement they stuck. While Napoleon set in hand the reconstruction of the Grande Armée, much of which had been demobilized at the Restoration, the Allies sealed plans for the concentration of four large armies of their own on France's eastern and north-eastern borders. An Austrian army of 200,000 was to enter France through Alsace-Lorraine, to be followed later in the summer by a Russian army of over 150,000; a Prussian army of over 100,000 was to march into southern Belgium; and an Anglo-Dutch army, formed on a British nucleus already in the Low Countries, was to concentrate in the north. When in position, the four armies were to advance simultaneously into France.

Napoleon could not hope to match these numbers. He had found only 200,000 men actually under arms on his return and, recognizing that Louis XVIII's abolition of conscription had been too popular a step to revoke, had to resort to exhortation and illegalities to add to them at all. His inferiority in strength limited him to a choice between only two strategies: a Fabian one of defence and delay which, if protracted long enough, might persuade the Allies to make peace with him out of sheer frustration; or a spoiling offensive against the already forming British and Prussian armies in Belgium which, if successful, might deter the Austrians and Russians from subsequently risking defeat in detail themselves. His natural inclination being for the offensive, and the British and Prussians scarcely outnumbering him, he decided for the second. It had the added attraction over the first of averting another foreign invasion of the national territory within eighteen months.

He was not, however, strong enough to tackle the British and Prussian armies combined. His plan required therefore that he should bring one of the two Allies to battle before the other and in such a fashion that the unengaged army should not come to its neighbour's assistance until it was too late. The difficulty was to choose which to attack first. A study of their lines of communication supplied the answer. The British army's base was in the Belgian ports, the Prussian's in the Rhineland. Whichever was attacked would tend to fall back on its base and hence away from the other, to whose lines of communication its own ran at right angles. An attack on the British might provoke the Prussians into coming to their assistance; but Napoleon thought it most unlikely that Wellington would risk endangering his communication with the channel ports to help Blücher. This line of reasoning determined that he should strike at the Prussians first.

Napoleon's reading of personal and national character was unfair to the British and to Wellington, for the Duke was determined to fight it out in harness. But the initial stage of the campaign seemed nevertheless to bear out Napoleon's view. So successful was he in assembling his army on the Belgian border without the Allies being able to fix its exact whereabouts that, on 16 June, he managed to concentrate the greater part of his force against the unsupported Prussians at Ligny, and to beat them. The British, rather ominously, did make an appearance late in the day on the extreme western flank of the battle at Quatre Bras, but were contained and repulsed by a detached French force and fell back, as Napoleon had predicted and desired, northwards. The Prussians meanwhile made their escape, apparently eastward.

So far, splendid. Strategically, the campaign was won. It now remained only to tie up the tactical loose ends by defeating the British; and a British army a great deal less formidable than that which had fought him in Spain. Less than half of it was British at all, the remainder being German, Dutch and Belgian, while many of the British regiments were composed of inexperienced troops.

Its cavalry nevertheless covered its retreat efficiently throughout 17 June and when darkness lifted next day Napoleon found it deploying for battle in a strong position across the main road to Brussels just south of the Forest of Soignes. Its front measured about 6,000 yards across and its flanks were well protected – to the east by the farm buildings and cottages of Papelotte, Frichermont and La Haye, to the west by the village of Braine l'Alleud. The centre of the position was reinforced by two strongly built farms, La Haye Sainte and Hougoumont. British possession of those farms made a frontal attack unpleasant to contemplate. But to manoeuvre his way round the position, with an unlocated Prussian army somewhere on his right, was perhaps even more dangerous, and certainly very time-consuming. Napoleon therefore decided to make his attack a frontal one.

There was no question of the British attacking him. Although the two armies were almost equal in number at about 70,000 each, so many of Wellington's German and Dutch-Belgian soldiers were politically or militarily unreliable that he could not contemplate using them in any other than a static role. The King's German Legion, an émigré force of Hanoverian regulars which had fought the campaign of the Peninsula, was stout-hearted enough to be trusted anywhere; British officers and soldiers willingly conceded K.G.L. regiments to be the equal of their own. But many of the Hanoverian regiments proper were undertrained and inexperienced, while the Dutch-Belgians were suspected, on sound evidence, of preferring Napoleon to their own recently restored Prince of Orange. Wellington therefore disposed them where they could get into least trouble, putting most of the Dutch-Belgians into Braine l'Alleud at one

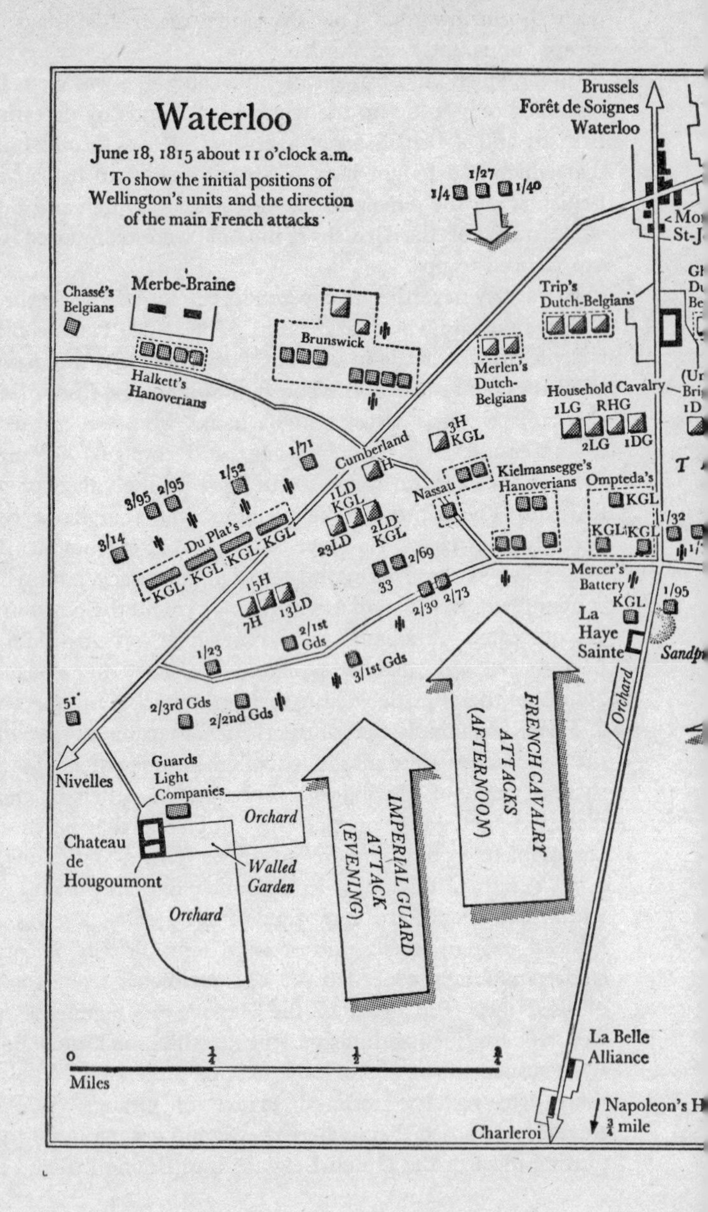
Waterloo
June 18, 1815 about 11 o'clock a.m.
To show the initial positions of Wellington's units and the direction of the main French attacks
Brussels
Forêt de Soignes
Waterloo
1/27
1/4
1/40
Chassé's Belgians
Merbe-Braine
Halkett's Hanoverians
Brunswick
Trip's Dutch-Belgians
Merlen's Dutch-Belgians
Household Cavalry
1LG
RHG
2LG
1DG
3H KGL
Cumberland
H
Nassau
Kielmansegge's Hanoverians
Ompteda's
KGL
1/71
1/52
3/95
2/95
3/14
Du Plat's
KGL
1LD KGL
23LD
2LD KGL
2/69
33
2/30
2/73
Mercer's Battery
KGL
1/95
La Haye Sainte
Orchard
15H
7H
13LD
2/1st Gds
1/23
3/1st Gds
51
2/3rd Gds
2/2nd Gds
Nivelles
Guards Light Companies
Orchard
Chateau de Hougoumont
Walled Garden
Orchard
IMPERIAL GUARD ATTACK (EVENING)
FRENCH CAVALRY ATTACKS (AFTERNOON)
0
¼
½
¾
Miles
La Belle Alliance
Napoleon's H
¾ mile
Charleroi

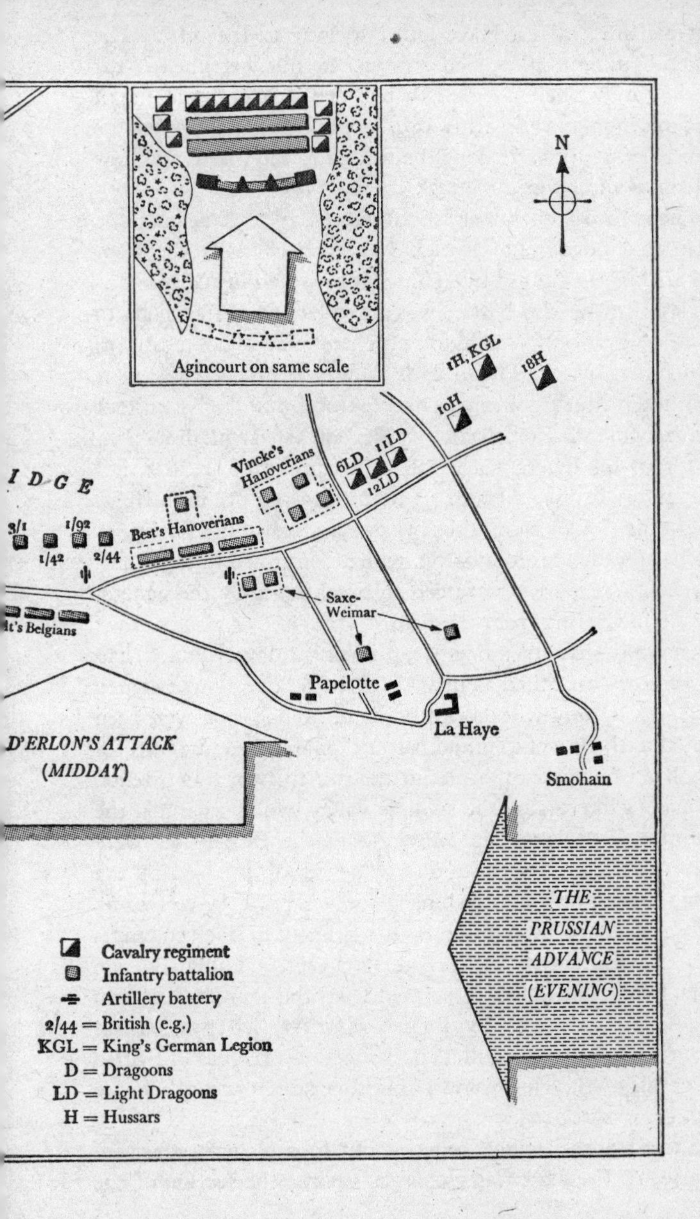
Agincourt on same scale
N
1H, KGL
18H
10H
11LD
6LD
12LD
Vincke's
Hanoverians
IDGE
3/1
1/92
1/42
2/44
Best's Hanoverians
Saxe-
Weimar
lt's Belgians
Papelotte
La Haye
Smohain
D'ERLON'S ATTACK
(MIDDAY)
THE
PRUSSIAN
ADVANCE
(EVENING)
Cavalry regiment
Infantry battalion
Artillery battery
2/44 = British (e.g.)
KGL = King's German Legion
D = Dragoons
LD = Light Dragoons
H = Hussars

end of his line and La Haye and Papelotte at the other. The irreducible minimum needed to thicken out his line in the centre he sandwiched between British or German regiments of dependable quality. His army thus deployed, generally on the crest of a gentle forward slope, he waited to see where and how Napoleon would open the attack.

Napoleon chose to attack about eleven o'clock against the Château of Hougoumont, which was garrisoned by the Foot Guards. This, the first of five phases into which historians conventionally divide the battle, was intended as a diversion, to draw reserves from Wellington's centre where he meant to make his main attack. The Guards, however, proved capable of holding the château – an immensely strong building – without assistance; while the French commander entrusted with the assault quite forgot his diversionary role and committed greater and greater numbers of soldiers in an attempt to capture it outright. The fight for Hougoumont thus became, as 'territorial' struggles often do, a battle within a battle, which continued to rage until the French attackers were forced to break it off by the general retreat of their army from the field.

The second phase of the battle, d'Erlon's infantry attack, had therefore to be launched against the British centre unweakened at any rate by any withdrawal of men. It had however been subjected to the fire of a 'grand battery' of about eighty guns for over half an hour when, at about quarter to two, four French divisions began crossing the shallow valley which separated the two armies. Two important outworks of the British line were quickly captured – Papelotte and 'The Sandpit', used by the British riflemen as a skirmishing-place – but La Haye Sainte, though by-passed, did not fall. As the French, in thick columns, approached the crest of the ridge, however, a Dutch-Belgian brigade, which had suffered heaviest from the cannonading, ran away. A counter-attack by British infantry, led by General Picton, restored the line and a charge by two brigades of British heavy cavalry – the Heavy and Union Brigades – then drove the French off in disorder.

The third phase, which began about four o'clock, consisted of a series of French cavalry charges against the section of the

British centre that had not been attacked by d'Erlon. The decision to charge was made by Ney, Napoleon's battlefield commander, who had misinterpreted movement behind the British line to mean that it was giving way. In fact the section of the British line which the French cavalry struck was well-prepared to receive them. It formed square and drove off charge after charge. The horsemen who survived this hour eventually retired, pursued by British cavalry, with whom they entered into a running fight. Napoleon, aware that the vanished Prussians were now approaching the battlefield, hastily sent Ney the armoured cavalry of the Imperial Guard and two other divisions of cuirassiers. They also were beaten by the British and Hanoverian squares, as too were some infantry, launched into the battle as an afterthought about six o'clock.

The artillery of both armies had played a vital attritional role in the second and third phases. The fourth phase, however, was almost wholly an affair of infantry. It was quite brief and stands out as a separate episode because it centred on a clear-cut French success – the first of the day. This was the capture of La Haye Sainte, abandoned by its King's German Legion garrison because they had run out of ammunition. Its loss put the section of British line behind it in great danger and Ney almost succeeded in breaking through with another infantry attack. But he was now running out of soldiers, the reserve being fully committed against the Prussians, while Wellington, a thriftier commander, could still produce sufficient to reinforce the threatened front. Soon after half past six the situation in the centre was restored.

The crisis now shifted to the French side. Napoleon was heavily engaged on two fronts and threatened with encirclement by the advancing Prussians. He had only one group of soldiers left with which to break the closing ring and swing the advantage back to himself. This group was the infantry of the Imperial Guard. At about seven it left its position at the rear of the battlefield and ascended the slope just to the east of Hougoumont. The British battalions on the crest fired volleys into its front and flank – the flank fire of the 52nd Light Infantry was particularly heavy and unexpected – and, to their surprise, saw the

Guard turn and disappear into the smoke from which it had emerged. On the Duke of Wellington's signal, the whole line advanced, behind the charging horses of the remaining British cavalry. The battle of Waterloo was over, almost – the Prussians were still locked in combat with the French on the east flank – and Napoleon had been beaten.

The Personal Angle of Vision

It is probably otiose to point out that the 'five phases' of the battle were not perceived at the time by any of the combatants, not even, despite their points of vantage and powers of direct intervention in events, by Wellington or Napoleon. The 'five phases' are, of course, a narrative convenience. But it is nevertheless important to emphasize, before turning to look at the battle in terms of the experience of the men in the line, how very partial indeed was the view most of them got of it. An extreme example is provided by the case of the 27th Regiment, the Inniskillings. They, having been employed on the line of retreat during the night of 17–18 June, did not get into bivouac until about eleven on the morning of the battle. There, around Mont St-Jean, about three quarters of a mile from the front, they lay down to sleep. Many were still sleeping when at about three o'clock, after the battle had been in progress for four hours, they were ordered forward to the La Haye Sainte crossroads. Near that spot they formed columns of companies and stood, occasionally having to form square, until the general advance was ordered over four hours later. During those four hours, over 450 of the regiment's 750 officers and men were killed or wounded, in almost every case by the fire of cannon several hundred yards distant or by the musketry of French skirmishers in concealed positions. So heavy were the casualties among the officers (only one out of the eighteen went untouched) that very little about those four hours was ever written down. But it seems unlikely that any Inniskilling had eyes or thoughts for much but the horror that was engulfing him and his comrades;

and though, when formed in square, the men in the rear face would have been protected from direct fire by those in the four ranks of the front face forty feet behind them, their view would have been of the one sector of the battlefield where fighting was *not* going on, while their nerves would have been taut with the expectation of a cannon-ball in the back. To have asked a survivor of the 27th what he remembered of the battle, therefore, would probably not have been to learn very much.

Many of Siborne's correspondents prefaced their replies to him with a warning of how incomplete or unbalanced their view of the day had been. Thornhill, A.D.C. to Uxbridge, the cavalry commander, wrote that he had been so busy with the 'prompt and direct transmission of his orders' that he had 'little time to contemplate passing events irrelevant thereto'; Robbins of the 7th Hussars noted that up to the moment when he was hit, at the very end of the day, 'we never scarcely saw an enemy', although the regiment changed position three times and was usually under fire; and during its one direct encounter with the French cavalry, he 'was too much occupied with my own men to have been able to pay much [attention] to what was going on around us'. Browne, a lieutenant of the 4th Regiment, a neighbour of the Inniskillings but less exposed than they were to the enemy's fire, developed this point explicitly: 'the smoke, the bustle, which I fear is almost inseparable to Regiments when close to the enemy, and more particularly the attention which is required from the company officers to their men, intercepts all possibility of their giving any correct account of the battles in which they may be engaged.' Pratt, the lieutenant commanding the light company of the Inniskillings – which skirmished to the regiment's front and was therefore spared its terrible ordeal of standing motionless to be cannonaded – makes the same point in a more graphic way: 'I think you will readily agree with me that a young Subaltern officer ... harassed and fatigued after two days' previous marching, fighting and starving ... was not likely to take particular notice of the features of the ground over which he was moving, or to direct his observations much beyond the range of what was likely to affect himself and the few soldiers immediately about him.'

There were other causes, besides the preoccupations of duty, which deprived men of a coherent or extended view of what was going on around them. Many regiments spent much of their time lying down, usually on the reverse slope of the position, which itself obscured sight of the action elsewhere. Mercer, whose set-piece description of the field in his *Journal* is one of the most enthralling passages in Waterloo literature, transmitted a flatter, if more convincing, recollection to Siborne, in a paragraph which describes the situation of his battery in the early afternoon: 'Of what was transacting in the front of the battle we could see nothing, because the ridge on which our first line was posted was much higher than the ground we occupied. Of that line itself we could see only the few squares of infantry immediately next to us, with the intervening batteries. From time to time bodies of cavalry swept over the summit between the squares, and, dispersing on the reverse of the position, vanished again, I know not how.' A few feet of elevation, therefore, made the difference between a bird's-eye and a worm's-eye view; indeed, Sir John Colborne, commanding the 52nd Light Infantry, was 'persuaded that none but mounted Officers can give a correct account of the Battle' (not an opinion, as we have seen, borne out by the experience of the cavalrymen). But even on the crest of the position, physical obstacles could limit the soldier's horizon very sharply. In many places, at least at the beginning of the battle, the crops of wheat and rye stood tall enough for the enemy to approach to within close musket shot undetected. At Quatre Bras, 'the rye in the field was so high', Llewellyn of the 28th Regiment remembered, 'that . . . the Enemy, even in attacking our Squares, were obliged to make a daring person desperately ride forward to plant a flag, as a mark, at the very point of our bayonets.' At Waterloo, a longer battle, the crops were eventually trampled flat ('to the consistency of an Indian mat', Albemarle of the 14th Regiment noted) but earlier in the day the light company of the 51st Regiment, on the far west flank of the position, was fired on by French infantry which had got unscathed to within forty feet of their line under cover of the standing grain; while Lieutenant Sharpin, of Bolton's battery, got his first inkling of the approach of the Imperial Guard in the 'crisis of the battle' when he 'saw the French bonnets just above the high corn, and within forty or fifty yards of our guns'. Again, the men in the rear or interior of dense columnar formations, of the type adopted by the Guard in their advance, would have glimpsed little of the battle but hats, necks and backs, and those at a distance of a few inches, even when their comrades at the front were exchanging fire with the enemy. And almost everyone, however well-positioned otherwise for a view, would for shorter or longer periods have been lapped or enveloped by dense clouds of gun-powder smoke.

Smoke not only limited visibility (though that was one of its primary effects, of which more later); it also played tricks with vision. Cathcart, one of Wellington's A.D.C.s, remembered the Guard as 'black looking columns' which 'loomed through smoke and fog'; to the 1st (British) Guards, who stood in their path, the Imperial Guardsmen looked 'with their high bonnets ... through the smoky haze ... like a corps of giants'; and an officer of Picton's division, observing the clash of the two corps d'élite from a distance, recalled that 'the slanting rays of the setting sun, reaching us through the medium of the smoke of the guns, rendered the atmosphere a *camera obscura* on a giant scale' in which individual figures appeared magnified and sooty black.

Yet despite distance, smoke and inconvenient features of terrain obscuring their line of sight, many combatants confidently recorded detailed accounts of longer or shorter episodes and even precise identifications of personality. Albemarle, of the 14th Regiment, saw Jérôme Bonaparte and his suite riding across the front before the opening of the cannonade; soon afterwards Vivian, who commanded the Hussar Brigade, was convinced he saw Napoleon 'with a large suite of Officers ... amongst the Columns forming in front of the British left'. A very large number of British officers and soldiers saw Wellington, often at close hand, heard him speak, or even exchanged words with him. Lieutenant Drewe of the Inniskillings passed under his balcony in Waterloo village, from which the Duke was watching his troops march up, some time after six o'clock in the morning. About ten he was west of Hougoumont, where some panicky Nassauers shot at him after he had ordered them back into position, and he remained in the vicinity until after the opening of the French attack on the château itself. In the early afternoon he was in the centre, during the succession of attacks launched by d'Erlon, sometimes near 'his tree' – an isolated elm on the crest of the ridge near the La Haye Sainte crossroads – sometimes in the interior of a square. After the repulse of d'Erlon's attack he visited the companies of the 95th Rifles in The Sandpit, just east of La Haye Sainte, having followed the 1st Battalion, King's German Legion, which he had sent to their support. Later in the afternoon he was mostly behind the right centre, while it was under assault from the French cavalry: Rudyard, of Lloyd's battery, saw him frequently near the square of the 33rd and 69th Regiments. Calvert, a major in the 32nd Regiment, saw the Duke east of the crossroads during the attack on La Haye Sainte – he estimated the time at five o'clock, but it would have been later – and he was quite close to the farmhouse itself (and 'much vexed' Cathcart, one of his A.D.C.s, noted) when its defence collapsed. During the final phase, that of the attack of the Imperial Guard, many soldiers recalled seeing him, perhaps because death and wounds had so reduced his entourage that he at times rode alone or with only a single companion: Gawler, of the 52nd Regiment saw him riding unaccompanied to the east of Hougoumont, and Hunter Blair, brigade major of the 3rd Brigade, exchanged words with the only staff officer then following the Duke (a Sardinian liaison officer who spoke no English). But the Duke was also more conspicuous at the end of the battle because it was at that stage that he mostly directly involved himself as a commander. Though 'Up Guards and at 'em' is a fictitious ascription, he certainly did give orders directly to Maitland, commanding the Guards Brigade, and later to the 52nd Light Infantry. After the retreat of the Imperial Guard, he rode east along the whole line, his approach being signalled by cheers which rolled from battalion to battalion. In the general advance which followed, he made his way behind the leading columns to La Belle Alliance, where he and Blücher met, at some time between nine and ten. Soon afterwards he rode back to his headquarters in the village inn at Waterloo, to go to sleep on a mattress on the floor because a staff officer was dying in his bed.

This chronology of the Duke's movements on the day of Waterloo, besides providing an index of the temperature of the battle at any time – for he always managed to be present where the fighting was hottest – also allows us to calculate what he did *not* see, and so in a sense to estimate how distorted *his* view of events was in comparison with that of other combatants less free than he to move about. Thus we can safely say that he did not see what was going on inside Hougoumont (though he issued a most pertinent order based on his observation of the spread of flames inside the building) and he can have known little of what was going on around Papelotte, since he scarcely crossed to the east of the battlefield. He did not witness the culmination of the Union Brigade charge, since that occurred deep within the enemy's positions, nor did he preside over the ordeal of the Inniskillings at the crossroads, being frequently occupied elsewhere. But to say that he did not experience close infantry combat, of the sort the Guards waged in Hougoumont, or a cavalry charge, of the sort led by Ney during the afternoon, or prolonged cannonading, of the sort undergone by the Inniskillings, is not to say that his view of the battle was any less 'real' than that of those who did. It was a view purchased at great personal risk (at least two officers – Gordon and de Lancey – suffered mortal wounds by his side) which was the price paid by almost everyone for the privilege of wearing a Waterloo Medal, and therefore typical. It was a longer view than most had: the Inniskillings, as we saw, missed the first four hours of the battle, while he had been busy about the field long before it began; it was much more varied than that granted to the majority: the 1st/95th Rifles, for example, spent all their time in action at or near The Sandpit, while he was constantly back and forward the thousand yards between Hougoumont and La Haye Sainte. In that respect his view, like his role on the battlefield, was highly individual. While junior officers and common soldiers naturally used his comings and goings as points of reference in their memory of the day, his personal chronology of the battle would have turned on quite different events. At what they were we cannot guess (though he recalled to an interrogator long afterwards that he had taken his cloak on and off 'fifty times' – an intriguing example of the sort of irrelevant detail which sticks in a mind subjected even to the greatest distraction). But by attempting to see how his view differed from his soldiers', how theirs differed one from another's, the cavalryman's from the gunner's, the man's in the rear rank of a square from the skirmisher's to his front, the wounded man's from that of the man left untouched, we offer ourselves the best chance of comprehending the character of the battle as a whole.

The Physical Circumstances of Battle

We ought to take account, nevertheless, as a prologue to consideration of individual experiences of the battle, several common factors which helped to determine its human context. The first of these was fatigue. It is a fairly safe generalization that the soldiers of most armies, at least before the development of mechanical transport, entered battle tired, if only because they had had to march to the field under the weight of their weapons and kit. The English army at Agincourt was certainly very tired, and hungry, cold and wet into the bargain. So too were both armies on the morning of Waterloo. Both had been on the march the whole of the previous day, carrying fifty to sixty pounds per man, had fought the day before that, and had been living on rations issued the day before that again. They had slept in the fields on the night of 17–18 June, when it had streamed with rain, and had woken to an overcast and breakfastless morning. For many of the British regiments we can calculate their state of deprivation with almost clinical precision. The 2nd Battalion, 30th Regiment had left Soignies at two in the morning of 16 June and marched the twenty-two miles to Quatre Bras, which they reached at 5 p.m. On the following day, having left forty men dead or wounded on the field – a comparatively light toll, for seven regiments suffered worse at Quatre Bras than at Waterloo and eight, including these seven, lost between 100 and

300 men on an average regimental strength of 600 – the 2nd/30th retreated to the Waterloo position. They went supperless and breakfastless, so that 'between the midday meal on the 15th and the morning of the 19th the men received somewhat over two days' bread rations (four pounds) and two days' meat (one pound) but had had no time to cook the latter. An attempt had been made to cook during the halt at Braine le Comte, but the march was resumed before the cooking was finished, and the soup and meat poured out on the roadside.'* Less hungry perhaps than the men of this regiment but more footsore were those of the 1st/40th. They had left Ghent, at half an hour's notice, very early in the morning of 16 June, marched thirty miles that day and twenty-one the next, to arrive at Waterloo at 11 a.m. on 18 June. Thus they had covered fifty-one miles in a little over two days and nights, with two brief halts of a few hours. The other two battalions of their Brigade had done likewise; since one of them was the Inniskillings, it prompts speculation whether the men of that regiment were not helped to endure the horrors of 'their' battle by the semi-anaesthesia of extreme physical tiredness.

These feats of endurance were not isolated. Adam's Brigade had spent nearly two days on the road; the 71st Regiment, which belonged to it, had left Leuze early on 16 June, without food, and marched for thirty-six hours, with no halt longer than thirty minutes, to reach Waterloo in time for the battle. The men had then sat on their packs throughout the night of the 17th–18th and the breakfast they got when the sun rose was the first meal they had eaten for two days. The soldiers of the 4th Regiment were so tired on the morning of the 18th that they could scarcely keep awake; they, brigaded with the Inniskillings and with the same march behind them, also slept through the first four hours of the battle, lying down in the open about a thousand yards behind the firing line.

They would have slept a great deal better than the soldiers of the regiments which, having fought Quatre Bras, got into their Waterloo bivouacs before darkness fell. For rain and cold

*From *A History of the Thirtieth Regiment* by Lieutenant-Colonel Bannantyne.

in the night of 17–18 June made sleep almost impossible. Captain Cotter of the 69th Regiment 'preferred standing and walking to and fro during the hours of darkness to lying upon ... mud through which we sank more than ankle deep'. Albemarle of the 3rd/14th noted that the rain lifted for an hour at sundown, was heavier again after dark, with thunder and lightning, and that, after much standing about, he eventually threw himself down 'on the slope of the hill ... it was like lying in a mountain torrent'. He nevertheless slept soundly until two in the morning when his soldier servant called him. Simmons, a lieutenant of the 1st/95th Rifles, 'smeared an old blanket with thick clayey mud' and lay down under it on some straw. He kept quite warm. But next morning in Macready's regiment, 'we were almost petrified with cold, many could not stand, and some were quite stupefied.' The Highlanders of the 92nd, Peninsula veterans, slept in fours under their 'united blankets', but were roused at midnight by a false alarm and stood to arms for some time. The troopers of the Scots Greys were constantly disturbed by their horses which, frightened by the thunder, kept stepping on their masters where they lay at their heads. Captain Wood, of the 10th Hussars, remembered that 'everyone was wet through. We had a shower that came down like a wall. Our horses could not face it and all went about. It made the ground up to the horses' fetlocks.'

The Quatre Bras regiments fared no better for food, moreover, than those which had had the long march. Leeke, an ensign of the 52nd Light Infantry, breakfasted on a 'half-mouthful of broth and a biscuit' which was all the food he got until the battle was over. Five officers of the 32nd, who had had no food since late on 15 June, shared a fowl and a handful of biscuits for supper on 17 June and appear to have got no breakfast. Mercer and his officers, who had also shared a fowl for supper on 17 June, made the serious mistake next morning of not eating 'stirabout', prepared by the gunners from some freshly delivered oats, but of waiting for meat to be cooked. Like the 30th's meal, two days before, its cooking had to be abandoned when the stand-to sounded, so that he and they fought the battle without food.

Besides being hungry and travel-worn the combatants at Waterloo were also rain-sodden. The regiments that had spent the night marching lay down to sleep in wet clothes and probably woke up to fight the battle still very damp. Those which passed the night in the fields, though they slept worse, or had no sleep at all, generally found means to dry out after sunrise. A young officer of the 32nd, who had woken wet through, managed to get into a shed where there was a fire and the men made large fires outside. The light company men of the 3rd Foot Guards, who had spent the night 'cramped sitting on the side of the wet ditch' south of Hougoumont, got a fire going 'which served to dry our clothing and accoutrements', and Leeke, of the 52nd, found a fire large and hot enough to get some sleep by. Wood, of the 10th Hussars, an officer whose Waterloo letter breathes the authentic cavalry spirit, 'got into a small cottage close to our bivouac ... most of us naked, and getting our things dry at the fire ... Old Quentin burned his boots and could not get them on.' Other cavalrymen, too, found their clothes spoiled by the wet. The Greys' scarlet jackets had run into their white belts overnight and Sergeant Coglan of the 18th Hussars attempted to dry his clothes by hanging them on the branches of trees. The Assembly was sounded before he had succeeded and he dressed in the saddle, 'crying out to those I had charge of to mount also'. Waterloo day was overcast, rather than sunny, so those who, like Coglan, failed to get near a fire at the beginning presumably stayed damp until well after midday. Houssaye's 'kaleidoscope of vivid hues and metallic flashes', his 'bright green jackets ... imperial blue collars ... white breeches ... breastplates of gold ... blue coats faced with scarlet ... red kurkas and blue plastrons ... green dolmans embroidered with yellow braid, red pelisses edged with fur' must have covered many limp stocks, sticky shirts and clammy socks.

How much better were the armies prepared for battle spiritually and mentally than they were physically? Of religious practice, which played such an important part in the English army's preparations for Agincourt, there seems to have been little or none before Waterloo. Chaplains, so numerous in medieval French and English armies, had almost disappeared from those of the nineteenth century (and were not to reappear until the more pious reigns of Victoria and Napoleon III). The great Napoleon's seems to have had none, its soldiers being among the last French citizens to parade the irreligiosity of the high Revolution. Wellington's army, indifferent rather than hostile to religion, had one chaplain per division, but as a group they were neither esteemed nor influential. He indeed had had chaplains appointed principally to combat the spread of Methodism, which he regarded as subversive of military hierarchy. But even of informal private prayer, to which Methodism in particular exhorted its followers, there is almost no mention in Waterloo memoirs. Leeke, a man of deeply religious temperament who subsequently took Anglican orders, recollected that 'my first thought of what would become of my soul in case I should be killed' did not occur until quite late in the battle. Bull, commander of one of the Royal Horse Artillery batteries, had made a habit in the Peninsula of holding prayers with his gunners, and may have done so before Waterloo, but it was not remarked upon; nor do any of the regimental histories record the holding of religious service before the battle, even though 18 June was a Sunday, and at least one of the regimental commanders – Colborne, of the 52nd – was well-known for his personal devoutness.

The most likely reason for this failure either corporately to observe the sabbath or to make private spiritual preparation was the uncertainty prevailing in two very tired armies and until late in the morning, as to whether there was to be a battle at all. The commanders, of course, were separately resolved to give battle – Napoleon from the outset, Wellington from shortly before dawn, when he received assurance of Blücher's intention to come to his support – but neither was certain of the other's frame of mind and until both made their intentions clear their subordinates could only speculate whether 18 June would be a day of fighting or marching. Mercer, who made a little promenade among the soldiers bivouacked near his battery position, heard a variety of opinions. 'Some thought the French were afraid to attack us, others that they would do so soon, others that the Duke would not wait for it, others that he would, as he certainly would not allow them to go to Brussels.' An officer of Picton's division recalled that his brother officers were 'generally gay and apparently thinking of anything but the approaching combat'; but it was hindsight which allowed him to write 'approaching', for the moment he described was still one of waiting for orders.

Gibney, assistant surgeon of the 15h Hussars, found 'waiting for orders ... tedious work'. 'We were anxious to be put in motion,' he wrote, 'if it were only to circulate our blood.' The English army before Agincourt had also found the prolonged wait physically tiresome and emotionally frustrating. To its soldiers, the decision for battle had come eventually as a welcome release, and the sounding of the Assembly may have evoked the same response from the men of both armies on 18 June. For in assessing their readiness for battle, it is important to remember that a very high proportion were experienced soldiers. While the appetite for battle grows with eating only in the most unusual individuals, most veterans would probably rather fight today than tomorrow if the intervening night is likely to be wet and the battle in any case unavoidable. Keppel, an ensign in the 14th Regiment, though not a veteran, probably spoke for many when he summarized his feelings during the long period of waiting as 'wishing the fight was fought'.

The 14th was alone among the British regiments of Wellington's army in being wholly unblooded. For the seven other non-Peninsula regiments had just fought Quatre Bras, where indeed three of them – the 2nd/1st Guards, the 2nd/69th and the 33rd – had suffered heavy or very heavy casualties. Of course, many of the Peninsula regiments now contained sizable contingents of young soldiers, but the majority of their officers and sergeants would have had experience in Spain. The raw Hanoverian regiments, however, for the most part lacked veteran leadership. Wellington had therefore so divisioned, and where possible, brigaded, the army that no very long section of his line would be held by inexperienced troops.

Napoleon's ingenuity had been less taxed than his opponent's. His army was nationally homogeneous and composed to a very high degree of professional soldiers. The Old Guard contained none but veterans of long service; but even in the line regiments the majority of men had seen action, and had handled their weapons under fire. They would thus have learnt not only how to bear the fatigues of campaigning but would also have been familiar with the two other most oppressive characteristics of the battlefield: smoke and noise. Smoke had been quite absent from the atmosphere at Agincourt, for the few cannon present had fired only once or twice, if at all. The black-powder weapons with which the artillery and all private infantrymen fought at Waterloo discharged smoke in dense, whitish-grey clouds, which hung low, needed a brisk breeze to disperse them and therefore usually obscured the front of any unit heavily engaged. We get some idea of how seriously smoke hampered visibility by a number of incidental remarks set down by combatants. Thus Lieutenant Wilson, of Sinclair's battery, which was in position four hundred yards north-west of La Haye Sainte, found 'the smoke so dense' during d'Erlon's attack that he could 'not see distinctly the positions of the French' (by which he must have meant those actually in combat with the British), 'being at that time ordered to direct my fire over the dead bodies of some horses in front'. Ingilby, another gunner whose battery stood on the extreme left of the English line and who enjoyed a good lateral view, interspersed his narrative to Siborne with such caveats as 'the thick rolling fire of the musketry, adding to the smoke from the Artillery, I could not perceive the further result' and 'it was only occasionally when the wind freshening and partially cleared away the smoke, that other charges . . . and movements in both Armies . . . could be distinguished.' Eventually his battery limbered up to follow the French off the battlefield: 'For some while we could see nothing whatever from behind the Infantry (which advanced slowly step by step) on account of the dense smoke from their musketry.'

Infantry memoirs certainly suggest that smoke clung more densely around them than around the artillery (an artillery salvo consumed more powder; infantry volleys followed each other more rapidly) and several regimental histories claim that their squares were enveloped in thick smoke for most of the day. The 1st/4th King's, in square near the La Haye Sainte crossroads, could not make out the farmhouse, though they were less than four hundred yards from it at the time of the French assault which carried the place. They were warned of the assault by sound, not sight; when they advanced, at the end of the battle, 'the movement carried them out of ... darkness' in which they had stood 'for a great part of the day'. The 18th Hussars' commanding officer, Murray, remembered that, at their advance at the end of the battle, 'we burst from the darkness of a London fog into a bright sunshine.' And Vivian, commanding the Hussar Brigade, described the smoke at the time of the great French cavalry attacks on the right-centre as 'literally so thick that we could not see ten yards off'. Smoke, moreover, had an effect on other senses than sight. Gronow described the interior of the square of the 3rd/1st Guards at about 4 p.m. as so thick with smoke and the smell of burnt cartridges that he nearly suffocated (Leeke also remarked upon 'a peculiar smell ... arising from a mingling of the smell of the wheat trodden flat down with the smell of the gunpowder'). Mercer, going into action at three, 'breathed a new atmosphere – the air was suffocatingly hot, resembling that issuing from an oven. We were enveloped in thick smoke ...'

But if smoke oppressed the senses, the noise of Waterloo assaulted the whole being. At Agincourt noise would have been chiefly human and animal and would have overlaid the clatter of weapon-strokes. There was still a good deal of perceptible human noise at Waterloo: an officer of Picton's division had remembered the noise of the army preparing for battle as similar to that of the 'distant murmur of the waves of the sea, beating against some ironbound coast'. Once the battle got under way there was cheering – Leeke, like several others, mentions hearing 'continued shouts of "Vive l'Empereur"' at the time of the Imperial Guard's attack, shrieking – the 32nd 'set up a death howl' when the French reached within forty yards of their line, and confused shouting – an officer of the 73rd describes a French advance as 'very noisy and evidently reluctant'. There were cries of pain and protest from the wounded – though here the testimony is contradictory, Mercer being pierced 'to the very soul' by the scream of a gunner whose arm had just been shattered, Leeke insisting that the wounded kept unnaturally silent. And there were of course shouts of command. There was also music: Gronow, Leeke and Standen recall hearing the beating of the *pas de charge* (which, one of Picton's officers says, was called by his men, who remembered it from Spain, 'Old Trousers') and there was piping in the squares of the Scottish regiments. The 71st's pipers played and re-played 'Hey, Johnnie Cope' and Piper McCay of the 79th stepped outside the square under French fire to play 'Cogadh na sith'.

But it was weapons which made by far the loudest and most insistent noise at Waterloo. Some of the sounds were incidental and unexpected. Lieutenant Wyndham of the Scots Greys remarked on the 'extraordinary manner in which the bullets struck our swords', a phenomenon which, as we know from an eighteenth-century memoir, set up a weird harmonic vibration; something of the same sort was produced by shot hitting bayonets, quite a frequent occurrence, though that could also sound like a stick being drawn along park railings. Leeke noted the 'rattle' which grape made when striking arms and accoutrements and Gronow, in an often quoted simile, likened the impact of his Guardsmen's musket-balls on the breastplates of Kellermann's and Milhaud's cuirassiers to 'the noise of a violent hailstorm beating upon panes of glass'. These were sounds which could only be caught at close range, however, for at any distance they would be drowned by the much louder and pervasive crash and rumble of firearms and artillery. Several witnesses nevertheless make a point of recalling the whistle and sigh of projectiles over their heads and above the noise of the cannonade: Mercer described it as a 'mysterious humming noise, like that which one hears of a summer's evening proceeding from myriads of black beetles'; to an officer of Picton's division it was a 'whistling' and 'familiar music', to a sixteen-year-old officer of the Scots Greys, hearing it for the first time, a 'whizzing' with 'really something rather grand about it'. Brave sixteen-year-old words! Mercer's medical officer, also for the first time 'hearing this infernal carillon about his ears, began staring round in the wildest and most comic manner imaginable, twisting himself from side to side, exclaiming, "My God ...

what *is* that? What *is* all this noise? How curious! – how very curious!" and then when a cannon-shot rushed hissing past, "*There! – there!* What *is* it all?"' But these upper-register notes penetrated the sound-storm only because they were intermittent, heard at close-hand and spelt danger. The sonic constant was the 'roar', 'rumble', 'crash', 'thunder', 'boom' of gunfire – few who were subjected to it attempt to define its quality precisely. But that is not surprising; for though the nearby explosion of shells and the firing of musket volleys were sonically different from each other, and both different from the more distant discharge of artillery, the differences tended to be drowned by the sheer volume of noise. That volume was very great indeed. Murray, a matter-of-fact cavalryman, described it simply as 'deafening'; Gibney, assistant surgeon of the 15th Hussars, said the noise was 'so loud and continuous ... that you could hardly hear what was said by the person next to you' (he was speaking particularly of the opening cannonade); Mercer, at the end of the day, was 'almost deaf' – and we may take him quite literally. As a battery commander whose guns fired about seven hundred rounds each (an astonishing figure) he had been at the focus of enough prolonged noise to have suffered damage to his hearing; so too had many front-rank infantrymen, whose ears had been only inches away from the muzzles of the rear-rank men during sustained bouts of musketry.

There were other circumstantial ingredients of battle beside fatigue, hunger, smoke and noise. Many combatants mention the wetness of the ground which, though doing much to reduce the effect of artillery by shortening the ricochet of solid shot and allowing shells to bury themselves, made for squalor underfoot. The 40th Regiment, on the ridge near La Haye Sainte, had trampled itself 'almost knee-deep' in mud by the end of the day, through the frequency with which it had formed from square to column on the same spot. We must also remember that the men, not being able to leave the ranks, would have had to relieve themselves where they stood. But all these circumstances, though intrusive enough to have been thought worth recalling by many Waterloo men, *are* of course in the last resort circumstantial.

What sticks in the forefront of survivors' memories is combat itself: their own and their comrades' behaviour, the action of the enemy and the effects of the weapons they faced. Is it possible, from the reams of testimony they have left, to discern in these dozens of transient individual experiences any pattern of human activity, any concrete 'reality' of battle in this, the apogee of black-powder warfare?

Categories of Combat

Even to begin to do so requires first that we separate out the various categories of man-versus-man and man-versus-weapon encounters which went to make up the totality of the conflict. Compared with Agincourt, the variety of encounter was greater; but not that much greater. Henry V's army had been composed of missile-firing infantrymen and armoured cavalrymen, most of the latter dismounted and fighting on foot, some in the saddle and bearing lances; there had also been a few cannon on the field. Napoleon's army consisted of missile-firing infantrymen and of cavalrymen, some of the latter armoured, and some lancers; he also had 250 cannon, and it was the presence of these weapons which explains – in crude terms – the altogether greater lethality of nineteenth- over fifteenth-century armies. It was also artillery which principally served to multiply the number of potential man-to-man and man-to-weapon encounters. At Agincourt there had been, in practice, only three types of encounter: single combat (hand-to-hand fighting between individuals, whether mounted or on foot); missile-firing infantry versus cavalry; and missile-firing infantry versus infantry (strictly, in their case, dismounted cavalrymen). At Waterloo there were seven sorts of encounter at least: single combat; cavalry versus cavalry; cavalry versus artillery; cavalry versus infantry; infantry versus infantry; missile-firing infantry versus missile-firing infantry; and artillery versus artillery (virtually a one-sided exchange, for Wellington had forbidden his gunners to fight artillery duels).

Single Combat

Single combat, which at Agincourt had generally occurred between dismounted men and often been deliberately sought out – as by the French noblemen who went to challenge Henry – was at Waterloo exclusively the affair of cavalrymen, and arose as a result of cavalry charges losing impetus and formation. This needs immediate qualification. Several instances of single combat between dismounted men are recorded. Gawler, of the 52nd, bluntly describes how one of his light infantrymen, challenged by a French officer, 'parried his thrust, closed with him, threw him on the ground and keeping him down with his foot reversed his musket in both hands'; despite 'a groan of disgust from his surrounding comrades' he then 'killed him with one thrust of his bayonet.' This took place in the closing stage of the battle, however, when the troops had left their defensive positions and were advancing in comparatively loose formation. During the really desperate passages, the demands of discipline denied individual infantrymen that freedom of movement within or from the ranks which is the basis of single combat.

The skirmishers, operating in front of the close-packed columns, had such an independent role. But its proper performance required them to avoid coming to close quarters with the enemy (the instructions to the 30th Regiment's skirmishers were quite precise on that point). So although numbers of cavalrymen, for example, were singled out for attack by sharpshooters (Colonel Muter of the 6th Dragoons saw a 'French soldier on his knees, deliberately taking aim at the Adjutant . . . and sending his bullet through his head'; one of Picton's officers 'could distinctly see a French soldier level his piece' at an officer of the Greys, 'fire, and bring him rolling to the ground'), they died without warning or the chance to defend themselves. Single combat – hand-to-hand, blow-for-blow, face-to-face – demands, by definition, equality of risk and fore-knowledge of the consequences. It also appears to presuppose consent by both parties (it was in that that chivalry saw its glory). Was consent usually or often given at Waterloo? And if so, was it given freely, or under the compulsion of a 'him or me' situation? Cornet Gage of the Greys wrote to his mother of their first charge, 'The men were only too impetuous, nothing could stop them, they all separated, each man fought by himself'; and the famous Corporal Shaw of the Life Guards certainly sought out opponents – he was 'very conspicuous, dealing deadly blows all round him'. But the Greys had not seen battle since 1794 and Shaw, a champion boxer in an age when boxing was a branch of the blood sports, was also probably crazed with drink. What seems to have happened in experienced regiments is that, their charge having failed to break (i.e., frighten away) bodies of cavalry in their path, and the men finding themselves intermingled with the enemy, individual soldiers struck out, in drill-book fashion, at those near enough to threaten them. Lieutenant Hamilton of the Greys provides a convincing account of what was probably a common experience:

one of the red lancers put his lance to my horse's head, I made a cut at his arm as I passed him; and as I did not look behind me to see whether I had struck him or his lance, I should not have known that I had struck his arm, had I not in recovering my sword thrown the blood on my white pouch belt. On inspecting my sword I saw that I had succeeded in wounding the lancer and possibly thus saved my own life. My fears were, when I saw him thrust at my horse's reins, that he would shoot me with his pistol, having heard of the red lancers sometimes doing so.

Self-defence in a moment of danger was not the only motive for individual fights. The sight of an enemy regiment's standard or a party bent on capturing one's own provoked men to extremes of ferocity. Sergeant Ewart of the Greys – probably the most famous of Waterloo heroes – found himself near the Eagle of the French 45th Regiment, during the confused fighting which followed d'Erlon's attack. He struck at its bearer who 'thrust for my groin – I parried it off, and ... cut him through the head ... one of their Lancers threw his lance at me but missed ... by my throwing it off with my sword ... I cut him through the chin upwards, which cut went through his teeth. Next I was attacked by a foot soldier, who, after firing at me, charged me with his bayonet; but ... I parried it and cut him down through the head.' And a sudden turning of tables could also lead to a deadly duel. Leeke, at the end of the battle, saw a French cuirassier chasing a German light dragoon. 'The latter was retreating at speed ... with his head down on his horse's neck and his sword over his own neck [but] watching his opportunity ... on finding himself near his friends [he] suddenly pulled his horse up upon his haunches, and dealt the cuirassier a blow across the face; he wheeled round and engaged the [German] in single combat, who managed to strike him again on his face, so that he fell over on one side and was pierced under the arm and killed.' Little chivalry there; even less in the experience of General Vivian who, with his right arm in a sling, was attacked by a cuirassier: 'I was fortunate enough to give him a thrust in the neck with my left hand ... and at that moment I was joined by my little German orderly, who cut the fellow off his horse.'

Cavalry versus Cavalry

Do these instances tell us anything about the character of mass cavalry combat at Waterloo? Both popular impressions and copy-book drill – and the initial charges in the two great series, British and French, were launched copy-book style – supposed cavalry versus cavalry charges to mean the meeting of dense formations at high speed. Moreover at least two British cavalry officers maintained that this was what happened. Waymouth, of the 2nd Life Guards, informed Siborne that 'the (Heavy) Brigade, and the Cuirassiers too, came to the shock like two walls, in the most perfect lines' and Wood of the 10th Hussars, writing to a friend, was at pains to refute '(what) the English papers say, "The Light Dragoons could make no impression on the French Cuirassiers." Now our regiment rode over them. Give me the boys who will go at a swinging gallop for the last seventy yards, applying both spurs when you come within six yards. Then if you don't go through them I am much mistaken.'

Wood, however, did not actually complete his charge, being badly wounded before it got under way, while Waymouth was really retailing the witness of a comrade. Common sense tells us, too, that cavalry coming 'to the shock like walls' and 'at a swinging gallop' will achieve nothing but a collapsed scrummage of damaged horses and men, growing bigger as succeeding ranks are carried on to the leading ones by their own impetus. A little inquiry reveals, in any case, that formations were much less dense and speeds much lower than casual testimony, and certainly than the work of salon painters, implies. The British cavalry were too few in number to cover much expanse of ground and though French cavalry formations were fairly dense at Waterloo, their leaders attempted, even during the charges into the 'funnel' between La Haye Sainte and Hougoumont, to keep an interval between the squadrons, while regiments and squadrons themselves were formed in line. This meant that the 120 men of the squadron were formed in two ranks, one close behind the other, but that the succeeding squadron rode, if possible, 100 yards behind. In theory the squadron could be manoeuvred at a gallop, say over twenty miles an hour, but it would very shortly lose cohesion if it was, as stronger horses outstripped weaker; and in any case distances and gradients on the Waterloo field make it seem unlikely that high speeds were achieved with any frequency. The 'classic' encounter of the 2nd Life Guards and the French Cuirassiers, described by Waymouth, was as near as anything seen during the battle to a straightforward collision, in that the two bodies met head-on and in motion. But the French had come a long way, over 1,500 yards and the last stretch uphill; while the British, though having a shorter distance to cover, had had to negotiate a succession of obstacles – first the road on the top of the ridge, 'too wide to leap, and the banks too deep to be easily passed' (Waymouth), and then 'the enclosure of the farm of La Haye Sainte' – before they could get to the French. Acceleration into a 'swinging gallop' by either side appears, under the circumstances, to have been an unlikely conclusion to their advance. Indeed, Waymouth reveals that the 'shock' took the form of a 'short struggle' with swords, and that it was success in the sword fight which allowed the British to penetrate the French line. In other words, the two lines must have been almost stopped dead when they met, and the British able to penetrate the French line because they found or created gaps in it.

Confirmation of this surmise can be found in the testimony of other witnesses. Morris, a sergeant in the 73rd, relating his view of events during the great French cavalry attacks of the afternoon, writes that 'the Life Guard boldly rode out from our rear to meet [the Cuirassiers]. The French waited, with the utmost coolness, to receive them, opening their ranks to allow them to ride in.' Consent – the vital precondition for single combat proper – is thus made to appear equally necessary if cavalry formations were to fight each other in any effective fashion. When they did so, of course, they did not fight as *formations*, but as individuals or small groups. Morris continues his account with a description of the sort of fighting to which this 'opening of ranks' and 'allowing entry' led. 'I noticed one of the Guards, who was attacked by two cuirassiers at the same time . . . he disposed of one of them by a deadly thrust in the throat. His combat with the other lasted about five minutes.' We are back with single combat again.

Indeed, unless cavalry action resolved itself into a complex of single combats, it was pretty harmless to the participants. Mercer recalls watching two lines of French and British light cavalry skirmishing with each other on ground between the armies, on the evening of 17 June. 'The foremost of each line were within a few yards of each other – constantly in motion, riding backwards and forwards, firing their carbines or pistols, and then reloading, still on the move . . . I did not see a man fall on either side; the thing appeared quite ridiculous; and but for hearing the bullets whizzing overhead, one might have fancied it no more than a sham fight.' He has an equally dismissive account of cavalry's occasional mutual harmlessness even in the performance of its true shock role. It refers to an encounter during the afternoon cavalry battle. 'A Regiment of Cavalry (I think of the German Legion) . . . formed up to attack a [French regiment]. The French, immediately aware of this danger, wheeled to the left into line, and, both advancing to the charge, literally came into collision at full gallop. The shock appeared tremendous, yet there was no check, each party passing through the other, and closing their files immediately on being clear.' In another account, his explanation of what occurred is still more revealing:

There was no check, no hesitation, on either side; both parties seemed to dash on in a most reckless manner, and we fully expected to have seen a horrid crash – no such thing! Each, as if by mutual consent, opened their files on coming near, and passed rapidly through each other, cutting and pointing, much in the same manner one might pass the fingers of the right hand through those of the left. We saw but few fall. The two corps re-formed afterwards, and in a twinkling both disappeared, I know not how or where.

The cavalry in both these cases, however, were fresh – so, too, more importantly, were their horses – and had kept their formation and their heads. Cavalry could, it must be emphasized, suffer very grievously at the hands of other cavalry when nerves failed, horses were blown or weapons markedly unequal. The French Lancers, armed with a weapon which gave them an advantage in reach of many feet over their British opponents, frequently killed or wounded opponents without being touched themselves. The Cuirassiers who gave way before the charge of the Life Guards near La Haye Sainte sought an escape down the sunken road and 'the 1st Life Guards made great slaughter amongst the flying Cuirassiers who had choked the hollow way' – a ready-made demonstration of Ardant du Picq's view that the most dangerous course in war is to retreat when in close contact with the enemy. It produces a situation the exact opposite of that obtaining in single combat by consent, and appears to stimulate an almost uncontrollable urge to kill among those presented with a view of the enemy's backs. It is this urge which made it so perilous for cavalry to overextend a charge, finding themselves at the end of it alone or scattered, on blown horses, and deep within the enemy's positions. Hence the heavy casualties suffered by the Scots Greys who, carried away by success and inexperience, rode right across the valley separating the two armies after their repulse of d'Erlon's attack. 'Our men were out of hand,' wrote one of the staff officers present.

Every officer within hearing exerted themselves to the utmost to reform the men; but the helplessness of the Enemy offered too great a temptation to the Dragoons, and our efforts were abortive. It was evident that [his] reserves of Cavalry would soon take advantage of our disorder ... If we could have formed a hundred men we could have made a respectable retreat, and saved many; but we could effect no formation, and were as helpless against their attack as their Infantry had been against ours. Everyone saw what must happen. Those whose horses were best, or least blown, got away. [Most of the rest] fell into the hands of the enemy ... It was in this part of the transaction that almost the whole of the loss of the Brigade took place.

The Greys in fact lost nearly 200 men and over 200 horses in this short space of time, chiefly through being ridden down by French Lancers, who spared no one, mounted, unhorsed or even disabled. Ponsonby, the Brigade Commander, was among those killed, and lost his life because of a false economy. He had left his best charger, worth far more than the government compensation fund would pay if it were killed, behind the lines and chosen to ride instead an inferior hack. The French Lancers caught him struggling to safety over heavy ground, easily rode him down, and speared him to death.

Cavalry versus Artillery

Cavalry were also vulnerable to other cavalry which happened to be accompanied by horse artillery, even if the two bodies of horsemen were otherwise evenly matched. The 7th Hussars, in the retreat to Waterloo, had charged a body of French light cavalry 'but could make no impression ... we did not give ground, nor did they move. This state of things lasted some minutes, when they brought down some Light Artillery'; these guns knocked over several of the British and swiftly persuaded their commander to order the rest away. On the battlefield itself, however, mobile artillery was not usually a threat to cavalry, being too valuable to risk in detached action, when it might easily be captured. At Waterloo, as in most other pitched battles of the period, the artillery draught horses were sent to the rear once the guns had been brought up, and the gunners then fought where they stood, usually among or slightly in front of the infantry. This static artillery, for all that its crews looked so exposed to charging swordsmen, was one of two agents of destruction from which the cavalry had most to fear (the other being steady infantry formed in square), and the British gunners at Waterloo were certainly responsible for a very large proportion of the deaths and wounds which the French cavalry suffered. Rudyard, an officer of Lloyd's battery of 9-pounders, wrote of the afternoon attacks to Siborne:

> The Cuirassiers and Cavalry might have charged through the Battery as often as six or seven times, driving us into the Squares, under our Guns ... In general, a Squadron or two came up the slope on our immediate front, and on their moving off at the appearance of our Cavalry charging, we took advantage to send destruction after them, and when advancing on our fire I have seen four or five men and horses piled up on each other like cards, the men not having even been displaced from the saddle, the effect of canister.

This extract invites a short commentary, since it is highly descriptive of what the artillery did during the battle. Rudyard is telling us that guns were posted some distance, perhaps twenty or thirty yards, *in front* of the infantry (a position unthinkable a hundred years before or after); that they were firing a multiple projectile, consisting of 'a number of small cast-iron spheres in a sheet metal can which disintegrated on discharge' (canister), at an operational range of about a hundred yards or less; that the French who survived the salvoes, of which they had to stand perhaps two or three as they approached, actually rode past and round the guns, whose crews fled before them to the shelter of the infantry squares; the cavalry, then subjected to the square's musketry, lost whatever impulsion it had left and, on seeing the British cavalry to the square's rear, turned and retreated, leaving the guns, which it had no means of removing, where they stood, to be re-manned by their crews and used against the backs of the cavalrymen as they withdrew. Why the artillerymen were able to stand by their guns so long is explained by Mercer in another place: he estimated the pace of the Horse Grenadiers' advance as a brisk trot ('none of your furious galloping'), noted that the impact of his battery's first salvo, fired at sixty yards, brought it down to a walk, and that the second and subsequent salvoes piled up such 'heaps of carcasses' that the survivors either could not get past them, or, if they did, fell individually victim to his fire and that of the squares to his rear. Nevertheless some escaped by spurring their horses between the guns and riding through the intervals between the squares and back again; while others died without coming within striking distance of the artillery, for Mercer's guns were double-shotted, and the round-shot which followed the canister in the same discharge smashed deep into the French formation, striking several horses or men in succession. Little wonder that 'the survivors struggled with each other' and that he 'actually saw them using the pommels of their swords to fight their way out of the *mêlée* ... pushing furiously onward, intent only on saving themselves ... until the rear of the column, wheeling about, opened a passage, and the whole swept away at a much more rapid pace than they had advanced.' Even so they were not at once out of danger. The gunners in front of the 14th Regiment who, 'at the [French cavalry's] approach, had thrown themselves at the feet of our front rank men, returned to their guns and poured a murderous fire of grape into the flying enemy ... When the smoke cleared ... the matted hill was strewed with dead and dying, horses galloping away without riders and dismounted cuirassiers running out of the fire as fast as their heavy armour would allow.'

Thus the cavalry versus artillery fight at Waterloo turned out to be almost wholly one-sided affairs. Even when the French horsemen notionally took possession of the British guns, they were unable to remove them, having with them neither harness nor limbers with which to tow them away. That they did not try, or succeed in, spiking them either (that is, driving a spike into the touch-hole by which they were fired) has always caused puzzlement; the probable explanation is that the act required a man to dismount, something which no cavalryman, whether out of braggadocio, stupidity, caste-pride or self-preservation, seemed prepared to do in the face of the enemy.

Cavalry versus Infantry

How much more successful were the cavalry's encounters with infantry? To this a clear-cut answer is more difficult to offer, for such encounters were more varied in character. Cavalry *could* do infantry very great harm, using 'harm' in a military rather than human context. The regiments of the Union Brigade which charged the flank of d'Erlon's Corps, at a moment when it was under fire and attempting to deploy from column to line, reduced it to a purposeless crowd in a few instants. 'As we approached at a moderate place,' wrote Evans, a staff officer of the Brigade, 'the fronts and flanks began to turn their back inwards; the rear of the Columns had already begun to run away ... In going down the hill the Brigade secured about 2,000 prisoners, which were successfully conducted to the rear ... The enemy fled as a flock of sheep across the valley – quite at the mercy of the Dragoons.' Shelton, an officer of the 28th Regiment who followed these Dragoons on foot, 'distinctly saw them charge the heavy Reserve Column, and break it. The greater number of the French threw down their arms when broken by cavalry.' (Tomkinson of the 16th Light Dragoons saw these muskets later 'in two lines nearly as regularly as if laid on parade'.) Some did not. 'Many,' recalled Marten of the 2nd Life Guards, 'threw themselves on the ground until we had gone over, and then rose and fired.' But in neither case did many of these infantrymen suffer personal injury. To lie down was usually enough to put one beyond a swordsman's (though not a lancer's) reach, and those who shammed were already safely behind the cavalry, whose attention was focused on the enemy lines to which their impetus was carrying them; those who offered a genuine surrender had it readily accepted, for this was early in the battle, when there was clearly much fighting ahead, and no time or motive for casual slaughter. At the end of the day, however, isolated bodies of infantry whose nerve had gone and who could no longer expect support from the rest of the army, suffered wounds and death when trying to escape or even surrender. Duperier, a ranker-officer of the 18th Hussars, came late in the evening on 'a regiment of infantry of the franch, nothing but "vive le Roy", but it was too late beside our men do not understined franch, so they cut a way all through till we came to the body of reserve when we was saluted with a voly at the length of two sords. We tacked about and had the same fun coming back.' Murray, commanding the regiment, got in among a mob of the fugitives, one of whom thrust at him with his bayonet: 'his orderly was compelled to cut down five or six in rapid succession for the security of his master'; not a story, one feels, which would convince a court of inquiry.

But even that late in the day French infantry which 'would stand' could see off British cavalry without trouble. Taylor, of the 10th Hussars, saw, at about the same time as Duperier was massacring the unfortunate French turncoats, 'about thirty of the 18th ... gallantly, though uselessly, charge the square on the hill, by which they were repulsed'. And indeed if the story of Waterloo has a *leitmotiv* it is that of cavalry charging square and being repulsed. It was not absolutely inevitable that horsemen who attempted to break a square should fail. The 69th Regiment, caught before it had properly formed square at Quatre Bras, had had three of its companies sabred by French cavalry, and lost one of its colours (the disgrace was the greater because it had also lost a colour at Bergen-op-Zoom the year before). And at Garcia Hernandez, in 1812, Bock's Dragoons of the King's German Legion had broken clean into a regiment of French infantry standing securely in square and delivering fire. What happened on that occasion, however, helps to explain why the event had no counterpart at Waterloo – was, indeed, one of the rarest occurrences in contemporary warfare. It came about because one of the dragoon horses, moving on a true course and at some speed, was killed in mid-stride, and its rider with it; continuing the charge for several paces, the pair of automatons did not collapse until directly above the bayonets of the front rank. Carrying these down, they opened a gap through which a wedge, and then the remainder, of the regiment followed. The dead horse had done what living flesh and blood could not; act as a giant projectile to batter a hole in the face of the square.

The feat of breaking a square was tried by the French cavalry time and again at Waterloo – there were perhaps twelve main assaults during the great afternoon cavalry effort – and always (though infantry in line or column suffered) with a complete lack of success. Practice against poorer troops had led them to expect a different result: a visible shiver of uncertainty along the ranks of the waiting musketeers which would lend their horsemen nerve for the last fifty yards, a ragged spatter of balls over their heads to signal the volley mistimed, then a sudden collapse of resolution and disappearance of order – regiment become drove, backs turned, heads hunched between shoulders, helot-feet flying before the faster hooves of the lords of battle: this, in theory, should have been the effect of such a charge. This process was more nearly realized in many places along Wellington's front than the magnitude of the ultimate cavalry debacle suggests. 'The first time a body of cuirassiers approached the square into which I had ridden' (it was the 79th Regiment's) wrote a Royal Engineer officer, 'the men – all young soldiers – seemed to be alarmed. They fired high and with little effect, and in one of the angles there was just as much hesitation as made me feel exceedingly uncomfortable.' Morris, a sergeant of the 71st, testifies to the power of the psychological shock-waves emitted by these mounted onsets. 'A considerable number of the French cuirassiers made their appearance, on the rising ground just in our front, took the artillery we had placed there and came at a gallop down upon us. Their appearance, as an enemy, was certainly enough to inspire a feeling of dread – none of them under six feet; defended by steel helmets and breast-plates, made pigeon-breasted to throw off the balls. The appearance was of such a formidable nature, that I thought we could not have the slightest chance with them.' In every case, however, almost exactly the same sequence of events served to break the impetus of the cavalry's advance and to transfer the psychological advantage from attackers to defenders. First of all, the cavalry changed direction or decelerated or even stopped as they came within effective musket-shot of the square. Sometimes they did so because the protective artillery, or a well-timed and well-aimed volley, had knocked down horses in the leading ranks.

Leeke, of the 52nd, describes them coming on 'in very gallant style and in very steady order, first of all at the trot, then at the gallop, till they were within forty or fifty yards of the front face of the square, when, one or two horses having been brought down, in clearing the obstacle they got a somewhat new direction, which carried them to either flank ... which direction they all preferred to the charging home and riding on to our bayonets'. Eeles, of the 95th:

kept every man from firing until the Cuirassiers approached within thirty or forty yards of the square, when I fired a volley from my Company which had the effect, added to the fire of the 71st, of bringing so many horses to the ground, that it became quite impossible for the Enemy to continue their charge. I certainly believe that half of the Enemy were at that instant on the ground; some few men and horses were killed, more wounded, but by far the greater part were thrown down over the dying and wounded. These last after a short time began to get up and run back to their supports, some on horseback but most of them dismounted.

Sometimes the stop happened because the leaders hoped to trick or panic the square into firing before its shots could take proper effect, meaning to ride in during the fifteen-second delay necessary for re-loading. The Duke himself recalled watching squares which 'would not throw away their fire till the Cuirassiers charged, and they would not charge until we had thrown away our fire'; but, as he knew, the trick would not work against soldiers, like the British, who were trained always to keep half the fire of the regiment in reserve. Sometimes the French stopped simply because they feared to go forward, often when they had already entered the narrow, deadly killing ground immediately in front of the square and the safer course would have been to go on rather than back. Reynell, commanding the 71st, refers to these 'repeated *visits* from [the] Cuirassiers. I do not say *attacks*, because these Cavalry Columns on no occasion attempted to penetrate our Square, limiting their approach to within ten or fifteen yards of the front face, when they would wheel about, receiving such fire as we could bring to bear upon them, and, as they retired, *en passant*, that from the neighbouring square.'

Injurious though it was for cavalry to flinch or turn away from squares which had fire in hand, the results of riding round them, Red Indian fashion, or loitering with intent to terrorize were worse. For the infantry's fear of the cavalry seemed dissipated by the smoke of their first discharge. The Royal Engineer, sheltering with the 79th, noted how quickly moral superiority shifted:

No actual dash was made upon us. Now and then an individual more daring than the rest would ride up to the bayonets, wave his sword about and bully; but the mass held aloof, pulling up within five or six yards, as if, though afraid to go on, they were ashamed to retire. Our men soon discovered they had the best of it, and ever afterwards, when they heard the sound of cavalry approaching, appeared to consider the circumstance a pleasant change (from being cannonaded)!

Macready, of the 30th, remembered that his men 'began to pity the useless perseverance of their assailants, and, as they advanced, would growl out, "here come those d—d fools again"'. Confident in, even elated by their ability to outface the French squadrons (at Quatre Bras, after their second dispersal of a French charge, there had been 'a good deal of laughter and handshaking' in the 30th's square), the British infantry began to inflict on them heavy casualties whenever they were foolish or badly enough led to linger within range. Saltoun, commanding the Guards' light companies, ordered them to fire at a group of French cavalrymen who then 'rode along the front of the 52nd with a view of turning their right flank, and were completed destroyed by the fire of that regiment.' The 40th Regiment, alerted by an experienced sergeant who called out, 'They are in armour. Fire at the horses,' brought down Cuirassiers in swathes. 'It was a most laughable sight to see these guards in their chimney armour – trying to run away, being able to make little progress and many of them being taken prisoner by those of our light companies who were out skirmishing.'

This reference to casualties among the horses should remind us that the French troopers were engaged in a dual battle of wills – not only with the British musketeers but also with their own mounts. Gronow, an intensely acute observer in one of the Foot Guards' squares, describes how 'the horses of the first rank of cuirassiers, in spite of all the efforts of their riders, came to a standstill, shaking and covered with foam, at about twenty yards distance ... and generally resisted all attempts to force them to charge the line of serried steel'; much the same thing happened in front of Mercer's battery, where a 'confused mass stood before us ... vainly trying to urge their horses over the obstacles presented by their fallen comrades.' As the casualties increased and the going on the slope up to the British positions deteriorated, and as the litter of carcasses grew to form a tide-mark around the edge of the squares' killing zones, it became more and more difficult to force the horses to face fire. The less resolute French units drew off a hundred or a hundred and fifty yards, leaving their skirmishers to loose off their pistols at the British infantry, or trot up and down firing their carbines. It was a perfectly fruitless, almost pathetic proceeding. Indeed, the question poses itself to the modern reader whether sympathies – given that sympathy is an appropriate emotion – over the conflict between cavalry and squares are not misplaced. On the face of it, the predicament of the storm-racked battalions (Mercer's analogy for the attack of the cavalry on Wellington's chequer-board of squares was that of 'a heavy surf breaking on a coast beset with isolated rocks, against which the mountainous wave dashes with furious uproar, breaks, divides and runs, hissing and boiling, far beyond up the adjacent beach') is breath-catching. But, as Jac Weller has shown by careful analysis of formation-widths, the number of cavalrymen in an attacking line was always much lower than the number of infantrymen with whom their onset brought them face to face. If the average strength of a battalion was about five hundred, it would, formed four deep, present in square a face about sixty feet across, opposing about 140 men to the approaching French cavalry. They, because of the greater bulk of their horses, could present no more than about eighteen men on the same width of front, with another eighteen immediately behind, and it was these thirty-six who would take the brunt of the square's fire. But even though they would suffer worst by the first volley, the full strength of the squadron to which they belonged was only a hundred and twenty; and if its moral power failed to disarm the infantry – as it always did fail at Waterloo – then each horseman theoretically became the target for four infantrymen. Viewed like this, 'Here come those d—d fools again' seems an appropriate judgement on the character of the conflict.

Artillery versus Infantry

Indeed, even the best cavalry could normally hope to break good infantry only with the help of artillery. Hence the existence of 'horse artillery' or 'galloping guns', whose task was to accompany cavalry to within charging distance of the infantry and, from just beyond musket-shot, to open gaps in the square so large that its members were either stunned into passivity or driven to flight. But Waterloo, at least so far as the use of artillery was concerned, was not a normal battle. The 'ratio of men to space', particularly of cavalrymen to space in the 'funnel' between Hougoumont and La Haye Sainte where, during the afternoon, about 10,000 French horsemen were milling about on a front only 800 yards wide, was so high that room for artillery to accompany the cavalry to within charging distance, let alone to unlimber when it got there, could not be found. The result was that the infantry versus cavalry combats were reduced to exactly no more than that – very much to the infantry's advantage and safety.

Furthermore, the near approach of cavalry caused the French gunners bombarding the British line from long distance to cease firing, for their own horsemen obscured the view as they breasted the slope on the British side of the valley, and risked becoming the recipients of their shots; thus, as Leeke wrote – and many other infantrymen expressed the same sentiment – 'the charges of cavalry were a great relief to us all . . . at least I know they were to me.' For though the eighty-odd guns in Napoleon's 'grand battery', 700 yards distant from the British line, could not do any particular infantry formation the same concentrated harm as could a 'galloping battery' firing grape or canister into it from close range, the arrival of their solid cannon-balls was so frequent, the effect of the balls on human flesh so destructive, the apprehension of those temporarily spared so intense that the cannonade came as near as anything suffered by the British at Waterloo to breaking their line. Wherever and whenever he could, the Duke positioned his battalions just on the reverse of the crest, in what soldiers call 'dead ground', often allowing them to lie down, so that most of the balls skimmed their heads. But many battalions had nevertheless to spend some of the day under direct fire. Bylandt's Dutch-Belgians, east of La Haye Sainte, were demoralized by it and decamped; the Inniskillings, who stood their ground, drew their wounded into the square, threw their dead out and closed their ranks, were destroyed.

Even in the shelter of the crest, some battalions suffered casualties. Reed, of the 71st, reported that French artillery on the left of Hougoumont 'were able to throw in a fire among us as we lay down under the slope of the hill, by which we suffered some loss, I think fifty men', and the 1/95th was 'annoyed' by cannon-shot which 'rolled over the hill behind which we were posted' (Leeke's sergeant prevented him from trying to stop one of these shot which came 'rolling down like a cricket ball' with the warning that it would have seriously injured his foot). When the approach of French infantry or cavalry forced them to their feet, making them solid targets, even a single hit by a French gun could cause awful injury. Leeke, who had watched mesmerized while a French gun-crew several hundred yards distant sponged out, loaded, rammed and fired, apparently straight at him, and had even glimpsed the ball leave the muzzle, saw the four men in file next to him fall dead or mutilated two seconds later. In the neighbouring square of the 71st a round shot killed or wounded seventeen; and the 40th Regiment, though in open column, suffered a succession of horrors, described by Lieutenant Hugh Wray: 'We had three companies almost shot to pieces, one shot killed and wounded twenty-five of the 4th Company, another of the same kind killed poor Fisher, my captain, and eighteen of our company . . . and another took the 8th and killed or wounded twenty-three ... At the same time poor Fisher was hit I was speaking to him, and I got all over his brains, his head was blown to atoms.'

When artillery of either side found the opportunity to 'co-operate' with other arms, that is, make its attack simultaneous with infantry or cavalry action against the same enemy formation – something difficult to achieve, as we have seen, because of the danger it ran of hitting its own men – the effect of its fire was magnified. For the threat offered by the presence of enemy soldiers close at hand forced a defending formation to stand up and stand still; and this 'standing to be cannonaded, and having nothing else to do, is about the most unpleasant thing that can happen to soldiers.' 'Take us out of this,' demanded some men in Picton's division of their officers, 'are we to be massacred? Let us go and fight them.' The French, who managed to send some guns forward with d'Erlon's Corps, unlimbered them within 120 yards of the 32nd Regiment and 'opened sad gaps in its square'. Bull's battery, supporting the Scots Guards against the Imperial Guard near Hougoumont, exploded howitzer shells 'to such an extent in the midst of those fine fellows that [Maitland] could distinctly see, above the smoke of these explosions, the fragments of men, Grenadier caps, muskets and belts.' In both these cases, the artillery wrought the slaughter it did because the infantry that formed its target, being at close quarters with other infantrymen, were unable to shelter from its fire.

Infantry versus Infantry

This conflict of infantry with infantry, though it occupied nearly everywhere at Waterloo a much shorter span of time, continuous or intermittent, than that between artillery and infantry or cavalry and infantry, was, in 'result' terms, the crucial element of the battle – a statement which can be made with fair safety of almost every battle fought in the period between the eclipse of the armoured horseman in the fourteenth century and the rise of the armoured fighting vehicle in the twentieth. For the action of cavalry and artillery against infantry was subsidiary and preliminary; during its course, the role of infantry was indeed 'to be massacred', if that could not be avoided. Naturally, it behoved a commander to shield his infantry as much as possible from cannonading or cavalry charge. But since infantry was (and is) the only force with which ground could (and can) be held (physical occupation being ten points of the law in war, and infantry the bailiff's men), it could never be withdrawn from ground whose possession was held vital simply to avert loss of life. (Wellington, asked by Halkett at a particularly critical moment that 'his brigade, which had lost two-thirds, should be relieved for a short time', sent the message 'Tell him what he asks is impossible: he and I, and every Englishman on the field must die on the spot we now occupy.') But, *per contra*, infantry which refused to yield ground required by the enemy, despite the menaces of his cavalry and the efforts at massacre by his artillery, had ultimately to be attacked by other infantry.

But 'ultimately' did not necessarily mean 'after every other method had failed'; it could also mean 'because trial and error had recommended it'. In point of fact, the attack of the Imperial Guard was essayed at the end of the day because every other means had failed. But the infantry attack on Hougoumont, at the beginning of the battle, was decided upon because other methods were judged unprofitable, while the constant skirmishing between the light infantry troops of both sides, who operated in front of the two battle lines, was an element in the fighting whose necessity was ordained by experience.

It was their extreme skill in skirmishing that had enabled the French, in the early battles of the twenty years' war, to inflict heavy loss on infantry without sending their own to close quarters. Eventually their enemies had grasped the need to oppose skirmishers with skirmishers, and the special formations raised by the British – the 95th Rifles and their own and the K.G.L. light infantry – had learnt to achieve the same standards as the French. Most of these regiments had, at Waterloo, to stand in the line which held the ridge; but they detached companies to their fronts to provide a covering screen, of which the specially trained light companies of the ordinary infantry regiments also formed part. Pratt, a lieutenant commanding the light company of the 30th Regiment, provides an account of his day at Waterloo which perfectly demonstrates the duties of skirmishing light infantry. His instructions from the adjutant were: '"To cover and protect our Batteries. To establish ourselves at all times as much in advance as [prudent]. To preserve considerable intervals ... for greater security from [artillery] fire. To show obstinate resistance to [enemy light] Infantry, but to attempt [nothing against] Cavalry, but to retire ... upon the Squares in our rear ... When the charge was repulsed, to resume our ground"!' He carried out these orders faithfully, 'creeping down the hill to nearly its foot' where he carried on 'a desultory fire' with the French light infantry, interrupted by 'frequent advances and retreats' to his own square or more often a K.G.L. or Hanoverian. 'Towards the close of the day I found myself for the last time near the bottom of the slope with the few Light Infantry troops that were remaining.' They were 'gradually retiring before the overwhelming force opposed to them'. The moment to which he refers was just after the loss of La Haye Sainte; he was then wounded and 'ceased to be an eye-witness of what took place afterwards'. But by that time, the importance of the role of light infantry was almost played out. The 'crisis' of the battle – the clash of heavy infantry with heavy infantry – was upon the armies.

But the battle had also begun with just such a clash: that between the Foot Guards and Jérôme's Frenchmen for possession of Hougoumont. Here is an example of a command decision for an immediate infantry attack in preference to other methods – one of our meanings of 'ultimately'. It was for Napoleon a necessary decision (though in 'result' terms probably a bad one), for the strong, loopholed walls of the château threatened death to any cavalry which came near and were fairly proof against even heavy field artillery. But, if necessary, it was also a desperate decision, the solidity, size and complexity of Hougoumont and its outbuildings, and the strength of its garrison making it almost impregnable to infantry as well. The French did, at one moment, manage to break open the gate to the central courtyard, but the British defenders succeeded in shutting it behind the small party which gained entry and then hunted them down until all, but a drummer boy whose life was spared, were dead. The extreme ferocity of this episode testifies to the special character of the Hougoumont Battle. It has often been called a 'battle within a battle' and, in that it was fought for the greater part of the day by two strong detachments which took almost no part or interest in anything else which was happening on the field, that description is accurate. Modern students of aggression theory would probably be more struck, however, by its intensely 'territorial' quality. And, tautological though the concept of 'territoriality' is in this context, the behaviour of the defenders of Hougoumont, and of the smaller La Haye Sainte, was indeed wholly directed throughout the several hours the fighting lasted in those places to preserving the absolute integrity of very precise boundaries: at La Haye Sainte, those of the farmyard and garden, at Hougoumont of the château, walled garden and orchard. Given, however, that men are going to fight for walled enclosures, the choice of what they will and will not defend is pretty narrowly determined by the configuration of those enclosures themselves. Moreover, intra-specific fighting (that is, fighting between members of the same species) for territory is in animals, on the observation of whose behaviour modern aggression theory is based, highly 'ritualized'. Title to property is all in nature: the title-holder usually has only to simulate an attack on the interloper, and he to rehearse a conventional gesture of submission and beat a retreat, for hostility instantly to cease. But title is not all in war. The practice of ritualized attack, defence and submission certainly goes on – one thinks of the demand of the German commander of the Cherbourg Arsenal on 27 June 1944, that a tank should be produced to fire a token shot at the main gate to make respectable his capitulation, and, for all their bloodiness, one would also recognize a strong ritual element in the cavalry attacks on the squares at Waterloo ('Now and then an individual more daring than the rest would ride up to the bayonets ... wave his sword about and bully'). It is probably also the case that human attackers concede to human defenders a certain claim – which one would call moral but for the ambiguity implied – to their territory, be it a mere shell-pocked hilltop or water-logged trench, that would provide an additional explanation of the tendency for the 'defensive' to prevail over the 'offensive' in warfare and a reason why surprise – which allows the claim-jumper to stoke the fires of his acquisitiveness while the defender drowses over the deeds – is so much valued as a tactical achievement by aggressive commanders. But that is about as far as 'territorial' theories of behaviour can be pushed in a battlefield context. The offensive usually fails in war only after real fear has been excited, real humiliation inflicted, real blood spilt. Indeed, almost the exact contrary of the situation which observers have perceived in nature prevails on the battlefield: there it is proprietorship which is fictive, combat which is in earnest.

One principle of animal behaviour which does seem to be applicable, however, to human combat is that which its promulgator, the zoologist Hediger, called the *critical reaction.* He derived it from his observation of the response of animals to threat, which he saw was determined by the distance at which the threat was offered. Beyond a certain distance – which varies from species to species – the animal would retreat, within it he would attack. He called the two distances 'flight distance' and 'critical distance' and, in the sort of instantly illuminating example after which all communicators strive, explained that 'lion-tamers manoeuvre their great beasts of prey into their positions in the arena by playing a dangerous game with the margin between flight distance and critical distance.' There is some evidence (besides ordinary self-knowledge) to suggest that instinctual judgements of critical and flight distances also impinge on human behaviour; it has, for instance, been found that some abnormally violent men consistently underestimate the distance separating them from other human beings, consequently investing inoffensive gestures with menace and subjecting those who make them to apparently unprovoked assault. Soldiers certainly play games with critical and flight distances. The advice of Sun Tsu, the ancient Chinese philosopher of war, that confrontations with the enemy should begin by the donning of fearsome masks and the uttering of dreadful threats, and that only after these preliminaries had failed to put him to flight should recourse be made to the use of weapons themselves, contains an implicit recognition of the *critical reaction*; and a great deal of primitive warfare – foot-stamping, spear-waving and drum-beating – clearly takes place safely outside the two sides' critical distance. Indeed, it is probably true to say that the more primitive the peoples involved in warfare, the less they will be prepared to violate critical distances. But even ruthless modern commanders have shown themselves ready, in certain circumstances, to respect critical distance when it served their purpose: the displays of overwhelming air- and sea-strength mounted by the Americans during the Second World War offshore of Japanese-held islands they intended to invade – the wheeling and massing of landing-craft, the circling and swooping of aeroplanes – had as part of their purpose, vain though it may have been, the intimidation of the defenders; and the grand review of his army organized by Napoleon on the morning of Waterloo itself, out of range but within sight of the Allies, a proceeding without parallel in his generalship, seems to have been intended to frighten the Belgians, perhaps also the British, into leaving their positions.

But, if we really want instances of the influence of critical and flight distances on the fighting at Waterloo, the place to look is in the records of the defence of the two strongpoints. At La Haye Sainte the garrison, when their ammunition ran out, were rushed by French tirailleurs who broke into a central passage-way in the buildings. At first many of the intruders were killed with the bayonet, and their bodies used to barricade the entry, but their comrades then managed to scale the roof, and by firing down into the mass of the defenders, force them to run. At Hougoumont, as we have seen, the French were less lucky when they broke in; the garrison still had ammunition, were able to hold off those who would have followed the storming-party until the gates had been closed, and then massacred those inside. There had been a similarly bloody struggle a little earlier across the wall of the formal garden, which had resulted in the death of every Frenchman who got inside. There are, no doubt, several ways of describing what went on during these moments at those three points; but one at least is to say that the French had triggered among the British a *critical reaction*, compelling them instinctively to strike to kill. If that were the case, it would help to explain why accounts of the fighting in the confined spaces of La Haye Sainte and Hougoumont, though in one sense so abhorrent, are in another so comprehensible, so easy to accept. Walls, passageways and corners bring men suddenly face to face with each other, restrict their room for manoeuvre and bar their line of retreat. If the stories of La Haye Sainte and Hougoumont are familiar to us, or seem almost *déjà vu*, it is not necessarily because we are reminded by them of more recent fights in the ruins of Stalingrad or Hué. The nature of the fighting at all these places would be as readily grasped by a time-traveller from Tancred's Jerusalem or Achilles' Troy, just as its exigencies will be understood by anyone whose stomach has jumped at a creak on the turn of a darkened stairway in an unfamiliar house.

The encounter which eludes the comprehension of the modern reader, though also between infantry and infantry, is a different one. It is the Queen's Move of black-powder warfare, the head-on clash of heavy infantry, at close-range, in close-order, over levelled musket barrels. Discounting the attack which led to the fall of La Haye Sainte, since it really took the form of a skirmish on a gigantic scale, supported by light artillery run forward for the event, there were only two of these Queen's Moves on Wellington's front. The first is known as d'Erlon's attack, the second as the 'Crisis' – the attack of the Imperial Guard near Hougoumont at the very end of the battle. In both, very large and dense masses of French infantry advanced across the whole width of the valley separating the two armies to within a few yards of the British line, exchanged fire with it for a very brief period, then turned summarily about and fled.

What makes episodes of this sort so difficult for the modern reader to visualize, if visualized to believe in, if believed in to understand, is precisely their nakedly face-to-face quality, their offering and delivery of death over distances at which suburbanites swap neighbourly gardening hints, their letting of blood and infliction of pain in circumstances of human congestion we expect to experience only at cocktail parties or tennis tournaments. The descriptions, nevertheless, are unequivocal. Mountsteven, an officer of the 28th Regiment, who remembered 'looking over the hedge . . . and admiring the gallant manner the French officers led out their Companies in deploying', estimated that they were at about thirty or forty yards when 'we poured in our fire, sprung over the fence and charged. The Enemy ran before we could close with them, and, of course, in the greatest confusion.' An officer of a neighbouring regiment, the 92nd, thought the distance was twenty yards and 'could hardly believe, had he not witnessed it, that such complete destruction could have been effected in so short a time'. Reports from the brigades on the other flank which defeated the Imperial Guard are similar in tone and content. Dawson Kelly was with the 73rd Regiment,

> when the last attacking column made its appearance through the fog and smoke, which throughout the day lay thick on the ground. Their advance was as usual with the French, very noisy and evidently reluctant, the Officers being in advance some yards cheering their men on. They however kept up a confused and running fire, which we did not reply to, until they reached nearly on a level with us, when a well-directed volley put them into confusion from which they did not appear to recover, but after a short interval of musketry on both sides, they turned about to a man and fled.

Powell, of the 1st Guards, standing directly in the path of the main French column, saw the Grenadiers

> ascending the rise *au pas de charge* shouting '*Vive l'Empereur*'. They continued to advance until within fifty or sixty paces of our front, when the Brigade was ordered to stand up. Whether it was the sudden and unexpected appearance of a Corps so near to them, which must have seemed as starting out of the ground, or the tremendously heavy fire we threw into them, *La Garde*, who had never before failed in an attack, *suddenly* stopped. Those who from a distance and more on the flank could see the affair, tell us that the effect of our fire seemed to force the head of the Column bodily back!

Another Guardsman, Dirom, confirms his account: 'The French Columns appeared staggered . . . convulsed. Part seemed inclined to advance, part halted and fired, and others, more particularly towards the centre and rear of the columns, seemed to be turning round . . . On our advance the whole of the French Columns turned round and made off.'

The facts then are not in dispute. The French approached to within speaking distance of the British, were halted by their fire, failed to overcome it with their own, and retired. In practice, there was a little more to their attacks than these eye-witness accounts reveal. Both were preceded by a good deal of heavy skirmishing, and accompanied by some close-range cannonading, from guns which the French infantry brought along with them. But in so far as 'pure' infantry encounters were possible on Napoleonic battlefields, that is what these two episodes were.

How are we to explain the suddenness and completeness of the collapse of French endeavour in each case? The French outnumbered the British and were fresh ('[They] showed no appearance of having suffered on their advance, but seemed as regularly formed as if at a field day', wrote Dirom of the Imperial Guard). Both formations were composed of excellent and deeply experienced soldiers. Both were well led – five generals marched in front of the Guard and Ney, *le plus brave des braves*, disentangled himself from the fifth horse he had had killed under him that day to walk beside them. Both had had their attacks more than properly 'prepared' – that is, preceded by multiple cavalry charges or prolonged bombardment or both. Both delivered something more than simple 'columnar' assaults, in that the commanders of each formation attempted to deploy their men, once British musketry struck, into the sort of linear arrangement which would allow them to fight fire with fire. Yet it was British, not French, fire which prevailed and, in prevailing, led to a collapse of the French which was not partial, but total. How so?

Some things have been left out of account. First, there was a heavy British artillery fire brought to bear on both d'Erlon's Corps and the Imperial Guard, which caused severe casualties and must have weakened resolve before they came within musket-shot of their infantry opponents. Second, the recoil of d'Erlon's corps was hastened and then completed by a very powerful British cavalry charge, while another, in weaker strength, accelerated the retreat of the Guard. Third, the two attacks were not without some initial or local success. D'Erlon's Corps unhinged Bylandt's Brigade of Belgians from their place in the line, while the 3rd and 4th Grenadiers of the Guard shook the British 30th and 73rd Regiments and broke the Brunswick and Nassau contingents, which had to be prevented from retreating by British cavalry. Nevertheless, the two attacks resolved themselves essentially into conflicts of infantry against infantry and culminated in a clear victory of British over French: at the focus of the Imperial Guard's attack, of five battalions (52nd, 33rd, 69th Regiments and 2/1st and 3/1st Guards) against five (1/3rd and 2/3rd Chasseurs, 1/4th and 2/4th Chasseurs and 2/3rd Grenadiers); at the focus of d'Erlon's attack, of seven battalions against twenty-four.

To say that it was done by superior musketry is not to explain very much, though the mechanics are easy enough to describe. The British battalions, formed two, in some cases four deep, were in line: thus a strong battalion, like the 52nd Light Infantry, presented a front of 250 men, with three equal rows behind them, to the enemy. Their fire would have been effective – that is, would have achieved a significant percentage of hits – at over 100 yards, but commanders, as was normal, reserved it until the enemy were much closer. When they had fired, they could reload in about twenty to thirty seconds; but, after the discharge of the first volley, it was habitual in British battalions to fire by platoons or by ranks, so that part of the unit was reloading while another fired. Overall, the result was the same: the projection at the enemy of about 2,000 heavy leaden musket-balls every minute.

The French, initially at least, were in column, a fact which makes their collapse easier for us to understand. For columns were, by definition, much deeper than lines. The men in the battalions of the Imperial Guard were probably formed nine deep, and the battalions ranked closely one behind the other; the same arrangement in d'Erlon's Corps did much to nullify the three-rank formation he had adopted for his battalions. Thus, whether broad like d'Erlon's or narrow like the Imperial

Guard's, the columns' fate was to be overlapped and outflanked by any British battalion; the main column of the Imperial Guard met five, two to its front, two on one flank and one on the other. D'Erlon, who had foreseen that his columns risked being literally engulfed in fire, had hoped to avert the danger by deploying his broad columns into lines of equal width to Wellington's as soon as they reached musket-shot. But the British beat him to the draw. The Imperial Guard, whose commander had taken no such precaution, was caught in narrow column and so deprived of even a theoretical chance of equalling the weight of musketry to which it was subjected. For columnar formation, of course, effectively disarmed the majority of soldiers confined within it. Only those at the very front and along the margins could use their weapons; those in the interior, even if they glimpsed the enemy, could not raise their muskets to fire.

Thus both d'Erlon's men and, six hours later, those of the Imperial Guard, were 'beaten in the fire fight', in that those at the front and along the flanks were outnumbered by the British soldiers opposite and suffered fearfully disproportionate casualties. Even so, the French suffered nothing like the total of loss to which they were mathematically liable. Each British battalion ought, at fifty yards, to have sent each of its shots home, which means for example that the whole of the Imperial Guard's main column should have been destroyed by their opponents' opening volley. But, far from this being the case, many of the foremost French soldiers survived the first blast: Shelton, of the 28th, remembered that one of d'Erlon's columns 'attempted a deployment to their right' after his regiment had given it 'a very steady volley' and Dawson Kelly, with the 73rd, credited the French he met in the 'Crisis' with returning the British fire at least for 'a short interval'. Not only does this tell us something about the marksmanship of the period – that even at fifty yards a large proportion of musketeers clean missed their target – it reinforces suspicions that many musketeers did not aim at all, or at least did not aim at a particular human target. This is borne out by the recollection of a Waterloo officer that the word of command generally used was 'Level' rather than 'Aim'. But the deliverance from seemingly certain death of so many Frenchmen at the head of the columns also draws attention to another and more significant phenomenon. Although it was they who had suffered most from the British fire, it was also they who did what little was done to counter or return it effectively. The men at the rear did nothing, or did nothing useful. Indeed, it seems safe to go further. It was at the back of the columns, not the front, that the collapse began, and the men in the rear who ran before those in the front.

We can assemble several hints that this was in fact what happened. Colborne, commanding the 52nd, 'observed the Enemy in great confusion, some firing, others throwing away their packs and running to the rear'; and unless the men at the front were prepared to run through the fire of their own rear ranks, we must suppose it was they who were firing, those behind who were making off. Dirom is absolutely specific: after the British Guards had delivered their volley, 'part [of the French column] seemed inclined to advance, part halted and fired, and others, *more particularly towards the centre and rear of the Columns*, seemed to be turning round.' Sir de Lacy Evans, who had charged d'Erlon's column with the Union Brigade six hours before, had noticed the same phenomenon: 'As we approached at a moderate pace the fronts and flanks began to turn their backs inwards; *the rear of the Columns had already begun to run away.*' In other words, those least immediately threatened were the soonest off. It was behaviour such as this, rather than direct British action, that rendered useless the most critical French attacks of the day, and led to Napoleon's defeat. How can we explain it?

It is tempting to apply the concept of 'critical' and 'escape' distances to the situation; but probably too mechanistic. 'Critical reaction' is an explanation of individual rather than mass behaviour. More rewarding is an attempt to visualize the difference in conditions prevailing at the open face and in the closed interior of the French columns. At the front were the officers – Mountsteven had seen and admired 'the gallant manner of the French officers' during d'Erlon's attack; Dirom remembered 'the Officers of the leading Divisions (of the Guard) in front waving their swords'; Dawson Kelly described 'the Officers being in advance some yards cheering their men on'. If there were officers in the heart of the columns, they were prevented by the press from setting any heartening example to their men, were indeed hidden from them and, like them, deprived of a view of events. The men at the front could see their officers, see the enemy, form some rational estimate of the danger they were in and of what they ought to do about it. The men in the middle and the rear could see nothing of the battle but the debris of earlier attacks which had failed – discarded weapons and the bodies of the dead and wounded lying on the ground, perhaps under their very feet. From the front came back to them sudden crashes of musketry, eddies of smoke, unidentifiable shouts and, most important, most urgent, tremors of movement, edging them rearward and forcing them, crowd-like, in upon each other. Crowdlike too, in their leaderlessness, in their lack of information, in their vulnerability to rumour, they would have needed very little stimulus and what that little was we cannot guess ('without any *very apparent* cause', Dawson Kelly remarked [his italics] of the fight on his front) to transform them from an ordered mass into a suddenly fugitive crowd, and so carry them off the battlefield. Canetti, in his weird, inchoate book *Crowds and Power*, provides a poetic vision of what may have happened next and why:

> The *flight crowd* is created by a threat. Everyone flees; everyone is drawn along. The danger which threatens is the same for all . . . They feel the same excitement and the energy of some increases the energy of others . . . So long as they keep together they feel that the danger is distributed . . . No one is going to assume that he, out of so many, will be the victim and, since the sole movement of the whole flight is towards salvation, each is convinced that he personally will attain it . . . Everyone who falls by the way acts as a spur to the others. Fate has overtaken him and exempted them. He is a sacrifice offered to danger. However important he may have been to some of them as a companion in flight, by falling he becomes important to all of them . . . The natural end of the flight is the attainment of the goal; once this crowd is in safety it dissolves.

This appeal to the action of the irrational is defensible on three grounds. First a physical one: the men in the front of a stricken formation *cannot*, as we saw at Agincourt, run away until those behind them have opened the road. Second, there is eye-witness evidence of crowdlike behaviour at the rear of the French columns during both d'Erlon's and the Guard's attacks; a 52nd officer, writing years after the event, used the analogy 'making off like a mob in Hyde Park when a charge is made *towards* them' to describe the Guard's flight. Third, crowds are implicit in armies. Inside every army is a crowd struggling to get out, and the strongest fear with which every commander lives – stronger than his fear of defeat or even of mutiny – is that of his army reverting to a crowd through some error of his making. For a crowd is the antithesis of an army, a human assembly animated not by discipline but by mood, by the play of inconstant and potentially infectious emotion which, if it spreads, is fatal to an army's subordination. Hence it is that the bitterest of military insults contain the accusation of crowdlike conduct – rabble, riff-raff, scum, *canaille*, *Pöbel* – and the deepest contempt soldiers can harbour is reserved for leaders whose armies dissolve between their fingers – Cadorna, Kerensky, Gough, Gamelin, Perceval.

Many armies begin as crowds, like Lincoln's militia of 'ninety-day volunteers' or the British 'New Armies' of 1914, and the transformation of such a crowd into an army is in itself enough to win a soldier a lasting title to fame. Kitchener, his reputation otherwise demolished, is still accorded respect for his triumphs of army-building in 1914–15 and Carnot and Trotsky, the latter even less of a general than the former, enjoy the posthumous title of military leader solely for having provided their respective revolutions with disciplined soldiers, recruited from the mobs which had destroyed the old order.

Many armies, beginning as crowds, remain crowdlike throughout their existence. The great medieval hosts, tenuously bound together by links of kinship and obligation, were formidable only by reason of their size and because of the very variable military skills of their individual members. Tactically quite unarticulated, they were vulnerable to the attack of any drilled, determined, homogeneous force. Clive and Gordon, at the head of quite tiny European, or European-style, armies were consistently able to disintegrate the vast oriental armies they met because the latter were really not very much more than feudal crowds of retainers and followers who not only outnumbered but actually impeded the quite small nucleus of genuine fighting men they contained.

The replacement of crowd armies by nuclear professional armies was one of the most important, if complex, processes in European history. Its complexities are such as to require a literature rather than a paragraph for examination; but one ramification demands emphasis: the singularity of the institution which the process produced. Whatever its origin – whether it was, like the British army, forged by civil war from a bumpkin militia, or, like the Russian, hammered out of a conscripted serfdom by foreign mercenary officers – the standing army which emerged in most European states during the seventeenth century stood alone and apart, both among the other components of the state's apparatus and in the experience and imagination of the people it policed. Over no other group of subjects did the state exercise so rigorously, so minutely, so continuously its power; within no other group – except the religious orders, the newest and most 'progressive' of which, the Society of Jesus, was itself deliberately military in organization – were actions and attitudes regulated so scrupulously by code and timetable. Nor would they be, until industrialization and compulsory education, bearing their tainted gifts of factory discipline and textbook learning, came to transform the life of urbanized populations two centuries later. Even then, the army would remain for civilians a model of conformity and purpose, particularly for the leaders of the movements which, fathered by industrialism, then became its foe. Educated by the steady failure of 'their' crowds to overcome armies in the streets after 1830 – a failure which became absolute after 1871 – the men of revolution, whether violent or gradualist, made it their ambition to give to their followers the same advantages of order, command, pliability, enjoyed by the forces on which their class-enemies – a new name for a new idea – regularly called to frustrate their aims. Their transformation of the fickle and spontaneous crowd into the disciplined, mass political party was to be as important an achievement for the future of states, as, in its time, had been the creation of standing armies. As important but less remarkable; for Maurice and Gustavus had nothing, except the fragmentary advice of classical writers and the unsavoury example of mercenary companies, on which to base their norms of military excellence. Bebel, Jaurès, Guesde, the creators of the *Sozialdemokratische Partei Deutschlands* and the *Section française de l'Internationale ouvrière*, had in the organization of their own national armies examples of the degree of centralization and quality of staff-work necessary for the mobilization of the latent power of the proletariat. The second generation of mass political parties, populist and anti-Marxist, like the German Nazis and the Italian Fascists, would actually adopt the structure and dress of armies (the language of the mass political party, as Baudelaire had remarked, had been militarized for half a century) and, in Germany at least, eventually precipitate the most fundamental of political crises by demanding that the army transfer its functions to the party-in-uniform. The idea would have been even more repellent than it was to the generals had they realized that it had respectable socialist roots in the pre-1914 programmes of the S.P.D. and the S.F.I.O.

But perhaps its roots stretched even further back than that. The evaporation of the Revolution in revolutionary France is one of the most puzzling vanishing tricks in modern European history. A great deal has been done to demystify it; and perhaps too much should not be made of the role of the Armies of the Republic in absorbing both the wild men and wild ideas of 1792. Nevertheless, the existence of those armies and their continued success abroad was a factor in reconciling the libertarians and perhaps even the radicals to the stultification of the revolutionary movement at home after 1794. For one of the great unexpressed ideas of the Revolution was that 'Militarism is Theft': by its very existence, the standing army deprived free men of their right to protest, to demonstrate, to heckle, to jostle, to intimidate, to riot – all rights which it could be imagined had been freely exercised before the king had possessed soldiers to repress them. Napoleon's appropriation of the army cap-stoned his seizure of power and made possible his inauguration of a regime more effectively repressive than any administered by the king. Yet Napoleonic repression did not appear to be the betrayal it was because the army, which was the Empire's ultimate guarantee, remained in mood and ethos a creature of the Revolution. To the end it was anti-Bourbon, anti-clerical, egalitarian, open to talents. Thousands of young Frenchmen might seek to avoid serving beneath its standards. But as long as the standards were tricoloured, as long as they proclaimed Liberty, Equality, Fraternity, those who still cared could console themselves with the belief that the Revolution lived. The army, that extraordinary organism, which marched with a million synchronized legs to a single word of command, rose and ate and slept by the clock, practised punctuality, moved in unison to the tap of a drum, spoke a private language of command and submission, owed a wider loyalty than to family and place, which resembled, in short, no other institution under the wide skies of France, had been, in its white coats, both symbol and agent of the power of kings; dressed in blue, it stood for the victories of the Bastille, of the Tuileries, of the Champ de Mars, and embodied, at however submerged a level, the principle of the sovereignty of the people.

What happened, therefore, on the lower slopes of the ridge beneath Hougoumont on the evening of 18 June 1815, was of crucial importance in more respects than one. The agonized incredulous cry, '*La Garde recule!*' did not only speak Napoleon's defeat – though that it certainly did: Wellington's order to the commander of the 52nd, 'Go on, Colborne! Go on! They won't stand. Don't give them a chance to rally,' demonstrated his recognition that the disintegration of Napoleon's last reserve sealed his victory. But the reduction of the Guard to a fugitive crowd was also the reversal of the most powerful current in recent European history. The Revolution had made itself manifest by the Parisian crowd's defeat or subversion of the royal army in July, 1789; the metamorphosis of the Guard into a crowd, its spirit crushed, its solidarity broken, its militancy extinct, its only motive self-preservation, its only purpose flight, marked, as effectively as anything else we can point to, the restitution of power to its former owners. Louis XVIII at least got the point. After his second restoration, daring what he had not risked in 1814, he disbanded every regiment in the army and remade it in a new and different style.

From a strictly military point of view, however, the crowdlike behaviour of the Guard at the end of the battle, even though we wish to explain it and cannot (for the concepts of *anomie* and 'collective neurosis' fashionably applied to crowd behaviour by social psychologists certainly have no place in a consideration of fugitive soldiery), is of less interest than that of those from whom they ran. For although the suddenness and completeness of the Guard's collapse implies a long and terrible preliminary ordeal, it had in fact suffered little in comparison with many British regiments, which had been under fire for five, six or seven hours before it issued from its place of shelter near La Belle Alliance.

What had made these regiments stand? And it is important to emphasize that 'stand' is used precisely. Regiments, sub-units, individuals were allowed to, and did, take cover; Saltoun, halted by the Duke while marching his Guards light companies out of Hougoumont orchard, ordered them to lie down 'according to an invariable custom'; the Royal Scots, at the time of d'Erlon's second attack 'were moved forward to the hedge ... ordered to form line and lie down'; the Life Guards lay down before their great charge; the 3/14th were ordered to lie down in square, the men lying 'packed like herrings in a barrel'; and the men of a regiment of Picton's division, lying down behind the ridge during the afternoon, spent the time reading letters scattered from the packs of Frenchmen killed during d'Erlon's attack. But unless so ordered, to lie down, or even to duck, was thought at best cowardly, at worst a dereliction. Leeke draws an illuminating distinction in the case of one of the 52nd's new sergeants, who escaped a cannon-ball 'by stooping just as he saw it in line with him at some little distance; this was quite allowable when his comrades were lying down at their ease.' Shortly afterwards faced himself by the same predicament, he 'thought, Shall I move? No! I gathered myself up, and stood firm, with the colour in my right hand' (the shot was the one which killed the four men next to him). And later still, when an explosive shell fell in the middle of the 52nd's square, one officer called out 'Steady, men!' another, 'I never saw men steadier in my life', 'the shell burst, and seven poor fellows were struck by the fragments.' Men who flinched were reproved: when a shell passed over a column of the 52nd, the men 'instantly bobbed their heads'; Colborne, the commanding officer, shouted, '"For shame, for shame! That must be the 2nd Battalion (who were recruits), I am sure". In an instant every man's head went straight as an arrow.' Mercer too had chided his gunners 'for lying down when shells fell near them until they burst' and found himself compelled to stand 'looking quite composed' when, some time later, one fell at his feet, luckily to explode harmlessly.

His attitude towards taking cover would have been dictated by the soldier's code of honour, whose tenets are implied by the disdain he reveals for the conduct of one of the military doctors: 'a shot, as he thought, passing rather too close, down he dropped on his hands and knees . . . and away he scrambled like a great baboon, his head turned fearfully over his shoulder as if watching the coming shot, whilst our fellows made the field resound with their shouts and laughter.' But the infantrymen, who would have shared his code, had a stronger motive in forcing their men, and themselves, to stand still. For the whole purpose of enemy artillery fire was to make men break formation. When, out of self-preservation, they did, it could have disastrous results. The 30th and 73rd, ordered 'in an evil moment' to march under cover of a bank from the fire of a pair of French guns accompanying the Guard's attack, became 'disordered by our poor wounded fellows clinging to their comrades thinking they were being abandoned' and by bumping suddenly into some other British soldiers. On this occasion, 'fortunately the enemy took no advantage of it.' At other times and places, sudden confusions of this sort precipitated a charge by horse or bayonets which could lead to the defeat of a whole army.

So what was it that, even during moments of disorder and peril like that just described, could prompt men 'jammed together and carried along by the pressure' to make an effort to stand, 'good-humouredly laughing ... struggling to get out of the *mêlée*, or exclaiming "By God, I'll stop, Sir, but I'm off my legs"'? Alas, the 'motivation to combat', individual or collective, of the private soldier of this period is almost impossible to analyse, for we know so little about him. Simple courage should not be discounted, nor the wish to stand well in the opinion of comrades. But the line infantryman, as opposed to the sharpshooter or horse-trooper, had little opportunity to display the initiative which would have called attention to his bravery. His was the unspectacular duty of standing to be shot at. What sustained him?

Not all did stand. The non-British troops of Wellington's army, in particular some of the Dutch-Belgian and minor German contingents, shirked more or less flagrantly; most cavalry of these nationalities refused to charge, or even ran away; a lot of the infantry drifted out of the battle or had to be kept in place by coercion. The men in the right face of the 14th Regiment's square, 'irritated by the ... conduct' of some Belgian cavalry which first refused to charge and then ran away, 'unanimously took up their places and fired a volley into them'. The Duke of Cumberland's Hussars, a volunteer regiment of rich young Hanoverians, galloped away from d'Erlon's attack to Brussels with the news that Wellington was beaten. The Brunswickers, who had fled through the night from the noise of Mercer's hoofbeats at their heels on the road from Quatre Bras, gave way at the sight of the Guard, but allowed themselves to be rallied by the Duke, who led them back into the line; earlier some Belgian infantry had been spotted by a 16th Light Dragoon 'firing their muskets in the air, meaning to move off in the confusion'; they were also steadied by the Duke. No British regiment actually ran away. But some gunners panicked during d'Erlon's attack and some regiments were occasionally less than perfectly steady. Individuals went absent. A troop shoemaker of the 16th Light Dragoons, an old soldier but 'deranged', disappeared on the morning of 18 June and reappeared in the evening. 'The men ... did not resent his leaving them, knowing the kind of man and his weakness.' Another from the same troop who, it was thought, 'had got away during the advance to plunder, was reported to me [Tomkinson] by the men and booted by them the morning following the action.'

It may not have been for slackness that he was kicked round the troop. Looting appears to have been so universal an activity, so energetically practised even during the battle itself, in the firing line and in advance of it, that it may have been for taking unfair advantage that his comrades punished him. Certainly we ought to consider the possibility that it was the prospect of loot which helped to keep men in place, handy, as it were, for the fruit as it fell. At the height of the battle an officer of Picton's division 'saw (the truth must be told) a greater number of our soldiers busy rifling the pockets of the dead, and perhaps the wounded, than I could wish ... with some exertion we got them in. Those of our own regiment the Colonel beat with the flat of his sword as long as he had breath to do so. The fellows knew they deserved it; but, they observed, someone else would soon be doing the same, and why not they as well as others?' Such a one was seen by Seymour, Uxbridge's A.D.C., who found himself unhorsed beside Picton at the moment he was killed, repelling d'Erlon's attack; 'from [Picton's] trousers' pocket a Grenadier of the 28th was endeavouring to take his spectacles and purse.' A little later, when the 44th Regiment was charged from the rear by a French straggler, one of the privates unhorsed him with a single shot in front of the regiment, ran forward to kill him and swiftly robbed his body before rejoining the ranks.

Soldiers have always looted; indeed, the robbing of the enemy, particularly an enemy killed in single combat, and, for preference, of an object worthy of display for its intrinsic or symbolic value – the finery or weapons of the vanquished – has always provided an important motive for fighting. But an economic motive operates too. The capture of a ransomable captive had offered the medieval warrior one of the few chances then available of making a sudden fortune. Ransom had long since lapsed as a practice and its institutionalized substitute, prize money, offered nothing like the same rewards, even though it accrued by right and not by hazard: Waterloo prize money for privates amounted to £2 11s. 4d. Very much larger sums than that – which equalled forty days' pay – were to be found, however, on the bodies of the dead and wounded, for the only safe storage for valuables in an army without bankers was about the person. Officers knew very well what would happen to their coin and watches once they were hit; hence the fund of stories – beloved, if misunderstood, by Victorian readers – of stricken officers sending for their best friends to receive their trinkets. Sir William Ponsonby was handing a locket to his A.D.C. when both were speared by the French lancers; Howard, of the 33rd, was 'sent for . . . repeatedly' by his friend Furlong 'who was wounded dangerously' and, when Howard could not be found, 'said he must die and therefore sent his watch'. Furlong recovered, and so could reclaim the keepsake; but that evening 'plunder was for sale in great quantities, chiefly gold and silver watches, rings, etc., etc. Of the former,' wrote an officer of Picton's division, 'I might have bought a dozen for a dollar a piece (but) I do not think any officer bought . . . probably reflecting (as I did) that in a few days' (they expected another battle) 'our pockets would be rifled of them as quickly as those of the French had been.'

This selling of loot at prices far below its value tends to demolish the notion that the hope of plunder sustained the ordinary soldier's steadfastness. Were there other factors?

Drink may certainly have been one of them. Tiredness, I have suggested, helped to inure the soldier to fear, and much of the army was tired. But many of the soldiers had drunk spirits before the battle, and continued to drink while it was in progress. Shaw, the slashing Lifeguardsman, was, in the opinion of Sergeant Morris who watched him guzzling gin at about noon, drunk and running amok when he was cut down by the French cuirassiers; Morris himself took three canteens full of gin 'for the wounded', but shared some of it with a friend; Sergeant Lawrence's officer, in the 40th Regiment, kept running to him during the battle for a swig at his spirits flask; and Dallas, the commissary of the Third Division, hailed by his general with the demand, 'My brave fellows are famished for thirst and support, where are the spirits you promised to send them?', managed to get a cart forward and rolled a barrel into the middle of a square, where it was broached, and the contents distributed, during the closing stages of the battle itself.

Almost every regimental memoir refers to drink being distributed. But we should probably not think of alcohol having more than an indirect effect in keeping the ranks unbroken. Much more positive, in the case of those soldiers who wanted or tried to run, was the simple mechanism of coercion. Most of the reports we have are of British soldiers, particularly British cavalry, acting to prevent the non-British contingents leaving position. The 10th Hussars stood behind some Brunswickers during the French cavalry attacks and 'kept their files closed' to prevent them leaving the field; the 11th Hussars did likewise, and the 16th Light Dragoons; Vivian stood his hussars '10 yards behind infantry which were running away. They returned to line, our cavalry cheering them, and began firing again.' Duperier, the ranker-officer of the 18th Hussars, had passed 'the Belgun troope, which I saw of my own eyes, officers behind them lethring away (as the Drover did the Cattle in Spain) to make them smell the gunpowder.' Later, during the Imperial Guard's attack 'We ... formed line close to our infantry' (these may have been British) 'close to their tails and them almost nose to nose with the french ... to pass the time away I done like the Belgum officers, every one that faced about I laid my sword across his shoulders and told him that if he did not go back I would run him through, and that had the desired effect for they all stood it.'

But other officers, beside the Belgians, could be brutal with their own soldiers. Mercer noticed that the ranks of the Brunswickers 'presented gaps of several file in breadth, which the Officers and Sergeants were busily employed filling up by pushing and even thumping their men together; whilst these, standing like so many logs ... were apparently completely stupefied and bewildered.' (Could they have been drunk?) 'I should add that they were all perfect children. None of the privates, perhaps, were above 18 years of age.' The French manhandled their men. Leeke saw 'a French officer strike, with the flat of his sword, a skirmisher, who was running ... to the rear' and an officer of the French 45th Regiment was 'thrusting a soldier

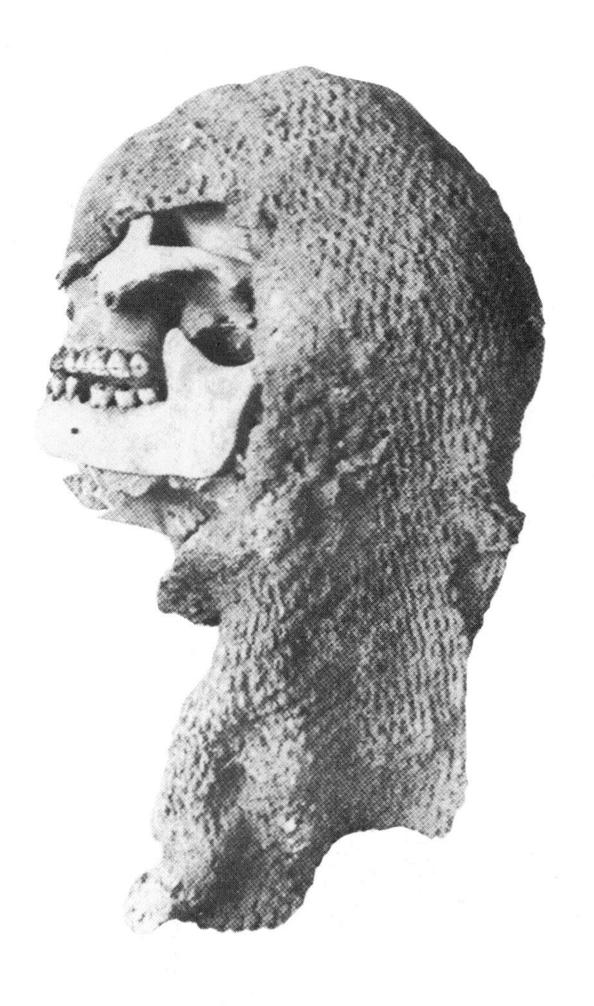

1. The skull of a Swedish soldier killed in the Battle of Visby, 27 July 1361. The fallen defenders of Visby were buried, most unusually, in their armour – it is supposed because hot weather and their great number (about 2,000 bodies were disinterred 600 years later) defeated the efforts of the victors to strip them before decomposition began. The mass grave yielded one of the most fearsome revelations of a medieval battle known to archaeologists. *Kungl. Vitterhets Historie och Antiqvetsakademien, Stockholm*

2. This miniature from a fifteenth-century *Chronicles* of Froissart is of the Battle of Poitiers, 1356, but the armour and weapons are contemporary: it conveys the effect of archery on cavalry at short range and suggests something of what happened on the flanks at Agincourt in the opening stage of the battle. *Bibliothèque Nationale, Paris*

3. These heaps of dead and wounded are from a miniature in the *Chroniques du Hainaut* of Jean de Guise (Mons, 1478); it may have been something similar which the Agincourt chroniclers described as the 'wall of bodies'.

4. This detail from a fifteenth-century miniature in a version of the *History* of Valerius Maximus may portray the fate of the French men-at-arms at Agincourt cut off by the English archers from the protection of their main body.
Harley 4374, f. 68v., Harley Collection, British Museum

5. This representation of a square of Highlanders (perhaps the 1/92nd) receiving cavalry at Waterloo – the time would be mid-afternoon – ruthlessly subordinates reality to artistic convention: the horsemen are arriving much too fast, charging the square head-on instead of lapping round the edges, and actually getting to weapon's reach with the infantry instead of stopping, falling or turning back several horse-lengths away.
Wellington Museum, Apsley House

6. *Scotland for Ever* (1881) is an imaginative reconstruction of the charge of the Scots Greys in the early afternoon at Waterloo. The artist, Lady Butler, a sister of the poetess Alice Meynell, was also the wife of a general and so able to persuade the then commanding officer of the regiment to form it up and charge her at her easel. Nevertheless, she has heavily dramatized her subject-matter: these riders (the right-hand half of the canvas is shown) are moving far too fast and on a collision course.
Leeds City Art Gallery

7. A German attack on an improvised British line of 'scrapes'. The Germans have probably reached as close as they have thanks to the cover of the hedge on the right. One of them appears to be a mounted officer, which would place and date the photograph – one of the very few to show both attackers and defenders – to Belgium or France, August-September 1914.
Ullstein

8. One of the few existing photographs of attacking troops taken from their front. These Russians, in battalion column, are charging a German or Austrian trench somewhere on the Eastern front in the autumn of 1914. Several have already been hit. At some places on the Somme the German defenders may have been presented with targets almost as dense.
Ullstein

9. A French counter-attack at Dien Bien Phu, probably by men of the *6ème Bataillon de parachutistes coloniaux* at strongpoint Eliane 1, 10 April 1954. The helmets in the foreground are of the Vietminh defenders.
Daniel Camus, Paris Match

forward' at the moment when the Union Brigade charged into them.

Even though we have only the most indirect references to British officers coercing their soldiers, we should not suppose that they did not do so. Indeed, the very formation of the square, merely tactical as it may seem, concealed a strong coercive purpose. Infantry in line, particularly if formed four deep, offered just as much fire to cavalry as when in square. In line, however, the ratio of officers to 'attacked length' was altogether lower than in square, for there all the officers were grouped in the centre and could turn in an instant to consolidate whichever face of the square was attacked; moreover the weapons they and the sergeants carried, swords and halberds, though of little offensive value, were exactly what was needed to keep individual soldiers, or groups of them, from running away. In one of General Lejeune's paintings of a Napoleonic battle in which he fought, he has actually portrayed a French sergeant pushing against the back of one of the French ranks, using his halberd horizontally in both hands to hold the men in place. It is not improbable to think of British sergeants having done the same at Waterloo.

But to see the square as a disciplinary device only is to underrate its overall importance, and to miss sight of two elements, perhaps the most important, acting on the British 'will to combat': group solidarity and individual leadership. Nothing better conveys the significance of the group to the individual under stress of battle at Waterloo than the pathetic little incident recounted by Albemarle of a bugler of the 51st Regiment. He had been out skirmishing and, returning, mistook the 14th's square for his own. 'Here I am,' he was heard to exclaim, 'safe enough'; at that instant of apparent homecoming, a cannon-ball took off his head. The point, of course, is that he had probably been safer out skirmishing, for the French did not waste cannon-shot on strays; indeed, the ball which killed him was followed by two more which disarmed six men and fatally wounded a sergeant. But though objectively more dangerous a post than a skirmishing line – at least in most circumstances – a square *felt* safer. Indeed, if one were wounded, it was altogether safer, for one would be dragged into the centre of the square and carried by the bandsmen to the rear when conditions permitted, something for which a skirmisher could not hope; it was the prospect of abandonment which caused panic among the 30th's wounded when their square moved. But even when a square was under fire, and men falling fast, those untouched seem to have drawn strength to stand from the proximity of their comrades and from the square's existence, tangibility, configuration, stepping sideways to close the gaps even though that improved their chances of being hit. 'What is that square lying down in front?' Sir Alexander Cadogan is said to have asked during the attack on La Haye Sainte, to receive the answer, 'That is the position from which the 30th and 73rd have just moved.' They had left three hundred dead and wounded on the ground where they had stood.

Symbolizing the square's integrity, and that of the regiment which formed it, were the colours. Each regiment carried two, a Union Jack as the King's Colour, another of the regimental facing colour – blue, yellow, green, white – as the Regimental Colour. The modern colours have shrunk to modest proportions. Those carried at Waterloo were enormous, six feet square, and requiring considerable physical strength to handle in any sort of wind. They were carried by the two most junior officers of the battalion, each escorted by two senior sergeants, and these posts were the most dangerous which could be held in action. Sergeant Lawrence of the 40th, ordered to the colours at four in the afternoon, recalled his reluctance: 'This ... was a job I did not at all like; but still I went as boldly to work as I could. There had been before me that day fourteen sergeants already killed and wounded while in charge of these colours, with officers in proportion, and the staff and colours were almost cut to pieces.' A contributor to the *New Statesman*, writing in October 1973, affected to believe that 'all the stories of deeds of heroism in defence of military colours can only have been so much myth-making.' The record of Waterloo certainly does not bear that belief out. It might today seem more promising – it would certainly be more fashionable – to look for an explanation of the square's rockfast steadiness in terms of their alignment on some territorial landmark or boundary. And there were indeed such features on the field, particularly in the centre, where the position was traversed by hedgerows and embanked roadways. Significantly, few memoirs make mention of these features, or, when they do so, make much of them. The colours, however, are mentioned frequently and their importance as a rallying-point and source of inspiration explicitly emphasized. More indicative of their importance, because the point is implicit, are the many accounts of extraordinary heroism displayed in defence of, or attempts to capture, colours. Several Frenchmen virtually committed suicide in hopeless and quite unnecessary efforts to carry British infantry colours back to their lines. Belcher, carrying the Regimental Colour of the 32nd Regiment, found himself close to a French officer who had been unhorsed during d'Erlon's attack. Instead of running off with his men, who were then retreating, the Frenchman, Belcher writes, 'suddenly fronted me and seized the staff, I still retaining a grasp of the silk. At the same moment he attempted to draw his sabre, but had not accomplished it when the Covering Colour-Sergeant, named Switzer, thrust his pike into his breast, and the right rank and file of the division, named Lacy, fired into him. He fell dead at my feet.'

There are yet more hair-raising stories of British officers' bravery in defending the colours they were carrying. Ensign Christie of the 44th was charged by a Frenchman whose lance 'entering the left eye, penetrated to the lower jaw ... Christie, notwithstanding the agony of his wound ... flung himself upon (the colour)' wrestled it away from the Frenchman and fell to the ground on top of it. He survived his terrible injury, dying of fever in Jamaica in 1833. Volunteer Clarke, carrying the new Regimental Colour of the 69th – it had lost its previous Regimental Colour at Bergen-op-Zoom the year before – saved the regiment from inextinguishable shame at Quatre Bras by a courageous tenacity which ought to have cost him his life. Isolated when the regiment was caught half-formed in square, he received twenty-two sabre wounds, but hung on to the colour and killed three French cavalrymen with his own sword. During this *mêlée*, the King's Colour was lost, so that the 69th was narrowly spared having to fight at Waterloo without colours at all. Clarke was only sixteen, having volunteered for the campaign as a Sandhurst cadet.

Important to both British endurance and French élan as were the actions of groups and symbols at Waterloo, it is vain to seek explanations of their motive power, for the solidarity of groups and the power of symbols is not inherent or self-made. They derive from the influence of those who lead and those who manipulate; in the case of armies, from the officers. To suggest that their example and leadership was crucial at Waterloo may seem a boringly conventional view to advance. The facts nevertheless seem to bear it out. What else are we to make of the experience of the 40th Regiment? They had arrived at Waterloo dead tired after a march of fifty-one miles in forty-eight hours; three weeks before that they had disembarked from America, having been six weeks at sea. During the day of Waterloo, they lost nearly two hundred soldiers dead and wounded out of seven hundred, and fourteen out of thirty-nine officers. 'The men in their tired state,' Sergeant Lawrence wrote, began to despair during the afternoon, 'but the officers cheered them on continuously.' When the French cavalry encircled them 'with fierce gesticulation and angry scowls, in which a display of incisors became very apparent' the officers would call out, 'Now men, make faces!' and at the very end of the day, when the men 'were dreading another charge', the officers kept up the cry they had been making throughout the afternoon, 'Keep your ground, my men,' adding the promise, 'Reinforcements are coming.'

This may not sound very original stuff – though 'Make faces' is – but the baldness of Lawrence's account implicitly makes the point that it was upon the officers' behaviour that the men's depended. We do not understand unfortunately the basis of the relationship between officers and men in Wellington's army (the power of corporal punishment which the former held over the latter certainly makes it different from that in the modern British army) and it may be that what is always at bottom an emotive tie would defy analysis. We can however infer from the way in which memoirists refer to their soldiers that the relationship was not a personal one of the closeness which the modern British subaltern is encouraged to establish with his platoon. Soldiers, in the memory of the officers who led them, are always simple surnames – 'my orderly (a man named Dwyer)', 'a young soldier . . . whose name Penn now forgets', 'statement by —— Aldridge, late Corporal', 'a Private (Penfold; I forget his Christian name)' – and neither their doings nor their sufferings appear to have required particular recollection. Macready of the 30th, describing the death of one of his soldiers, does recall that the man 'uttered a sort of reproachful groan', at which he 'involuntarily exclaimed, "By G—, I couldn't help it"', but that sense of intense responsibility for their soldiers as individuals, which becomes so characteristic of the British officer's attitude later in the nineteenth century, is quite lacking. Given the social distance then prevailing between classes, and the extreme class difference between officers and men, it is perhaps foolish to expect to find anything like it. But if leadership was not founded upon personal sympathies, upon what did it then depend?

Modern theorizing on military leadership makes much of the officer's need to impress his men by his 'professional' and 'technical' competence. And several Waterloo officers do mention well-judged exercises in military technique by themselves or others; Maitland recalls his withholding the fire of the Guards brigade until the French were within twenty paces; Macdonald, of the Royal Scots, relates a conversation with his brigade commander – '"Do you think you can hit those fellows out there?" "No, but more to the right I think I could"'; and Mercer describes gauging the speed of the French cavalry charges so as to unleash his salvo with maximum effect. But officers are equally ready to admit to lapses or improprieties in military technique. Saltoun, writing of the handling of 1st Guards in the 'Crisis', says 'The word of command passed was "Halt, front, form up", and it was the only thing that could be done,' though it was not an order in the drill book; Davis, of the same regiment, complacently describes its most un-regulation formation of line from square at a moment of danger in the afternoon; and Mountsteven of the 28th, relating how his regiment responded to d'Erlon's attack, wrote 'as to "the right wing being wheeled by sections to the left, etc", I can assure you nothing half so regular came within my notice' – a comment which many modern military historians whose battle-narratives read like choreographic notations might think on with profit.

Mere technical competence, then, undoubtedly ranked lower in the officer's system of values than other attributes. What were these other attributes? Courage, of course, stood at the head of the list. But we should be careful about judging what the contemporary officer thought courageous and what not. Participation in single combat, the apogee of achievement for the medieval warrior, seems to have lost much of its glamour. We cannot yet imply, as we can of the officer 100 years later, that he thought killing almost degrading of his rank. But it is significant that he had begun to carry weapons of very little lethal value; and the infantry officer at least seems to have looked on himself as a director rather than agent of violence. Captain Wyndham, inside Hougoumont, on seeing a French Grenadier climbing the gate, 'instantly desired Sergeant Graham, whose musket he was holding while the latter was bringing forward another piece of timber, to drop the wood, take his firelock and shoot the intruder.' Death, indeed, is something about which some Waterloo officers freely admit feelings of disgust or remorse. Ensign Charles Fraser, 'a fine gentleman in speech and manner', could raise a laugh when a French cannon-ball, beheading the unhappy bugler of the 51st, 'spattered the whole battalion with his brains, the colours and ensigns in charge coming in for an extra share', by 'drawling out, "How extremely disgusting"' but that was a deliberate display of hauteur. By contrast Albemarle of the same battalion, recounting a somewhat similar incident, comes close to making the sort of revelation one would not be surprised to find included in a modern psychiatric case-book:

> As I was rising from the ground, a bullet struck a man of my company named Overman ... He, falling backwards, came upon me with the whole weight of his accoutrements and knapsack, and knocked me down again. With some difficulty I crawled from under him. The man appeared to have died without a struggle. In my effort to rejoin my

regiment, I trod upon his body. The act, although involuntary, caused me a disagreeable sensation whenever it recurred to my mind.

In a more conventional, but still revealing vein, Leeke recalls shedding tears at the sight of the first two soldiers of the 52nd to be killed on 18 June, while some of the most common ingredients of officers' post-battle letters are very tender and concerned listings of the wounds received by their fellows. Wray, of the 40th, wrote:

Poor Major Heyland (who commanded) was shot through the heart, and poor Ford was shot through the spine of his back but did not die for a short time after he was carried away. Poor Clarke lost his left arm, and I am much afraid Browne will lose his leg, he is shot through the upper part of the thigh and the bone terribly shattered. There are eight more of our officers wounded, but all doing well except little Thornhill, who was wounded through the head. Anthony ... got his eighth wound and is doing well.

Here we approach perhaps as close as we are going to get to the officer's central motivation. It was the receipt of wounds, not the infliction of death, which demonstrated an officer's courage; that demonstration was reinforced by his refusal to leave his post even when wounded, or by his insistence on returning as soon as his wounds had been dressed; and it was by a punctiliousness in obeying orders which made wounds or death inevitable that an officer's honour was consummated. Officers, in short, were most concerned about the figure they cut in their brother officers' eyes. Honour was paramount, and it was by establishing one's honourableness with one's fellows that leadership was exerted indirectly over the common soldiers. 'Two of our officers,' wrote Albemarle, 'were not on terms; the one saw the other behaving gallantly, he ran up to him and cried, "Shake hands and forgive all that has passed; you're a noble fellow".' The criteria of honour are best conveyed by the cases of Major Howard of the 10th Hussars and Lord Portarlington of the 23rd Light Dragoons. Howard, at the very end of the day, was ordered to charge a French regiment. He asked another officer what he thought of his chances, 'who said that without the cooperation of infantry it was better not as the Square was well formed . . . Major Howard said that having been ordered to attack he thought it a ticklish thing not to do it, and gave the order accordingly.' Grove, of the 23rd Light Dragoons, saw him ride forward: 'we nodded to each other . . . and a very fine handsome fellow he was; but he evidently looked as if his time had come.' A few moments later, 'he gave the order [to charge] and did it with effect, though the enemy stood well, the [British] Officers being wounded close to their bayonets and Major Howard falling so that a man in the ranks [stepped forward and] struck him with the butt end of his musket' (in fact he beat his brains out). Howard's open-eyed 'going upon his death' seems to have epitomized for most Waterloo officers what honourable conduct was, for he is picked out for mention more than any other British soldier present and his kinsman, Byron, who made a pilgrimage to his grave, wrote a funerary ode for him. Portarlington, by contrast, attracted obloquy – even if much of it was of his own imagining. The commanding officer of the 23rd, he had left his regiment for Brussels, probably to enjoy himself, the night before the battle and was not back by the time the battle started. Very late in the day he arrived on the field to find his regiment already heavily engaged, and in a frenzy of shame joined in a charge by the 18th Hussars, in which he lost his horse. Numbers of excuses were made for him – that he had been '*dangerously* ill with spasms and a violent bowel attack', that he had been 'prevented from joining his regiment in time to command it' – but gossip could not be stilled and he was obliged to resign his colonelcy in September 1815. Pathetically, he repurchased a commission as an ensign but the army would not forget, and he died, unmarried, penniless and broken in spirit, in a London slum in 1845.

And what could any excuse be worth when Portarlington's part is compared with that of Picton – painfully wounded at Quatre Bras but concealing his suffering to die at the head of his division near La Haye Sainte; or Uxbridge, who, in Duperier's words, 'got a ball in the lage [leg] which fracted the bone so much that he was forced to leave us, but he don it so well that nobody saw it – I suspected it from his slow pace and his shaking hands with Lord Vivian;' or some unnamed officers of

Picton's division who, having 'been wounded on the 16th but refused to retire' could not be persuaded to do so until the evening of the 17th; or Bull, the gunner officer, who 'feeling much pain and losing a good deal of blood ... went to the rear to have (his) arm tied up' but was back within half an hour; or the six officers of a regiment in the Fifth Division who 'had been wounded, gone to the rear, been bled, bandaged and returned in time for the final advance'; or the wounded officers of the 10th Hussars whom a comrade remembered meeting as they rode up from the rear through the twilight, with bandaged heads and splinted arms, to take their places again at the head of their men? Beside these displays of constancy and disdain for preserving a whole skin, Portarlington's frivolous expedition to Brussels and his recklessly ill-disciplined effort to recover his good name do indeed look tawdry. The others demonstrated that they were men of honour. He had not.

In its starkness, the concept of honour acted out by the Waterloo officers had about it much of what we think of as classically Heroic. This is particularly true in that, to apply Professor Finley's test, it contained no 'notion of social obligation'. Eighty or 100 years later, the British officer's principal motivation would be defined in terms of 'duty to the regiment', the regiment into which many had been born and to which all were attached for the length of their service lives. But the modern regimental system had not been invented by Waterloo. Officers were still independent gentlemen, holding rank by cash purchase, which provided a rough measure of their family status, and swapping regiments almost at whim. They demonstrated their fitness to hold whatever rank they enjoyed by their conduct in battle, or course, but their behaviour, good or bad, reflected on themselves, not on the regiments to which they belonged. Yet the Waterloo officers' concept of honour differed from that of Homer's heroes in two important respects. As Professor Finley explains, for the Homeric hero 'there could be no honour without public proclamation, and there could be no publicity without the evidence of a trophy.' Worth-while trophies could be won only in single combat, and single combat could be concluded only by violent death – and a death in which the victor exulted.

There is little or nothing of this at Waterloo. The facts of death in battle are invested, by those who recount what they witnessed, or even perpetrated, with a tinge of Romantic regret, caught at goodness knows how many removes from *Young Werther* and the poetry of Schiller; while on the acquisition of trophies, which meant the personal possessions of the dead, there was something approaching a taboo. Honour, so absolutely concrete in Homer, was for the British officer of 1815 an almost wholly abstract ideal, a matter of comportment, of exposure to risk, of acceptance of death if it should come, of private satisfaction – if it should not – at having fulfilled an unwritten code.

Hence, in a way, it is that the most perceptive of all the comments about Waterloo is the best known and apparently the most banal; that it was 'won on the playing-fields of Eton'. The Duke, who was an Etonian, knew very well that few of his officers were schoolfellows and that football bears little relation to war. But he was not speaking of himself, nor was he suggesting that Waterloo had been a game. He was proposing a much more subtle idea: that the French had been beaten not by wiser generalship or better tactics or superior patriotism but by the coolness and endurance, the pursuit of excellence and of intangible objectives for their own sake which are learnt in game-playing – that game-playing which was already becoming the most important activity of the English gentleman's life. Napoleon had sent forward each of his formations in turn. They had been well led; many of the British speak with admiration of the French officers' bravery. But they had not been able to carry their men with them the final step. Each formation in turn had swung about and gone back down the hill. When at last there were no more formations to come forward, the British still stood on the line Wellington had marked out for them, planted fast by the hold officers had over themselves and so over their men. Honour, in a very peculiar sense, had triumphed.

Disintegration

In the last hour of the battle which followed the Imperial Guard's flight, the order which the tension of combat had imposed on both armies dissolved. Among the Allies, disorder manifested itself in a rash of accidental killings and woundings as units stumbled unrecognized upon each other in the gathering dark. Among the French, it took the form of a panicky or craven capitulation as regiments found their line of escape from the field impeded or threatened.

Accidental wounding is one of the major hazards of battle, and the desire to avoid it one of the principal reasons underlying the professional soldiers' much derided obsession with drill. For among close-packed groups of men equipped with firearms, one's neighbour's weapon offers one a much more immediate threat to life than any wielded by an enemy. Lieutenant Strachan, of the 73rd, had been killed by the accidental discharge of a musket in the ranks on the retreat from Quatre Bras; and, without strict obedience to the sequence of 'Load – Make Ready – Level – Fire', many others would have met the same end on the field itself. As it was, Colonel Hay of the 16th Light Dragoons was shot by British Infantry during the repulse of d'Erlon's attack (the 10th Hussars suffered several casualties from a battery of British artillery when riding in pursuit of some French cavalry who were the gunners' real target), and the 52nd fired by mistake on the 23rd Light Dragoons following the repulse of the Guard (they managed to kill their own Colonel's horse). Lieutenant Anderson overheard the colonel of the 23rd complain, at the sight of a 'considerable number' of his troopers lying dead or unhorsed around the 52nd, 'It's always the case, we always lose more men by our own people than we do by the enemy.' This was an exaggeration. But there are numerous authentic accounts of losses by 'friendly fire' – or even 'friendly' swordcuts – at Waterloo. Mercer describes at length how he suffered from a Prussian battery which mistook his men for French, inflicted on them more casualties than they had suffered throughout the day's fighting and were at length only silenced by the arrival of a Belgian battery – 'beastly drunk and ... not at all particular as to which way they fired' – who in their turn mistook the Prussians for the enemy. Among the cavalry, the 11th Hussars nearly charged the 1st King's German Legion Hussars, who were forming up to charge them (until they 'recognized them by their cheer') while the 10th and 18th Hussars did 'exchange cuts' with a regiment of Prussian cavalry, killing or wounding several. Tomkinson, of the 16th Light Dragoons, reveals in an aside of his own how woundings could occur even between people well known to each other: a Frenchman had feigned surrender and then fired; 'Lieutenant Beckwith ... stood still and attempted to catch this man on his sword; he missed him and nearly ran me through the body. I was following the man at a hand gallop.'

This incident occurred among a crowd of Frenchmen, most of whom were trying to surrender or who had given way completely to panic. Their behaviour, which Tomkinson describes in detail, was remarkable, shedding light on one of the most obscure of all battlefield transactions – how soldiers get their offer of surrender accepted. Leeke remarks gnomically, 'soldiers of a defeated army can never feel quite sure that their lives will be spared by any of their enemies whom they may fall in with'; and certainly by this time, on the other side of the battlefield, the Prussians were enthusiastically bayoneting whatever French wounded they came across. Earlier in the battle, with much of the French army still at hand and full of fight, the French had surrendered easily when compelled to do so – a simple throwing down of weapons or cry of '*Prisonnier*!' being token enough – and had felt free to make a break back to their lines if opportunity offered (though this could result in their being killed, as some escaping French Cuirassiers were by a company of the 51st at two in the afternoon). However, at the end of the battle, Tomkinson and others found that isolated Frenchmen, whether individually or in groups, and presumably because they knew they could hope for no succour from their own side, abandoned every vestige of soldierly bearing in their anxiety to be taken captive. Murray, of the 18th Hussars, described how the infantry of the Guard 'threw themselves down, except two squares, which stood firm, but did no good. The sneaking prisoners we had taken holloaed, "Vive le roi" ... On charging, not only did the infantry throw themselves down, but the cavalry also from off their horses, all roaring "*pardon*", many of them on their knees.' Tomkinson, with the 16th, also found that 'many of the infantry immediately threw down their arms and crowded together for safety ... We were riding in all directions at parties trying to make their escape.' He goes on next to describe a phenomenon which may help to explain those extraordinary 'piles of dead' at Agincourt: 'The enemy were lying together for safety – they were a mass some yards in height – calling out from the injury of one pressing upon the other, and from the horses stamping upon their legs.' Can it be that, in extremes of fear, men will not only press together for protection – or its semblance – but actually fall together to the ground in heaps? Do we not, in our memories of childhood, recall the sense of immunity we derived from burrowing together in scrums at a parent's simulated rage, those deepest inside feeling safest? And if men *in extremis* call, as they are so commonly reported to do, for their mothers, why should not their actions, as well as their cries, revert to those of infancy?

Aftermath

The collection of prisoners occupied the last minutes of twilight. The pursuit of the remnants of Napoleon's army, a work chiefly done by the Prussian cavalry, lasted into the night. The British took almost no part in it; 'having been nearly twelve hours under arms, ten under fire and, perhaps, eight hours hotly engaged in some shape or other' as one officer puts it – and his was the common experience – most of Wellington's men were too tired, as much by the nervous as physical strain of the day to do anything but slump down to sleep. Many of the private soldiers, like

Lieutenant Keowan's servant, slunk away to loot – there are several accounts by wounded British officers of their being plundered, and might be more if the looters had not killed a number of their victims; but sleep was what the survivors wanted most, often more than food. Keowan, before making a bed of straw for himself and another officer so that they 'should not be taken for dead by plunderers', found a 'hind leg of some animal' and washed down its shreds with 'some water tinged by blood', all that could be found: 'such was the wine we drank at our cannibal feast.' But the majority, under whatever cover they could find – an officer of the 52nd sent his servant back to take a blanket from one of the packs shed by the Imperial Guard near Hougoumont – simply dropped wide to the sky. One of Picton's officers fell asleep the instant the halt was sounded and did not think of food until later in the night, when he woke to eat some chops cooked in the breastplate of a dead cuirassier (meat fried in a breastplate was very much *à la mode* in the Waterloo campaign, rather as rats spitted on a bayonet were to be in 1871 or champagne exhumed from château gardens in 1914).

'About four o'clock,' his account goes on, 'we sat up and conversed. We talked of the battle, our minds more and more filled with *what they would say about us at home* than anything else. There was no exaltation! None! We had, many of us, when in the Peninsula, tried the mettle of French soldiers – we concluded the campaign *just begun*, and looked forward to have another desperate fight in a day or two, therefore we determined not to holloa until we got out of the wood.' Others beside were re-living the battle. Keowan, asleep under the bloody overcoat of a dead French dragoon, was disturbed by 'the shrieks of the dying and the agitation of our minds, for the waves will roll high, after the storm has ceased, and as much of the fight recurred to me as I had time to dream of.' This manifestation of battle shock has a parallel in René Cutforth's description of nights spent in a hut full of newly captured prisoners during the Second World War, which were disturbed for several weeks by the shrieks and gabblings of men dreaming through their experience of recent combat. And an expression of battle shock in a different form is conveyed by Lieutenant Hamilton's account of how he spent his evening:

> Upon entering a house at Waterloo, we found every room in it filled with the dead or dying. I was glad to get a chair, and sat down at a table in a large room, in every corner of which were poor creatures groaning. The master of the house having brought us a piece of bread and a bottle of wine, we began to talk over the events of the day; and as he had for years been a soldier of Bonaparte's himself, we found no lack of subjects ... after we had finished our bread and wine, which we enjoyed very much, notwithstanding the room was full of poor wounded human beings, we retired to a hayloft for the night, which we passed in perfect repose.

Such indifference to the sufferings of the wounded all about him can only be explained by the action of some mental defence-mechanism; and it may be that the instant and almost universal slumber which overtook the army was itself a collective defence-mechanism. For, tired as the army was, it cannot have been so exhausted as to be unable to offer even a little first aid to the wounded. Yet there is something amounting almost to a universal and specific insistence in the accounts of survivors that nothing was done at all until daylight, or in many cases for some time afterwards. Heartless as this sounds, it accords with what we know of much human behaviour in disaster situations, where the greater the scale of the devastation and loss of life, the more profound is the survivors' feeling of helplessness and frustration, from which they seek escape by inactivity. All battles are, in some degree, and to a greater or lesser number of the combatants, disasters. Waterloo was a disaster of very considerable magnitude. Within a space of about two square miles of open, waterless, treeless and almost uninhabited countryside, which had been covered at early morning by standing crops, lay by nightfall the bodies of forty thousand human beings and ten thousand horses, many of them alive and suffering dreadfully. The French, who might have helped in their relief, had fled; many of the Prussians were hot on their heels; those British who were left contemplated the spectacle and closed their eyes. They knew how little a regiment which had entered the battle with only three surgeons, had lost a third of its strength and had no wheeled transport to evacuate the graver cases could do to alleviate distress. Not until next morning were they prepared to put their inadequacy to the test.

The Wounded

The less seriously wounded of Quatre Bras had been conveyed from the field to Brussels on the horses of the 7th Dragoon Guards, which had been detached on ambulance duty at the Duke's orders. Those who could not stand the jolting had eventually been collected in carts. The Waterloo wounded were, from the following morning onwards, collected up by their own regiments, the slighter cases treated by the regimental surgeons, the more serious brought into Brussels for hospital care. Extra transport, most of it local, was drafted to evacuate the wounded the regiments could not handle. In general, the British were evacuated first. Some of the French were evacuated as promptly as some of the British; but there were still Frenchmen left when all the British had gone. Some lay out two days and three nights, not being collected until 21 June. Shock and loss of blood had by then killed most of the seriously wounded and – where water had not been available – dehydration even the lightly wounded.

Some of the wounded had been evacuated during the battle: Howard, of the 33rd, wrote home that 'we were charged so furiously that we could scarcely send our wounded officers to the rear and much less the men' – a neat revelation both of contemporary military medical practice and class distinction. The wounded were taken off by the regimental bandsmen who, being non-combatant, had no other duty on the field; they were notorious as plunderers. The less seriously wounded were expected to walk back; the more seriously wounded might be carried on a door, as was the colonel of the 15th Hussars; mounted officers rode, as did Ellis, commanding the 23rd Regiment, when hit in the chest by a musket ball. He asked the rear rank of the square to open, rode out, was thrown when his horse jumped a ditch, then picked up and put in a shed, which caught fire. 'Exhausted by so many shocks, he soon after expired.'

The character of the wounds presented to the surgeons when the sufferers, sooner or later, were brought to them was fairly monotonous. A few of the patients were suffering simply from shock: Leeke described how 'a young lad . . . of our company was struck by a cannon shot and borne off motionless and white as a sheet. [We] concluded he was dying. Two or three days afterwards I could scarcely believe my eyes, when I saw him walk into the bivouac. The shot had carried away his pouch . . .' and the fright shocked him insensible. But Leeke also testifies to the effect of a real cannon wound: 'Woods . . . was struck down by a ball full on the knee. He was removed into the centre of the square. I observed the limb above the knee quickly swell till it became the size of the body.' Leeke himself suffered a freak wound, when a piece of skull from a man killed in front of him struck his left thumb where it rested on the staff of the Regimental Colour, so hard that next morning it was 'black and sore'; and he witnessed the result of another: two soldiers walking rearward appeared to have been hit by a cannon-ball passing between them 'for they were both struck in exactly the same place, about four inches below the shoulder, the wounded arm being attached to the upper part by a small portion of skin and flesh, and being supported by the man taking hold of the hand of that arm by the other hand'. There were numbers of sword and lance wounds to be treated and some bayonet wounds, though these had usually been inflicted after the man had already been disabled, there being no evidence of the armies having crossed bayonets at Waterloo (or in any other battle, come to that). Most of the sword and lance wounds were suffered by cavalrymen, though not all cavalrymen's wounds by any means were by edged weapons. The list of wounds suffered among the twenty-two officers of the 13th Hussars bears this out: two were mortally wounded by musket-balls, one killed by a cannon-shot; one was wounded by a shell splinter in the hip, two wounded by musket-balls in the head, of whom one was also bruised in the groin by a grapeshot which flattened his watch, one was wounded in the arm by a musket-ball, wounded in the hand (cause unspecified) and bruised on the side by a sabre (he did not leave the field), one was thrown from his horse and stunned, one hit by a spent ball on the jaw, two wounded – probably in cut-and-thrust – with sword-cuts on the hands.

In the infantry regiments, the majority of the wounds suffered were by missiles. Cannon-ball wounds were by far the worst; they took off heads, killed and wounded several men in line, killed a man and his horse; to be hit by a cannon-ball, unless in the limbs, almost certainly meant death. Grapeshot ranked next in lethality, but were not necessarily lethal: Lieutenant Doherty was 'struck by a grapeshot in the stomach, and instantly afterwards by a musket ball through the head' but lived to write of it in 1834. Musket balls, though much the lightest of missiles, killed easily: Canning, one of the Duke's A.D.C.s, 'received a musket shot in the centre of the abdomen, and, although perfectly collected, could hardly articulate from pain ... Raised to a sitting position by placing knapsacks around him ... a few minutes terminated this existence.' Because of the low velocity of musket-balls, men could be seriously wounded by them without being knocked over. Hence the reports we have of wounds suddenly appearing on the body of whoever the reporter was talking to or looking at: Dawson Kelly was talking to General Halkett when the latter 'received a wound in the face, the ball passing through his mouth' and Hamilton, of the Greys, caught sight of a Royal Dragoon 'whose cheek, just as I looked at it, opened, while I felt a ball pass close to my lips.'

To be wounded in any of these ways was horrible enough. What made the plight of the wounded man doubly pitiable was that his wounding, unless he could be swiftly evacuated, made his subsequent wounding more, not less, likely. There was a great deal of re-wounding of the wounded at Waterloo, a lot of it mortal and often deliberate. Jackson, one of Wellington's staff officers, found the Prussians bayoneting the French wounded near Rossomme on the evening of the battle and saved a British Light Dragoon 'over whose fate they were hesitating ... by calling out "*Er ist ein Englander*"' The French lancers, whose weapons made it so easy for them to stick a man recumbent on the ground, struck again and again at the unhorsed survivors of the Union Brigade. Many were brought in with about a dozen lance wounds in their bodies, one with eighteen, who lived. But the British cavalry too were guilty of cutting at the French wounded. An officer of Picton's division wrote of what he saw in the final advance: 'Selfish and hardened as men become ... we could not look back on the sabre wounds made by our cavalry without regret. Defenceless men ... were cut down in the wantonness of triumph. A poor French soldier [was] holding up his cheek nearly sliced off by a sabre-cut ... which he was trying to re-unite. He was wounded and disabled besides.'

That much of this wounding was by mounted men of infantrymen or unhorsed cavalrymen prompts one to speculate if some 'extra-specific' factor were not at work – if men on horseback may not feel superior to and different from men on the ground, and so feel a reduced compunction about killing them out of hand. Certainly there is little evidence from Waterloo of infantrymen killing defenceless fellow infantrymen, while the traditional and well-known antipathy of infantrymen for cavalrymen – in part the product of the customary use of cavalry in putting down mutiny in infantry regiments – may rest on this cavalry habit of spurning the underdog.

Speculation on what prompted the wounding of the disabled also raises the question of whether it is profitable to apply the concept of 'cruelty' to acts committed in the course of combat at all. Surely it is. For although combat subjects human beings to extreme stress, and although much military procedure *compels* men to kill, as in the 'load – fire' sequence, neither the strains nor the circumstances of battle completely extinguish free will, or the possibility of recognition between enemies of mutual humanity. Corporal Dickson's description of his behaviour during the Union Brigade charge is a straightforward, if remarkably honest, account of cruelty: 'Then we got among the guns ... Such slaughtering! I can hear the Frenchmen yet crying *Diable* when I struck at them, and the long-drawn hiss through their teeth as my sword went home ... The artillery drivers sat on their horses weeping aloud as we went among them; they were mere boys . . .' To demonstrate that men can behave better than this during combat, showing a true and quite voluntary generosity, one has only to read of Murray, commanding the 18th Hussars, ordering his men in exactly similar circumstances *not* to harm some French gunners: of Hughes, the adjutant of the 39th, taking an unhorsed French officer of the 6th Cuirassiers from under his men's bayonets into the centre of the square for protection; of the sergeant of the 14th Regiment, whose men were about to fire on a French cavalryman who had turned back to lend an unhorsed comrade a stirrup, crying, 'No . . . don't fire. Let him off. He is a noble fellow'; of a French cuirassier dropping the point of his sword on detecting the youth of a Life-Guard boy-trumpeter and another, shown by a Major of the King's German Legion that he had an empty right sleeve, saluting with his sword and riding on. The story of Waterloo, indeed, is full of instances of quite neutral and normal human contact between people who happened to be wearing different uniforms – of British infantry lying for cover among Frenchmen wounded in the most recent attack and asking their opinion of how the battle would end, of men making faces at each other, exchanging glances, coming to recognize each other as individuals as the tide of battle carried formations rhythmically together and apart. If neutral behaviour and generous action is possible in the heat of battle, so too are outright acts of cruelty. Hence the paradox that among all the suffering at Waterloo, some of it appeals more urgently to our pity than the rest.

Once the shock of the day itself had worn off, the spectacle of the battlefield deeply affected the survivors. The suffering of the horses, which had affected men during the battle when human anguish did not – Albemarle describes how the sympathies of the soldiers of the 14th 'were excited by the sufferings of some wounded horses which seemed to seek the protection of their square' – were even more distressing on the morrow. They had suffered the same range of wounds as the human combatants but most could be brought no relief except by destruction. Even for that many had to wait while the wounded soldiers were collected. 19 June was also distressing for another reason: in some regiments, there were almost as many deaths as there had been the day before. For at work now were the principal enemies of the wounded soldier of the period: shock, peritonitis, dehydration, loss of blood – much of it by artificial bleeding. In the 32nd Regiment, which had lost twenty-eight killed in the battle, another eighteen men died on 19 June; between 27 June and 28 July, twenty-three more were to die; and there were to be seven deaths later, the last on 16 January 1816. Other victims were to succumb even farther on; an officer of the Foot Guards, who had lost his jaw and tongue in the battle, was to die two years later of malnutrition.

Few detailed case-histories of the care of the wounded survive – if they were made (though Sir Charles Bell, who came out post-haste to Brussels with other English doctors and worked heroically in the makeshift hospitals, has left some extraordinary water-colour sketches of the wounds he had to treat). The best is by one of the wounded themselves, Lieutenant George Simmons of the Rifle Brigade. Wounded in the back at the exact moment he was congratulating himself on his apparent invulnerability, he was carried to the rear and had the ball, which had lodged under his right nipple, cut out by the surgeon who then bled him of a quart of blood. He was transferred to Brussels riding on a led horse in acute pain, and billeted on friends. For the next three weeks he was bled several times daily, until he had his servant kill the leeches he was supposed to apply, meanwhile falling for several days at a time into a feverish stupor, then waking to agony for several more. At last on 14 July his wound burst, releasing an enormous quantity of pus, and he began at once to recover. His case has been diagnosed by a modern doctor from the symptoms described as 'a subphrenic abscess with a swinging fever'; the treatment he received was almost exactly the converse of what would be applied today, when he would be transfused instead of bled, and medicated with antibiotics as a preparation for operating to drain the abscess. In the light of his experience, it does not perhaps seem so unfortunate that Doctor Bell could write on returning from the hospital for French prisoners, 'The second Sunday [i.e. 25 June], many [wounded] not yet dressed.'

By then, the armies were far away. The young officers who,

like Macready, had wondered on the evening of Waterloo if the day would be dismissed merely as an 'action' or dignified with the name of a 'battle' had had their minds set at rest; so too had those who had expected to have to fight again within a day or two. 18 June had clearly brought victory in plenitude, and the leaders were already discussing what the victory should be called. General Vivian had written home to express the hope that every British combatant should receive a medal with Mont St-Jean inscribed upon it. Blücher had proposed the appropriateness of La Belle Alliance. Wellington had decided upon Waterloo, since it came more easily off English tongues. Ensign Howard, of the 33rd, writing home from Paris on 3 July, announced, 'I have often expressed a wish to see a general engagement. I have – and I am perfectly satisfied.'

4 The Somme, 1 July 1916

The Battlefield

Somme – French department formed from part of Picardy; prefecture: Amiens; sub-prefectures: Abbeville, Montdidier, Péronne; 4 arrondissements, 41 cantons, 835 communes; court of appeal and episcopal seat at Amiens; the department takes its name from the river which waters it.

'Lethargically,' one feels like adding. For the Somme is a slow-moving river, winding its weed-choked way through a peat-bottomed valley below beech woods and bare chalk downland. The countryside, too, is slow-moving, under the gaze of the traveller who takes one of the long Roman roads which radiate north-east, east and south-east from Amiens. Low plateaux and ridges, separated by the shallow valleys of the Somme's tributaries, the Aire, Ancre, Noye, Avre, Buce, succeed each other monotonously, devoid of hedges and almost of woodland, thickly populated, intensely cultivated. Between the Noye and the Somme itself, in the plain of Santerre, the land is completely flat and yields the most characteristic of the department's crops, sugar beet, dull to contemplate, heavy to work, rich to harvest. In September the roads of the Santerre are slippery with mud ('*Attention! – Betteraves*') when the clay-smeared beet are hauled to the little refineries whose tall, single chimneys mark the villages of the plain; in October the refinery-owners summon their neighbours to shoot partridge and hare among the furrows and sit down in the evening a hundred strong to eat the bag; in November enormous ploughs emerge from the machinery sheds to be dragged on cables between stationary traction engines across the mile-wide fields. The pace of life on the Somme is as slow as its rivers, as regular as its natural features.

But as ploughing proceeds, little dumps of foreign objects appear along the verges of the roads. Rusty, misshapen, dirt-encrusted, these cones and globes reveal themselves, at a closer look, to be the fruit not of agriculture but of war; trench mortar bombs, howitzer shells, aerial torpedoes – eight-inch, seventy-fives, seventy-sevens, eighteen-pounders, 210mm, Jack Johnsons, coalboxes, whizz-bangs – veterans identify them in the language of sixty years ago with unhesitant certitude, though they approach the dumps with respectful caution. At the end of the ploughing season, bomb disposal officers of the French army arrive to remove the relics to a spot where they can be safely detonated. Occasionally an officer is killed. Even after sixty years, the fuses of these 'duds' and 'blinds' remain activated, the charges they contain explosive.

Bomb-disposal officers are killed also in Belgium, along the banks of the Yser and on the low crescent of heights which ring Ypres to the east; and, on French soil, south of the Somme, in the vineyards of Champagne and in the wheatlands of Lorraine. But the dumps which accumulate on the Somme differ in two respects from those with which bomb-disposal officers have to deal elsewhere. First, the heaps are bigger. The Somme was not the most heavily shelled of the Western Front battlefields. In terms of shells per square yard, that cachet belongs to Verdun; in duration of shelling it belongs to the heights of the Aisne and the Chemin des Dames; in remorselessness it belongs to the Ypres salient. But for a variety of reasons, dud shells do not resurface in anything like the same numbers on other battlefields as they do on the Somme. Around Verdun and in the Argonne, another heavily shelled sector, little of the ground is worked; as on the slopes of the Vosges, it was covered with forest before the war and has been replanted since. In the Champagne, historically the principal training ground of the French army, large areas have always been used for artillery practice, and the duds of the First World War, merely adding to the existing hazards, attract little attention. The high chalklands of Artois, up whose slopes the French, and later the British and Canadians, struggled towards the crests of Vimy and Notre Dame de Lorette, are now back again under pasture; while in Flanders, the glutinous soil swallows the jetsam, as it has swallowed the concrete pillboxes with which Hindenburg sought to fence the British and

Belgians into the Ypres salient and the water-logged valleys of the Lys and Yser. It is the Somme, therefore, with its busy agriculture and light, friable soil, which most plentifully throws up the dangerous debris of the great offensives.

Most of this debris is British – and this is the second respect in which the Somme dumps differ from those accumulated north or southwards. For the Somme was, in a way true of no other battlefield of the First World War, British territory. Ypres, of course, became during the war almost the corner of a native field; and, with its British church, English-speaking pubs ('Bass on draft'), English school for the children of the Commonwealth War Graves' gardeners and plethora of county regimental memorials, remains so. But it was always a tiny battlefield into which Haig, even at the height of his offensive obsessions, found it difficult to squeeze more than half a dozen divisions for an attack. The British, moreover, won no victories at Ypres, except that curious victory of the spirit which, over half a century later, still plucks back the survivors of the Salient to stand in silence beneath the tomblike arches of the Menin Gate and hear the evening Last Post blown, or sit under the willows in the Ramparts cemetery, where lawns grow over the roof of the Lille Gate dressing-station and bodies carried from it lie ranked beneath the turf. The Somme, by contrast, offered an immensely long front of attack on which twenty divisions could assault side by side. And, for all the miseries suffered there, it was also a front which brought its triumphs. On it the British drove the first tanks into action, in the ruined village of Flers, on 15 September 1916. Two years later, they organized before Amiens the first great armoured breakthrough of modern warfare. And earlier in the year of 1918 they had, after the terrifying and almost total collapse of one of their armies, brought to a halt near the city the greatest of Hindenburg's 'war-winning' offensives. It was these battles, together with the long periods of garrison duty intervening between them, which made the Somme a British, rather than a French, Belgian or American sector. The Americans would eventually come to think of the Argonne, that awful wilderness of shredded woods and choked-up streams, as 'their' battleground, as did the Belgians the Yser, on which they were to wage a semi-aquatic war for nearly four years. The French, who had poured out their blood along each mile of the five hundred between Nieuport and Switzerland, considered half a dozen points on the Western Front as *champs d'honneur* particularly their own – Verdun, of course, but also the Chemin des Dames (it was for the vistas of the Aisne it offered that Louis XV had had built for his daughters the road which gave the ridge its name), Tahure and the Main de Massiges in Champagne, Les Eparges in the St Mihiel salient (later also the scene of an American offensive) and Sainte Marie aux Mines and the Hartmanns-willerkopf (*le vieil Armand*) in the high Vosges, over which *Chasseurs alpins* and *Jäger* squandered their special mountain-warfare skills during 1915. Mort-Homme, Côte 304, Neuville St-Vaast, Somme-Puy, Malmaison, Moulin de Laffaux – it was the struggles for such unregarded corners of the homeland as these which made them 'French' in a way they had not been before. In the same way it was the battles of 1916 and 1918 – Bazentin, Pozières, Morval, Thiepval, Transloy, Villers Bretonneux – which made the Somme 'British'. But none more so than the first battle, the Battle of Albert, and its first day, 1 July 1916.

The Somme, in 1916, was new territory for much of the British army. In mid-1915 a small sector north of the river had been taken over by the embryo Third Army, as part of the Allies' agreed policy of reducing the length of front held by the French, but the bulk of the British Expeditionary Force had remained in Flanders. As its size increased, it had extended its front southwards, but only far enough to cross the wet levels of the River Lys into the dreary coalfields east of Lille, where during 1915 it fought a series of minor, murderous trench-to-trench battles – Neuve Chapelle, Aubers Ridge, Festubert, Givenchy – and mounted one major, miscarried offensive at Loos. All had been characterized by the extreme ferocity of the fighting and the miserable physical conditions which the terrain imposed. Between Ypres and Armentières, water is found everywhere close beneath the surface and much of the line had to be constructed of sandbag barricades instead of trenches. Almost everywhere, too, the Germans occupied what commanding heights there were: near Ypres, the Passchendaele and Messines ridges; in the coalfields, most of the slag-heaps and, until they were destroyed, the pithead towers. Compelled to struggle for possession of the higher, drier ground, the British had driven their lines in many places almost to within conversational distance of the Germans'; at Spanbroekmoelen, south of Ypres, the trenches were separated by a single fence of 'international' wire, which each garrison mended from opposite sides under cover of darkness. But despite the heavy and continuous toll of losses – about 300 a day – which this physical intimacy and constant attack cost the British, Flanders had become a sort of home for the B.E.F. Behind the lines – and battalions left the trenches at regular and quite frequent intervals for 'rest' – the villages provided roofs, straw, beer, *pommes frites*, fields for football. As peasants learnt to profiteer, the army and churches erected canteen huts, where the beer was both cheaper and stronger, and tea, something the locals never got the hang of brewing, was on tap. Farther back, in the little Flemish market towns, Poperinghe – 'Pop', Bailleul, Béthune, cafés and restaurants thrived, collecting clienteles of young officers who would ride over in groups to celebrate a spell out of the line on comfortable chairs at well-laid tables. Other little calling places had begun to make a name: clubs where concert-parties, over from England or got up by divisions, performed; spiritual ports of call, like the famous Talbot House (Toc H, in the war's phonetics) in Poperinghe where all visitors shed their rank; profane addresses which officers entered furtively or not at all. The geography of West Flanders, Nord and Pas de Calais, had thus, by the end of 1915, become extremely familiar to the B.E.F. Its network of roads, turnings, crossings (already much engineered and improved), was imprinted on the army's mental map, and its place-names had been Tommified over a large area, not only those of the larger spots whose soldier-equivalents were well-known at home – Armenteers, Wipers (Ypres), Vlam (Vlamertinghe). Eetaps (Étaples) – but of many quite tiny features of purely local tactical importance. The Tommy's names for a few of them had a wide currency – Plugstreet (Ploegsteert) Wood and Whitesheet (Wytschaete) – but most formed a private code: Tram Car

Cottage, Battersea Farm, Glencorse Wood, Beggar's Rest, Apple Villa, White Horse Cellars, Kansas Cross, Doll's House. The origin of most of the names, bestowed during the B.E.F.'s moment of epic in October and November 1914, had already been forgotten by the following summer. By the end of 1917, the places which they signified would themselves have been obliterated. But the names would still pass from mouth to mouth as new battalions relieved old among the slag-heaps, along the stream-bottoms, in the vanished woods.

The Somme, therefore, was country to be mistrusted by the divisions which came south to it in March 1916. These short sectors which had been held since the previous summer by the Third Army looked familiar to the new Fourth: the trenches were properly house-kept and laid out as per regulation, with continuous belts of wire in front, a parapet and a parados, traverses – those regular kinks in the line which prevented an attacker gaining possession of one stretch from shooting down its whole length, and a support and a reserve line at the proper distance – the one, 200 yards back; the other, 400 back again – connected to the front by communication trenches. Thus accommodated, the brigades of the Fourth Army could look forward to organizing a proper trench routine, rotating each of their four battalions between front, reserve and rest on a sixteen-day cycle. But the divisions of the Third, which, by a sideways move, inherited a sector vacated by the French, were at a loss. Their predecessors, by all the evidence, had not been conducting trench warfare at all, not at least as its rules were understood in the B.E.F. Flowers had been allowed to overgrow the parapets of the trenches – the 1/4th Ox and Bucks took over a 'Marguerite Trench' from the French – which in many places were not truly continuous, but organized as independent 'positions', comfortable to occupy, at least for a small garrison, with their little scrapes in the trench walls and wattle-shelters in the corners of the traverses, but ill-adapted for raiding or for solid battalion defence.

And the truth indeed was that the French had not been waging trench warfare on the Somme. For them, it was an 'inactive' sector which they were content to hold with the minimum of infantrymen in the front line, using their plentiful corps and divisional artilleries of seventy-five millimetre guns to warn off the Germans should they menace an attack. But the Germans opposite had never put them to the trouble. For their high command also was content to regard the Somme as a quiet sector, leaving it in the hands of reserve divisions which might be allowed an easy life as long as they improved their front – by digging and wiring – and kept the French on the right side of no-man's-land. Some of the German divisions on the Somme had, in consequence, been there since September 1914 with the loss of scarcely a man, except among the contingents which they were occasionally obliged to detach when the Allies attacked in Flanders or Champagne.

The Plan

All this was about to be changed. Since December 1915, the French and British had been planning a great offensive on the Western Front and the sector they had chosen for it was that 'athwart the Somme'. This was not to be the first offensive seen in the west. Indeed the S-like trace of the Western Front was itself the product of a series of offensives in 1914. The earliest, mounted by the French into Lorraine, had fixed the lower loop of the S along the line of the Vosges and the River Meurthe; the second, mounted by the Germans in conformity with the dead maestro Schlieffen's plan, and hinging on Verdun, had planted the centre of the S on the Aisne – though it had hovered for a week on the Marne; the third, a running battle between French and Germans, known as the Race to the Sea and fought up the rungs of the parallel main railway lines of north-eastern France, across the departments of the Oise, Somme, Pas de Calais and Nord, had ultimately planted the upper loop of the S on the River Yser. There, in October and November, in the only gap of open front remaining in the west, the French, British, Belgians and Germans had worried the last gasps of life out of their hopes of quick victory. Henceforth, if they were to seek a decision, it would be found on the far side of the trench line.

But much of the terrain which the trench line crossed was unsuitable for decisive operations. North of Ypres, both sides were prisoners of the floods; south of Verdun the proximity of rivers, forests or mountains threatened a rapid check to any advance, even if it could be initiated; between the Oise and the Aisne, in the dead centre, steep river valleys likewise offered little prospect of breakthrough; and there were a number of other unpromising stretches – the valley of the Lys, south of Ypres, and the forest of the Argonne, west of Verdun. The narrowness of suitable attacking front left after these sectors had been subtracted from the whole mattered little to the Germans, for they, once fully persuaded of the failure of the 'war-winning' Schlieffen plan, were content to stand on the defensive in the west while they won victories over the Russians. But it mattered a great deal to the French, whose national honour was in pawn to the Germans, together with much of their national wealth; and to the British, who, as the numbers in their armies grew, needed a battlefield on which to make their strength – and their commitment to the common cause – felt.

By any sensible strategic reckoning, there were only three sectors where the lie of the land and the direction of the railways so ran as to favour an Allied attack; the Somme, Artois and the Champagne. During 1915 it was upon the last two that they concentrated their forces. Both offered high, dry, chalky going and a hinterland across which an advance might be carried at speed if the trench line could be broken. But it was less the nature of the terrain than their relationship to each other which recommended these battlefields to French high command. Grasping still after the decisive battle of which he had been cheated in 1914, Joffre insisted on seeing the Western Front as a 'theatre of operations' rather than the fortified position which it truly was, and its central section therefore as 'major salient'. In that the front between Verdun and Ypres ran in a rough semicircle, forming the upper loop of the whole S, a salient it was, with its northern root in Artois and its southern in Champagne. It was an illusion, however, soon to be demonstrated as such, that an attack on the major salient was the right strategy or that the conventional method of dealing with a salient – by simultaneous attacks at the roots – would work in the circumstances of 1915. There were many reasons why not. Some were to be glimpsed in the French spring offensive of 1915, known as the Second Battle of Artois, which failed to capture Vimy Ridge. More were to be revealed, and more fully, in the September battle, a truly joint offensive between the French, attacking in Champagne, and a pair of British and French armies, which attacked side by side in Artois; the French again failed to carry the Vimy position, the British just got possession of the village of Loos (having hoped to create the conditions for a cavalry breakthrough), while the major French offensive in Champagne – the last in which they assaulted behind colours and bands – dissipated itself in blood and misery on the slopes of Tahure and the Main de Massiges.

Allied strategy for 1916 required, therefore, a new offensive plan. Joffre decided, moreover, that it needed a new front. This was due in part to his growing recognition that a front which had been attacked was so 'thickened up' in the process – cauterized and criss-crossed with a scar-tissue of new and old trenches – that a renewal of the assault on the same spot carried a diminishing prospect of success; but in greater part to his desire to involve the British in a major offensive effort in 1916. In suspecting their disinclination to be involved, he did them an injustice; in supposing that their choice of location for an offensive might not serve his Grand Strategy, he was on to something. Haig, who had made his reputation by his defence of Ypres in late 1914, had, as soon as he assumed command of the B.E.F. in December 1915, set his staff to plan for the next great British offensive to take place there again. In selecting the Somme front, which was where the French and British sectors touched, as the focus of Allied efforts for 1916, Joffre was at least to ensure that those efforts would be jointly directed towards the defeat of the German army on French soil and under his hand – even if the method by which it would be defeated was, as he was coming privately to accept, that of *usure* – attrition – rather than the break-through in which the British still hoped and believed.

Attrition is a game at which two can play. Both the British and French had too long and easily taken for granted that the German posture on the Western Front was a defensive one. It was almost as great a psychological as physical shock, therefore, when in mid-February 1916 the Germans opened a major and quite unexpected offensive at Verdun, the lower hinge of the Western Front. From the outset the French rightly grasped that its object was to impose upon them the necessity either of making a humiliating withdrawal or of bearing a prolonged butchery. The French settled for butchery; but from the date of the offensive's outbreak, their discussions with the British lost their academic, almost reflective pace and took an urgency which became more and more desperate as the numbers of French lives lost at Verdun grew. Death or wounds had taken 90,000 Frenchmen by the end of March, only six weeks after the offensive had begun. In May, when Joffre came to visit Haig in his headquarters, it was calculated that losses would have risen to 200,000 by the end of that month. Haig conceded the need to fix an early date of the opening of the Somme offensive; he indicated the period from 1 July to 15 August. At the mention of the later date, Joffre, extremely agitated, burst out that 'the French army would cease to exist' if nothing had been done by that date. On the spot, the two generals settled for 1 July. The British would attack with a dozen divisions north of the river, the French with twenty to the south.

The Preparations

Haig's planners now busied themselves with fixing the final details. Throughout the earlier part of the year, they had been creating behind the Somme front the infrastructure of roads, railway spurs, camps, hospitals, water-pumping stations, supply dumps and transport parks without which a deliberate offensive could not be mounted in an industrial age. The most important end-product of this labour was the accumulation of artillery ammunition, of which 2,960,000 rounds had been dumped forward. (By way of comparison, Napoleon probably had about 20,000 rounds with his guns at Waterloo.) Consequently, the most important element of the attack plan was the artillery programme. It was divided into two. The first instalment was to be a week-long bombardment of the German line, concentrating on the trenches occupied by the garrison but also reaching back to 'interdict' – deny the use of – the approach routes to those trenches, where those routes could be reached. The second instalment was to be the barrage. This word, the meaning of which has since been smothered in English by the weight of historical allusion attaching to it, was new to the British in 1916. Borrowed from the French, who use it to signify both a turnpike barrier and a dam (whence it has been taken into English by another route), it literally has the force of meaning 'preventing movement'. And such was the desired function of the *tir de barrage*; 'barrage fire' was a curtain of exploding shells which preceded the advance of the infantry, preventing the enemy infantry from moving from their positions of shelter to their positions of defence until it was too late to oppose the attackers' advance. In strict artillery theory, the barrage, by carefully timed 'lifts', could take the body of infantry it was protecting clean through an enemy position without their suffering a single loss from enemy infantry fire.

The only *theoretical* limit on the protection the barrage could offer was imposed by the range of the guns firing it – which meant that beyond about 6,000 yards from the gun-line, say 5,000 from the front trench, the infantry could not count on the artillery's fire reaching ground they wished to traverse. In practice 'effective' ranges were regarded as rather shorter – 'effective' having changed its meaning since Waterloo. There, as we saw, it meant 'making an effect on the enemy', something shot would not do at long range because it quickly lost its killing velocity and accuracy. By 1916, when better technology had improved velocity and accuracy ten-fold, and the bursting charge had made shells lethal even at extreme range, 'effective' fire really meant 'observable' fire, fire which fell within sight of an observation officer who could communicate with his battery and correct its guns' deflection and elevation. He was expected to keep close behind the attacking infantry – he was also expected to be able to keep his telephone cable to his battery intact, a much more doubtful expectation – and the limit on the effectiveness of fire, therefore, was that imposed by his ability accurately to spot the fall of shot. If we estimate his effective range of vision at a thousand yards, and his distance from the leading infantry also at a thousand, we arrive at an 'effective' range for the barrage of about 4,000 yards from the front trench.

This was almost exactly the maximum distance set for the advance of each of the infantry formations on 1 July, which is not surprising, for 'objectives' were arrived at by exactly these mathematics. On some divisional fronts, final objectives were closer to the British front line if the German front position was closer. On less than half the front, however, did objectives fall on the German second position, the siting of which the Germans had, of course, determined by the same calculations of artillery ranges which underlay the planning of the British attack. The British infantry were, therefore, being asked to commit themselves to an offensive of which the outcome, even if completely successful, would leave the Germans still largely in possession of a second and completely independent system of fortification untouched by the attack. Its capture would require the hauling forward of all the impedimenta of bombardment and the repetition of the opening assault on another day, at another hour. That they were not daunted by this prospect is explained in part by the briefing that the staff had given to the regimental officers, and the officers to their men: that the real work of destruction both of the enemy's defences and men, would have been done by the artillery before zero hour; that the enemy's wire would have been scythed flat, his batteries battered into silence and his trench-garrisons entombed in their dug-outs; that the main task of the infantry would be merely to walk forward to the objectives which the officers had marked on their maps, moderating their pace to that of the barrage moving ahead of them: finally, that once arrived there, they had only to install themselves in the German reserve trenches to be in perfect safety. Had anyone yet coined the phrase, 'Artillery conquers, infantry occupies' it would have been on everyone's lips. Or would it? For the better explanation of the army's optimism was that it was a trusting army. It believed in the reassurances proffered by the staff who, to be fair, believed them also. It believed in the superiority of its own equipment over the Germans'. It believed in the dedication and fearlessness of its battalion officers – and was right so to believe. But it believed above all in itself.

The Army

The British Expeditionary Force of 1916 was one of the most remarkable and admirable military formations ever to have taken the field, and the Fourth and Third Armies, which were to attack the Somme, provided a perfect cross-section of the sort of units which composed it. Four of the thirteen attacking divisions were regular, were wholly or largely formed, that is, of long-service volunteer soldiers. The 4th Division demonstrated what type of formation this was. All its twelve battalions of fighting infantry were old-sweat units, two Irish, one Scottish, five Midland or North Country, two West Country, one East Anglian, one London; and, despite continuous action since the Battle of Mons in August 1914, many of their experienced pre-war officers and N.C.O.s still survived. The 7th and 8th Divisions were less completely regular, containing as each did a war-raised 'Kitchener' brigade (three brigades of four battalions made a division) but were distinctively regular in spirit. This was true, too, of the 29th Division, which contained two war-raised units – the so-called Public Schools Battalion and the Newfoundland Regiment – but was composed otherwise of the toughest old-sweat battalions, those which had been overseas on imperial garrison duty in August 1914. Three of the 'Kitchener' divisions also contained regular battalions, the 21st, 30th and 32nd, having one in each of its brigades; the rest of their infantry, like all that in the 18th, 31st, 34th and 36th Divisions, was 'Kitchener' or 'New Army'. What made these battalions – 97 out of the 143 destined for the attack – so worthy of note?

First, that they were formed of volunteers. The regular

battalions were also raised by voluntary enlistment, but the impulsion which drove a pre-war civilian to join up for 'seven and five' – seven years with the colours, five on the reserve – was most often that of simple poverty. 'I would rather bury you than see you in a red coat' were the words his mother wrote to William Robertson, a ranker who became a field-marshal, on hearing of his enlistment, and they tell us all we need to know about what a respectable Victorian working-class family felt at a son joining the army. Almost any other sort of employment was thought preferable, for soldiering meant exile, low company, drunkenness or its danger, the surrender of all chance of marriage – the removal, in short, of every gentle or improving influence upon which the Victorian poor had been taught to set such store. It is against this background that we must review the extraordinary enthusiasm to enlist which seized the male population of the British Isles in the autumn of 1914 and provided the army, in a little under six months, with nearly two million volunteer soldiers.

Among the first hundred thousand – for administrative convenience, the volunteers were called for in batches of that number – many who joined up were without work, there being, for example, a serious slump in the building trade in the summer of 1914. Some might, therefore, have been eventually impelled into the army, while others perhaps used the pretext of a national emergency to camouflage a personal one and to justify the breaking of a taboo. But, from the outset, many surrendered well-paid, steady employment to join up, coming forward in such numbers that they overwhelmed the capacity of the army to clothe, arm and train them. Kitchener, hastily appointed Secretary of State for War, had originally called for a single increment of 100,000 men to the strength of the regular army. He was, by the spring of 1915, to find himself with six of these 'hundred thousands', from which he formed five 'New Armies', each of six divisions. The two original 'hundred thousands' provided two series of six symmetrical divisions, reflecting and to a large extent corresponding with the regional division of the country: 9th and 15th were called Scottish, 10th and 16th Irish, 11th and 17th Northern, 12th and 18th Eastern, 13th and 19th Western and 14th and 20th Light (formed from Londoners and other southerners into battalions of the rifle and light infantry regiments). But the sheer weight of the recruiting flood soon washed away the very flimsy framework of organization within which the War Office tried to contain it. The facts of demography, too, worked against their scheme, for the population of the British Isles did not neatly divide into six. The great reserves of manpower were in the northern and midland cities and in London, and it was this pattern which began to tell in the third, fourth and subsequent 'hundred thousands'. The men who had come forward in these waves chose their own titles for their units, in some cases their own officers, in almost every case their own comrades. These were the men who formed the 'Pals' Battalions'.

Perhaps no story of the First World War is as poignant as that of the Pals. It is a story of a spontaneous and genuinely popular mass movement which has no counterpart in the modern, English-speaking world and perhaps could have none outside its own time and place: a time of intense, almost mystical patriotism, and of the inarticulate élitism of an imperial power's working class; a place of vigorous and buoyant urban life, rich in differences and in a sense of belonging – to work-places, to factories, to unions, to churches, chapels, charitable organizations, benefit clubs, Boy Scouts, Boys' Brigades, Sunday Schools, cricket, football, rugby, skittle clubs, old boys' societies, city offices, municipal departments, craft guilds – to any one of those hundreds of bodies from which the Edwardian Briton drew his security and sense of identity. This network of associations offered an emotional leverage on British male responses which the committees of 'raisers', middle-aged, and self-appointed in the first flush of enthusiasm for the war, were quick to manipulate, without perhaps realizing its power. First among these men was the Earl of Derby who, in his role as caudillo of the commercial north-east, called in late August 1914 on the young men of Liverpool's business offices to raise a battalion for the New Army, promising that he had Kitchener's guarantee that those who 'joined together should serve together'. The numbers for the battalion were found at the first recruiting rally, and the overflow provided two others. The clerks of the White Star shipping company formed up as one platoon, those of Cunard as another, the Cotton Exchange staff, banks, insurance companies, warehouses contributing other contingents, so that between Friday, 28 August and Tuesday, 3 September a whole brigade of four infantry battalions – 4,000 young men – had been found. They called themselves the 1st, 2nd, 3rd and 4th City of Liverpool Battalions; later the War Office would allot them the more official title of 17th, 18th, 19th and 20th Battalions, King's (Liverpool) Regiment, thus accommodating them within the conventional regimental structure of the peace-time army, and designate them the 89th Brigade of the 30th Division. But they would continue to think of themselves as the Liverpool Pals.

And the Pals idea at once caught hold of the imagination of communities much smaller, less self-confident, less commercially dominant than Liverpool. Accrington, the little East Lancashire cotton town, and Grimsby, the North Sea fishing port, shortly produced their Pals, Llandudno and Blaenaw Festiniog, the Welsh holiday resorts another, the London slum boroughs of Shoreditch, Islington, West Ham and Bermondsey theirs. Artillery brigades were raised in Camberwell, Wearside, Burnley, Lee Green, Lytham St Anne's; Royal Engineer field companies in Tottenham, Cambridge, Barnsley, Ripon, and units arose with sub-titles like North-east Railway, 1st Football, Church Lads, 1st Public Works, Empire, Arts and Crafts, Forest of Dean Pioneers, Bankers, British Empire League, Miners. Miners, who numbered 1.2 million in 1914, about six per cent of the employed population, and whose places of work were concentrated almost exclusively in the West Midlands, South Wales, the North-East and the Scottish Lowlands, provided a disproportionately large number of the recruits and of the units they would eventually form. So many were physically stunted that at first they failed the army's height requirement; but being otherwise robust were later formed into special 'Bantam' units, for which the height requirement was reduced to between 5 ft and 5 ft 3 ins. The spectacle of these uniformed midgets in training touched the lowest strain of sentimentality in Hun-hating journalists, while many of the recruits, sharing nothing with the miners but their lack of stature, turned out poor fighting material and their units with them. But these were the exception. In physique, in subordination, in motivation, in readiness for self-sacrifice, the soldiers of the Kitchener armies, 'citizen soldiers' as the propaganda of the period, for once getting its categories right, called them, were unsurpassed, and were matched in quality only by the magnificent volunteer contingents provided by the white Dominions, and by the Ersatz Corps of German university and high-school students who had paid the price of going untrained to war in the *Kindermord** at Ypres in October and November 1914.

The *Kindermord*, had the Kitchener soldiers grasped its import, offered them an awful warning, for the Ersatz Corps, which outnumbered the tiny B.E.F. of 1914, had been beaten by the superior military technique of war-hardened soldiers. The Kitchener battalions had on formation, and for many months afterwards, no knowledge of military technique whatsoever. Indeed 'battalions', which implies an irreducible minimum of military organization, is a misnomer. Some 'battalions' entered into military existence when a train load of a thousand volunteers was tipped out on to a rural railway platform in front of a single officer who had been designated to command it. Few of these battalions, beyond those of the first two 'hundred thousands', were allotted more than three officers and three regular N.C.O.s, and those were often second-raters – retired Indian cavalrymen, militia colonels, disabled pensioners. Occasionally the choice was more promising (though 'choice' of course was sharply limited by the need to keep every fit and able officer in France) and the more intelligent of these instant commanding officers would send the men off in small groups for a few minutes to elect their own junior leaders, or would call for those with some experience of supervising others to accept probationary rank. Egalitarian though the mood of the Kitchener armies very distinctively was, appeals of this sort generally produced

**Kindermord* – 'Massacre of the Innocents'.

candidates, often ones whose authority was readily accepted by their fellows and could eventually be confirmed.

But, although this method yielded N.C.O.s, it did not do much to officer the new armies. The War Office was unwilling to grant commissions unless aspirants could prove their suitability, and although it devolved the power to adjudicate on to the local 'raisers', it and they shared common criteria of what 'suitable' meant. Officers had to be gentlemen. But just as the distribution of manpower failed to mesh with the regimental organization of the British army, so too did the social with the human geography of the country. Britain in 1914 was as sharply Two Nations as it had been seventy years before, so that throughout the industrial North, the West Midlands, South Wales and Lowland Scotland existed populous and productive communities almost wholly without a professional stratum and so without an officer class. Young men with the necessary qualifications – possession of the Certificate A or B granted by an Officer Training Corps was usually stipulated, though education at one of the public or better grammar schools which ran an O.T.C. was in practice often found sufficient* – were concentrated in the south and west and in half a dozen major cities. Thus there came about, during the first two years of the First World War, one of the most curious social confrontations in British history and in its long-term political implications, one of the most significant. It was almost always a meeting of strangers. It was sometimes a meeting of near foreigners. John Masters, in his description of his joining the 4th Gurkha Rifles of the old British Indian Army in the nineteen-thirties, has marvellously evoked the mutual incomprehension, good-humoured but absolute, which took hold

*R. C. Sherriff, author of *Journey's End*, describes his first attempt to become an officer in August 1914: '"School?" inquired the adjutant. I told him and his face fell. He took up a printed list and searched through it. "I'm sorry," he said, "but I'm afraid it isn't a public school." I was mystified. I told him that my school, though small, was a very old and good one – founded, I said, by Queen Elizabeth in 1567. The adjutant was not impressed. He had lost all interest in me. "I'm sorry," he repeated. "But our instructions are that all applicants for commissions must be selected from the recognized public schools and yours is not among them." And that was that. It was a long, hard pull before I was at last accepted as an officer. Only then because the prodigious loss of officers in France had forced the authorities to lower their sights and accept young men outside the exclusive circle!'

of a platoon and its new officer, fresh from England, when first they met. Something very similar fell upon the Kitchener armies in the winter of 1914 when nicely raised young men from West Country vicarages or South Coast watering-places came face to face with forty Durham miners, Yorkshire furnacemen, Clydeside riveters, and the two sides found that they could scarcely understand each other's speech. It was only the ardent desire on the one hand to teach, to encourage, to be accepted, on the other to learn and to be led which made intercourse between them possible. In this process of discovery, both of each other and of the military life, many of the amateur officers were to conceive an affection and concern for the disadvantaged which would eventually fuel that transformation of middle-class attitudes to the poor which has been the most important social trend in twentieth-century Britain.* Many of the Kitchener Tommies were to perceive in their officers' display of fellow-feeling an authenticity which would make attendance on that transformation tolerable. But by what strange communion did these feelings transmit themselves! Siegfried Sassoon has described how his own life was changed by the expression of total trust and self-surrender visible in the face of his men, looking up at him as they squatted cross-legged, while he inspected their feet after a route march.

Inspecting sore feet was one of those rituals of the regular army into which the Kitchener officers were earliest initiated, partly because its dotty dissimilarity from anything they had known in civilian life convinced their seniors that it was the right thing to make the subalterns do – as indeed it was in those days of unmechanical warfare, when tactical mobility depended upon marching endurance and untended blisters could cripple a whole battalion – and partly because route-marching was, in the first months of their existence, almost the sole form of training of which the Kitchener divisions could get their fill. For many months rifles, even uniforms were lacking, so that the

*'What a lesson it is to read the thoughts of men, often as refined and sensitive as we have been made by the advantages of birth and education, yet living under conditions much harder and more disgusting than my own.' Letter of 2/Lt Stephen Howett, Warwickshire Regiment (Downside and Balliol College, Oxford), written after censoring his own soldiers' letters home.

Pals' battalions could neither learn the trade of soldiers nor simulate their appearance. Only by endless drilling and marching in formation were these thousands of unblooded volunteers, still clad in civilian tweed, or a little later in postman's serge, of which 1915 yielded a strange surplus, able to remind roadside spectators, at times even themselves, that they were votaries of the Great Sacrifice. Many divisions received sufficient rifles to issue one to every man only within weeks of going to France in the autumn of 1915; and the equipments of the artillery, whose management was a great deal more complicated, were even slower to arrive. At least three divisions which were to attack on 1 July 1916, came to the Western Front in a state of training which must be described as quite deficient. The 30th, 32nd and 34th Divisions (all belonging to the fourth 'hundred thousand' – K4 in the jargon of the period) had been raised only in December 1914, been allotted the meagrest cadre of experienced officers and N.C.O.s, had received their proper complement of weapons as late as the autumn of 1915 and yet were all shipped overseas between November 1915 and January 1916. The promise of tragedy which loomed about these bands of uniformed innocents was further heightened by reason of their narrowly territorial recruitment; what had been a consolation for the pangs of parting from home – that they were all Pals or Chums together from the same close network of little city terraces or steep-stacked rows of miners' cottages – threatened home with a catastrophe of heartbreak the closer they neared a real encounter with the enemy. Grave enough in the case of the 30th, with its three Liverpool or Manchester brigades, the threat bore even more heavily on the 34th, containing not only the so-called Tyneside Irish and Tyneside Scottish Brigades – 8,000 young men all domiciled in or around Newcastle-on-Tyne – but also a Pioneer battalion, the 18th Northumberland Fusiliers, raised by the Newcastle and Gateshead Chamber of Commerce from the shop assistants of the city: the notion of a regiment of Kippses and Mr Pollys fine-tunes the poignancy of the Pals idea.

From this it might be thought that the composition of a division not yet mentioned – the 36th – was potentially tragic;

it was the most close-knit of all the Kitchener formations, its infantry having been raised wholly in Ulster (Ulster was the divisional subtitle). Its very existence, however, testified to the extreme militancy, and living military tradition, of the Protestant people from which it exclusively sprang. Half-Catholic though the nine counties of Ulster were, there were no Catholics in the 36th. Indeed its parent body – the Ulster Volunteer Force – had been raised well before the outbreak of the war, as a weapon of Protestant opposition to the grant of Irish Home – 'Rome' – Rule; and its leaders' offer to the British Government of its most able-bodied members as soldiers had come as a considerable relief to Westminster in the autumn of 1914, when the kingdom had seemed threatened by an Irish civil as well as a foreign war.

Completing the composition of the army which was to attack on the Somme were a number of formations whose foundation also antedated 1914, though by much longer than the U.V.F.'s. These were the Territorial Force divisions – the 46th North Midland, 48th South Midland, 49th West Riding and 56th London – whose existence recalled an earlier, mid-Victorian craze for amateur soldiering, brought on by a panic scare that Louis Napoleon's *navy* threatened the impregnability of the white cliffs and sustained by a simple bourgeois pleasure in the wearing of uniforms and the bandying-about of military titles. These Volunteers (the Ulstermen had pinched their emotive title) had eventually become an accepted, if slightly comic, feature of the Victorian social fabric – accepted by the respectable classes in a way the regulars were not, because they aspired after the military virtues without indulging in the military vices, found comical because their aspirations generally fell a little short of the mark. Some of the Territorial battalions, however, particularly the London ones, had latterly become very good, drawing on a stock of well-educated, games-playing young men to supply their ranks; Conan Doyle has Sherlock Holmes characterize a stockbroker's clerk who appears in one of his cases as 'representative of the type found in one of our better London Volunteer regiments', probably meaning the London Rifle Brigade or Queen Victoria's Rifles, both in 1916 forming part of the 56th Division which would attack at Gommecourt.

And after the reorganization of 1908 even the more rustic Volunteer units had been brought to a standard of training nearer that of the regulars. At that date, indeed, and on swapping their title of Volunteers for that of Territorials, they had lost their old battalion numbers and been hitched on to the series of the regular county regiments. In consequence, a roll-call of battalions in the 46th (North Midland) Division, to take one example, made it sound like a segment of the regular army: it contained the 5th and 6th Battalions, South Staffordshire Regiment, 5th and 6th North Staffordshires, 4th and 5th Lincolns, 4th and 5th Leicesters, and 5th, 6th, 7th and 8th Sherwood Foresters (the Nottinghamshire and Derbyshire Regiment). But in human terms, each of these battalions represented a voluntary, part-time, peace-time effort at soldiering by the menfolk of one or another midland town, rural or industrial: Walsall, Wolverhampton, Burton, Hanley, Leicester, Loughborough, Lincoln, Grimsby, Newark, Nottingham, Derby, Chesterfield; and in practice their domestication was even more intimate, many of them having their component companies located in a suburb or outlying township, where the drill hall was as much part of the fabric as the nonconformist chapel – and run almost as soberly – and some having platoons centred on a single village. The same went for the artillery, engineers and services of the division, the 46th having its batteries at Louth, Boston, the Lincolnshire market town, Stoke-on-Trent (Arnold Bennett country), West Bromwich, one of Birmingham's drearier appendages, and Leek (whither Winston Churchill had *not*, trade union legend to the contrary, sent troops to cow the miners in 1910). Its engineer field companies were located at Smethwick, another charmless outlier of Birmingham, and Cannock, a tiny country town, overwhelmed by its surrounding coalmines, on the edge of the magnificent countryside of the Chase.

In a more diffuse and traditional manner than the Kitchener divisions, therefore, those of the Territorial Force were a military embodiment of the regions from which they hailed. By the summer of 1916, however, many of their originals were invalided or dead. For, following the destruction of the regular Expeditionary Force in the 1914 battles, it was the Territorials who, arriving in France in early 1915, had held the line until Kitchener's men could arrive. By the spring of 1916 a regular battalion like the 2nd Royal Welch Fusiliers (that extraordinary battalion of poets, wartime home both of Siegfried Sassoon and Robert Graves) still had left about 250 of the men who had accompanied it to France eighteen months before, demonstrating a drain of about ten soldiers a week – sometimes more, of course, sometimes less. The Territorials, though 'out' less long, had suffered a similar total loss, for they had started with weaker battalions and had had to detach men as cadres to their 'second line', on which new battalions were formed. In almost no battalion among those earmarked to attack on 1 July, therefore, had more than a quarter of the men, of whatever rank, memories of peacetime soldiering. Some of the regulars, by pulling in their Reservists and Special Reservists, could still field an almost complete turn-out of long service men. But among the remainder there was only a very little to choose, in terms of collective military experience, between the first and last joined.

The Tactics

Awareness of this lack of experience was strong at General and Fourth Army Headquarters, where the staffs had, in consequence, framed plans of stark simplicity for the infantry. The Fourth Army's eleven front-line divisions, of which six had not previously been in battle, were, on the cessation of the artillery preparation, and following behind its barrage fire, to leave their trenches and walk forward, on a front of about fifteen miles, for a mile and a half. In the centre of the front, a walk of a little less than that distance would give them possession of the German second line of entrenchments; on the northern sector, the walk to the German second position was a good two miles; on the southern sector, the German second position was judged to be too far back for it to be taken in a single day and the objectives were accordingly set somewhat closer. Next to the British on the southern sector a French force, of which the insatiable demands of the Verdun battle progressively reduced the size until on 1 July it numbered thirteen divisions, was to attack up both banks of the River Somme, behind a great weight of artillery. French small-unit tactics, perfected painfully over two years of warfare, laid emphasis on the advance of small groups by rushes, one meanwhile supporting another by fire – the sort of tactics which were to become commonplace in the Second World War. This sophistication of traditional 'fire and movement' was known to the British but was thought by the staff to be too difficult to be taught to the Kitchener divisions. They may well have been right. But the alternative tactical order they laid down for them was over-simplified: divisions were to attack on fronts of about a mile, generally with two brigades 'up' and one in reserve. What this meant, in terms of soldiers on the ground, was that two battalions each of a thousand men, forming the leading wave of the brigade, would leave their front trenches, using scaling-ladders to climb the parapet, extend their soldiers in four lines, a company to each, the men two or three yards apart, the lines about fifty to a hundred yards behind each other, and advance to the German wire. This they would expect to find flat, or at least widely gapped, and, passing through, they would then jump down into the German trenches, shoot, bomb or bayonet any who opposed them, and take possession. Later the reserve waves would pass through and advance to capture the German second position by similar methods.

The manoeuvre was to be done slowly and deliberately, for the men were to be laden with about sixty pounds of equipment, their re-supply with food and ammunition during the battle being one of the things the staff could not guarantee. In the circumstances, it did indeed seem that success would depend upon what the artillery could do for the infantry, both before the advance began and once it was under way.

The Bombardment

The artillery fire plan was as elaborate as the infantry tactical scheme was simple. Artillery now comprised a great variety of weapons, firing several different sorts of ammunition: field artillery, the lightest and most plentiful variety, composed of 18-pounder guns and 4.5-inch howitzers, which fired small shrapnel or high explosives or (more rarely) gas shells out to a range of about 6,000 yards; medium artillery – 60-pounder and 4.7-inch or 6-inch guns which fired high explosive shells out to 10,000 yards; and a variety of heavy howitzers, 6-, 8-, 9.2-, 12- and 15-inch calibre which dropped 100- to 1400-pound shells from a high angle at ranges between five and eleven thousand yards. In addition, the infantry brigades controlled their own 'trench mortars', simple smooth-bore tubes which lobbed 2-inch, 3-inch or 4-inch bombs in a very steep trajectory from one trench to another across no-man's-land.

Range, weight of shell and trajectory determined what the different tasks of these weapons should be. Trench mortars, having the shortest range and firing a projectile without any penetrative power, were turned against near-by surface targets, the enemy's trenches, which they were intended to collapse, and his wire, which they were expected to help cut. Wire-cutting was indeed the most fundamental of the artillery's duties, for should the German entanglements remain intact on the morning of Z-Day (the day of the attack), the infantry advance would terminate on the far side of no-man's-land. The belts were very thick. Accordingly the 18-pounders of the divisional field artilleries were also assigned almost exclusively to wire-cutting – though their fire, with the shrapnel shell of the period and its slow-acting fuse, tended to waste itself in the ground *under* the entanglements (instead of bursting on 'graze' against the wire). Some of the 18-pounders' fire was also allocated, however, to 'counter-battery' – firing, that is, at the estimated position of the enemy's guns, in the hope of knocking them out before the infantry had to advance through the barrage which those could put down on to the British parapet and into no-man's-land. What little gas shell was available was to be chiefly reserved for last-minute counter-battery fire, the British artillerymen understanding how difficult their German opposite numbers would find it to work guns while wearing gas-masks.

Howitzers and the heavier guns had the task of material destruction – of communication trenches, approach roads, railway spurs, of anything which aided the movement of men and supplies into the trenches which were to be attacked, but above all of strongpoints and machine-gun posts. These were of different sorts. In several places, notably where the German front crossed the site of a former village, the defences were notably stronger than in the open fields between. For although the Germans had excavated thirty-foot-deep dug-outs at regular intervals all along their front, which were proof against a direct hit by any weight of shell, and had thus assured that their trench garrisons would be alive even at the end of a prolonged British bombardment, these 'field' positions could not be given, without enormous extra labour, the complex illogicality presented to an attacker by the ruin of an inhabited area. In some spots, like the Leipzig salient between the devastated villages of Thiepval and Pozières, sporadic local attack and counter-attack had produced a maze of trenches as impenetrable as any ruin; and elsewhere, as at the Schwaben Redoubt, the Germans had thought it worth devoting the necessary spadework to building an artificial fortress-entrenchment. But the villages were the most important revetments of the German line; and the most important ingredient in the Germans' scheme of defence for these strongpoints was the fire of machine-guns. It was to the destruction of their emplacements, or the entombment of their crews in their positions of shelter, that the British heavy artillery was to devote its bombardment during the six days of 'preparation'.

The machine-gun was to be described by Major-General J. F. C. Fuller, one of the great *enragés* of military theory produced by the war, as 'concentrated essence of infantry', by which he meant his readers to grasp that its invention put into the hands of one man the fire-power formerly wielded by forty.

Given that a good rifleman could fire only fifteen shots a minute, to a machine-gunner's 600, the point is well made. But, as Fuller would no doubt have conceded if taxed, a machine-gun team did not simply represent the equivalent of so many infantrymen compressed into a small compass. Infantrymen, however well-trained and well-armed, however resolute, however ready to kill, remain erratic agents of death. Unless centrally directed, they will choose, perhaps badly, their own targets, will open and cease fire individually, will be put off their aim by the enemy's return of fire, will be distracted by the wounding of those near them, will yield to fear or excitement, will fire high, low or wide. It was to overcome influences and tendencies of this sort – as well as to avert the danger of accident in closely packed ranks – that seventeenth- and eighteenth-century armies had put such effort into perfecting volley fire by square, line and column. The result was to make an early-nineteenth-century – Waterloo – infantry regiment arguably more dangerous to approach than a late-nineteenth-century – Boer War – one. For though the latter had better weapons than the former, and ones which fired to a much greater range, these technical advantages were, if not cancelled out, certainly much offset by the dispersion of the soldiers which the very improvement of firearms itself enjoined – dispersion meaning lack of control, which in its turn results in poor musketry. Hence the wonder with which the machine-gun was viewed when Maxim first made it a practicable weapon of war. For it appeared to have put back into the hands of the regimental commander the means to inflict multiple and simultaneous wounding by the giving of a single word of command. But the appearance of the machine-gun was, of course, very much more than a reversion to a former order of things. For the most important thing about a machine-gun is that it is a *machine*, and one of quite an advanced type, similar in some respects to a high-precision lathe, in others to an automatic press. Like a lathe, it requires to be set up, so that it will operate within desired and predetermined limits; this was done on the Maxim gun, common to all armies of 1914–18, by adjusting the angle of the barrel relative to its fixed firing platform, and tightening or loosening its traversing screw. Then, like an automatic press, it would, when actuated by a simple trigger, begin and continue to perform its functions with the minimum of human attention, supplying its own power and only requiring a steady supply of raw material and a little routine maintenance to operate efficiently throughout a working shift. The machine-gunner is best thought of, in short, as a sort of machine-minder, whose principal task was to feed ammunition belts into the breech, something which could be done while the gun was in full operation, top up the fluid in the cooling jacket, and traverse the gun from left to right and back again within the limits set by its firing platform. Traversing was achieved by a technique known, in the British Army, as the 'two inch tap': by constant practice, the machine-gunner learned to hit the side of the breech with the palm of his hand just hard enough to move the muzzle exactly two inches against the resistance of the traversing screw. A succession of 'two inch taps' first on one side of the breech until the stop was reached, then on the other, would keep in the air a stream of bullets so dense that no one could walk upright across the front of the machine-gunner's position without being hit – given, of course, that the gunner had set his machine to fire low and that the ground was devoid of cover. The appearance of the machine-gun, therefore, had not so much *disciplined* the act of killing – which was what seventeenth-century drill had done – as *mechanized* or *industrialized* it.

It was this automatic and inhuman lethality of the machine-gun which determined that the posts from which it would operate must be the principal target of the heavy artillery between 25 and 30 June. Unfortunately for the British infantry, the heavy howitzer of 1916 was a piece of technology very much less developed towards perfection, relative to its potential, than was the machine-gun. The desirable characteristics of the machine-gun, besides those of functional efficiency, were portability, concealability and compactness. The Maxim met the first fairly, the other two very well. The desirable characteristics of the heavy howitzer were pin-point precision and intense concussive effect. These neither the 6-, 8-nor 9.2-inch howitzer achieved (the larger calibres were too few in number to matter). Their shells had an aiming error of at least twenty-five yards and an explosive power insufficient to collapse the very deep dug-outs – 'mined' dug-outs, the British called them, for they were driven by mining technique thirty feet below the surface – in which the machine-gunners sheltered, with their weapons, during a bombardment. Thus the British could not destroy the kernel of a German strongpoint. The best they could hope to do was to trap the crews below ground by choking the entrance shaft with spoil from the collapsed trench; but to hit the shaft, unless by luck, required either an altogether more revealing sort of air photograph than the Royal Flying Corps' cameras could supply or else constant, life-wasting raiding across no-man's-land to locate precisely where the dug-out entrances lay.

If we look, then, at the preliminaries to the attack of 1 July as a struggle between competing technologies, between the manifest power of the British artillery and the latent power of the German machine-guns, it will be seen clearly as a struggle the British waged on unequal terms – and terms which they failed to reverse, despite achieving the appearance of terrible devastation. The bombardment opened on 24 June. It was intended to last five days, but a postponement of Z-day extended it to seven. Over the period, about 1,500,000 shells from the stocks which had been dumped were fired – 138,000 on 24 June, 375,000 on 30 June. Much the greater number – about a million – were 18-pounder shrapnel shells; the 6-inch howitzers fired about 80,000, the 8- and 9.2-inch about 50,000 each. These are impressive totals. To achieve them the artillery crews had to labour, humping shells or heaving to re-align their ponderous weapons (the 8-inch howitzer weighed thirteen tons), hour after hour throughout the day and for long periods of the night. At the receiving end, the noise, shock-waves and destructive effect were extremely unpleasant. At first the Germans in the trenches opposite thought the bombardment heralded an attack and stood to arms in their dug-outs. Then, as the shelling continued, waxing and waning in strength, they realized that they were in for a long ordeal and settled down to bear it as best they could.

During 25 June ... the fire of the British ... batteries increased, and whereas on the previous day nine-tenths of the fire had been shrapnel or from guns of small calibre [shrapnel was disregarded because its scatter of man-killing pellets was of very little effect against entrenchments], the heavy batteries seemed now in the majority. Their shells crashed into the German trenches, the ground shook and the dug-outs tottered. Here and there the sides of a trench fell in, completely blocking it. Masses of earth came tumbling into the deep dug-outs, obstructing all entrances [which of course faced *away* from the direction of the shelling] to many of them. By evening some sectors of the German front-line were already unrecognizable and had become crater-fields.*

The British next began to mix gas with their shelling, using primitive projectors and the prevailing wind-stream to carry it across no-man's-land.

In the early hours [of 26 June] clouds of chlorine gas ... reached the German position [near Fricourt] and, being heavier than air, filled every crevice on the ground. The dense fumes crept like live things down the steps of the deep dug-outs, filling them with poison until sprayers negatived their effect ... during the afternoon aerial torpedoes, fired from heavy mortars in the British front-line, made their first appearance. Coming down almost perpendicularly from a great height, these monsters bored deep into the ground and then burst. [This reference is almost certainly not to mortar bombs but to the shells of the super-heavy howitzer, fortunately for the Germans very few in number.] Tons of earth and great blocks of chalk and rock were hurled into the air, leaving craters, some twelve feet deep and fifteen feet in diameter. Only deep dug-outs of great strength could stand the shock ... The Germans, who up till now had endured the inferno almost with indifference, began to feel alarmed. Every nerve was strained as they sat listening to the devilish noise and waited for the dull thud of the next torpedo as it buried itself in the ground, and then the devastating explosion. [The similar experience of listening to the Krupp 420mm siege-howitzers 'walking' their shells up to the target had driven men hysterical inside the Liège forts in August, 1914.] The concussion put out the candles and acetylene lights in the deepest dug-outs. The walls rocked like the sides of a ship and the darkness was filled with smoke and gas fumes ... The 27th and 28th June brought a similar picture of continuous devastation ... The bombardment continued to appear without method, an intense and apparently wild shelling, then carefully observed heavy artillery fire by individual batteries, then trench-mortar bombs and aerial torpedoes or gas

*From *If Germany Attacks* by Captain G. C. Wynne.

attacks, or again a sudden tornado of shells, with occasional periods of complete quiet.

June 30th was a repetition of the previous six days. The German front defences no longer existed as such ... [But] in spite of the devastation and chaos on the surface, the defenders in those of the deep dug-outs still intact (the majority) had ... survived the ordeal. For seven days and nights they had sat on the long wooden benches or on the wire beds in the evil-smelling dug-outs some twenty feet and more below ground. The incessant noise and the need for constant watchfulness had allowed them little sleep, and ever-present, too, had been the fear that their dug-outs might at any time become a living tomb from which escape would be impossible. Warm food had seldom reached them ... so that they had had to live on [iron rations].

But they were alive.

At 6.30 a.m., however, [on 1 July] a bombardment of an intensity as yet unparalleled suddenly burst out again along the whole front. At first it was most severe in the centre, about Thiepval and Beaumont, but it spread quickly over the entire line from north of the Ancre to south of the Somme. For the next hour continuous lines of great fountains of earth, rocks, smoke and debris, played constantly into the air ... The giant explosions of the heaviest shells were the only distinguishable noises in the continuous thunder of the bombardment and short, regular intervals of their bursts gave it certain rhythm. All trace of the front-trench system was now lost, and, with only a few exceptions, all the telephone cables connecting it with the rear lines and batteries were destroyed, in spite of the six feet of depth at which they had been laid. Through the long periscopes held up out of the dug-outs could be seen a mass of steel helmets above the British parapet ... The Germans in their dug-outs, each with a beltful of hand-grenades, therefore waited ready, rifle in hand, for the bombardment to lift from the front trench to the rear defences. It was of vital importance not to lose a second in reaching the open before the British infantry could arrive at the dug-out entrances.

The battle was about to begin. And its first, and indeed decisive, act was to be the 'race for the parapet' – a race which for the British ran from their own front trench to the other side of no-man's-land, for the Germans from the bottom to the top of their dug-out steps. Whoever first arrived at the German parapet would live. The side which lost the race would die, either bombed in the recesses of the earth or shot on the surface in front of the trench. Every British effort had been directed to ensuring that the Germans lost the race – that they would indeed lack the runners to make it a contest. But, as we have seen, the majority of the German trench garrisons still lived at zero hour on Z-Day. How had the British artillery effort been expended to such little purpose?

The greater part of the answer is revealed by isolating the proportion of active ingredient in the British bombardment; that is, of explosive delivered to the German-occupied area. The weight of shells transported to the British guns was about 21,000 tons, excluding propellant (the explosive needed to drive the shell up the barrel at the moment of firing). It had taken the efforts of about 50,000 gunners (almost the number of Wellington's army at Waterloo), working for seven days, to load this weight into their pieces and fire it at the enemy – or, more precisely, into the area, 25,000 by 2,000 yards square, which the British infantry were to attack. In crude terms, this meant that each 2,500 square yards had received a ton of shells; or, if numbers of shells are used for the calculation – and about 1,500,000 had been fired – that each 1,000 square yards had received 30 shells. However, about a million of the shells were shrapnel, fired by the 18-pounder field guns of the divisional artilleries, and these could do very little damage to earthworks, since they were filled only with light steel balls, and only a little more to wire, though it was their alleged wire-cutting capability which justified the firing of the enormous number used. In fact, the 18-pounders were set to firing shrapnel because the ammunition factories in England could not yet produce high-explosive shell for them in any quantity, though almost everyone in the B.E.F. from G.H.Q. officer to simple gunner had now come to realize that it was high explosive alone which did serious damage to an entrenched enemy.

Discounting the shrapnel, therefore, we are left with the output of the howitzers and heavy guns – about half a million shells of 12,000 tons weight. The lightest and most plentifully expended shell was that of the 4.5-inch field howitzers of the divisional artilleries, which weighed 35 pounds; the heaviest, that of the 15-inch howitzers, which weighed 1,400 pounds – but of which there was a strictly limited ration, there being only six of these monster guns on the battlefield. Nevertheless, their contribution to the bombardment – about 1,500 shells, weighing a thousand tons – is impressive, and all the more so if we recall that Napoleon had with him at Waterloo only about 100 tons of artillery projectile in all. Comparisons between the artillery efforts of 1815 and 1916 are pointless, however, for Napoleon's gunners had had the fairly simple task of firing solid shot from close range at dense and immobile masses of soldiers upon whom a hit meant a kill; Haig's gunners, by contrast, could not see their target and could not be sure that, even if they hit it, their fire would have a lethal effect. That this should be so was due to the very small proportion of explosive contained within the casing of the shell. The 1,400-pound shell of the 15-inch howitzer, for example, contained 200 pounds of explosive (Ammatol, a mixture of TNT and Ammonium Nitrate); the 35-pound shell of the 4.5-inch howitzer contained only four pounds ten ounces. The explanation of this disparity between total weight of shell and weight of filling was twofold; the stresses to which the shell was subjected during firing required that it have a very strong, and therefore heavy, casing, if it were not to disintegrate inside the gun with disastrous effect; while the purpose of the shell, as conceived by its designers, was to produce a large number of steel splinters, travelling at man-killing speed, as a by-blow of its explosion. For that reason, most shells were fused to explode on impact, their detonation producing those enormous fountains of earth and smoke which are the staple feature of First World War battlescapes.

It is these fountains which give the game away. Out of the 12,000 tons, weight of shell delivered on to the German-occupied area, only about 900 tons represented high-explosive. And the greater part of that small explosive load was dissipated in the air, flinging upwards, to be sure, a visually impressive mass of surface material and an aurally terrifying shower of steel splinters but transmitting a proportionately quite trifling concussion downwards towards the hiding places of the German trench garrisons. Each ten square yards had received only a pound of high-explosive, or each square mile about thirty tons. Twenty-eight years later, the Allied air forces would put down on German positions in Normandy, and in minutes not days, something like 800 tons of bombs to the square mile, most of that tonnage consisting of high-explosive, for free-falling bombs, being unsubjected to stress, can be given the thinnest and lightest of cases. Today, NATO tactical doctrine would regard the Somme position as a suitable target for several small nuclear warheads, each of which would yield many thousand tons of TNT-equivalent to the square mile. But some of the defenders, if properly dug in under overhead cover, would still be expected to survive – as many German soldiers who cowered under the aerial preparation for Operations Goodwood and Cobra in July 1944 survived to man their weapons against the British and American tank columns which emerged through the dust of the bombing.

We can see now, therefore, that the great Somme bombardment, for all its sound and fury, was inadequate to the task those who planned it expected of it. The shells which the British guns had fired at the German trenches, like those which a month earlier had broken up on the armoured skins of the German battleships at Jutland, were the wrong sort of projectile for the job, and often badly made. And while the British naval gunners had been able to see, and knew how to hit, their targets, the British field and garrison gunners, many of them amateurs, had largely to guess at where their real targets, the German machine-gun crews, were hidden, and then very often lacked the skill to put a shell where they wanted it to fall. Hence, despite the precision of the fire plan, that haphazard cratering of the battlefield, sometimes on, sometimes beyond, sometimes short of the German trench line and wire entanglements, which all observers of the Somme front mention.

The Final Preliminaries

The infantry, fortunately, remained largely unaware of the random and unsatisfactory result of the shelling which had filled their ears with sound for the last week, during every hour of the day and many of the nights. There was a good deal of individual apprehension. 'It was the Division's first battle,' wrote the historian of the 18th, 'and the solemnity of the occasion affected everyone.' Private Gilbert Hall, of the 1st Barnsley Pals (13th York and Lancs) was not feeling quite himself and had got a headache from the bombardment. Capt. E. C. T. Minet, machine-gun officer of the 11th Royal Fusiliers, felt himself 'sweating at zero hour. But that, I suppose, was nervous excitement.' Private Frank Hawkings, of Queen Victoria's Rifles, had found since 29 June 'the suspense very trying and everyone ... very restless'. But the long notice of the battle which everyone who was to be in it had been given – a new development in warfare and a function of the complex preparation which battles of the industrial age require – had allowed men the chance to make what personal accommodation with their fears they could. Most had written home, made out their wills, shaken hands with their pals. Many had gone to church. Each battalion of the B.E.F., the army of a church-going age and nation, had its own chaplain of the appropriate denomination, and they had held services behind the lines a day or two beforehand. Second-Lieutenant John Engall, of the 16th London Regiment, wrote home 'the day before the most important of my life ... I took my Communion yesterday with dozens of others who are going over tomorrow and never have I attended a more impressive service. I placed my body in God's keeping and I am going into battle with His name on my lips, full of confidence and trusting implicitly in Him.' Like so many other subalterns of the London division, Engall was to die outside Gommecourt. His explicit piety, which would have jarred with most of Wellington's ensigns, came as naturally to him as to them their stylish indifference. But it would not necessarily have surprised them;

the attendance at (Anglican) Communion of 'the dozens of others' – private soldiers of his regiment – most certainly would have done. The irreligiosity of their private soldiers was part and parcel of an altogether rougher persona than even the most hardened old-sweat regiments of 1914 could show.

It was a help, too, in calming fears that the last hours before zero were filled for most infantry soldiers with a carefully time-tabled programme of activity. The attacking battalions, which were out of the line, had to march up to the trenches from the villages where they had been billeted, first along the roads, then in the communication trenches which covered the last mile. On the way the men accumulated a growing load of kit. Starting with 200 rounds of ammunition and two days' rations, they successively picked up new empty sandbags (to fortify the positions they were to take), a wiring stake (for the same purpose), grenades, shovels, rockets and sometimes pigeon baskets, the two last items to help their officers communicate with the rear once they had passed beyond cablehead in the front trench. All this took a great deal of time, and the columns had also to press forward against the flow of men coming down the trenches from the battalions which were being relieved. When they arrived at their jumping-off places, the men were glad to huddle under a blanket or greatcoat on the floor of the trench and sleep.

Most awoke early, to find a light rain falling through white morning mist. In places it lingered even after the bombardment had struck up at 6.25 a.m. for the regular morning session, so that Lieutenant Chetwynd-Stapleton, on air patrol above the front, saw 'a bank of low cloud' on which 'one could see ripples ... from the terrific bombardment that was taking place below. It looked like a large lake of mist, with thousands of stones being thrown into it.' Across the greater breadth of the front, however, the mist quickly cleared, giving way to bright sunshine from a brilliant cloudless sky. Into it little plumes of smoke rose here and there from the British front where men were furtively cooking breakfast. Orders were for soldiers to be fed with food sent from the rear, and in the best organized battalions it arrived, prompt and hot. Lieutenant-Colonel Crozier, commanding the 9th Royal Irish Rifles (the West Belfast Battalion of the Ulster Volunteer Force), congratulated his cook-sergeant on having bacon rashers, fried bread, jam and tea ready for his riflemen, and a mixture of cold tea and lemon to go into their water-bottles for the trip across no-man's-land. Officers were being brought hot water in which to shave, and were tidying their uniforms, still conspicuously different from the soldiers', unless they belonged to battalions in which it was not thought bad form to don rough Tommy serge. Major Jack, commanding a company of the 2nd Cameronians, put on his silver spurs for the occasion, and his soldier servant gave him 'a final brush'. Few, if any, were to wear swords (though even temporary officers were still buying swords on commissioning) but all carried sticks, polished blackthorn with a silver band in the Irish regiments, malacca canes or ashplants with a curved handle, of the sort sold by seaside tobacconists, in others. Some carried nothing else, not even a revolver, thinking it an officer's role to lead and direct, not to kill – the need for which, in any case, they believed would have been nullified by the bombardment.

Between 6.30 and 7.30 a.m., the noise of the bombardment reached a level not yet touched, as weapons of every calibre and sort put down their final ration of shells on the German front trenches. Hawkings, of Queen Victoria's Rifles, had been watching a lark climb into the sky opposite Gommecourt, when the artillery, which had hitherto been firing spasmodically,

> suddenly blazed out in one colossal roar. The dull booms of the heavy guns in the rear could just be discerned amidst the sharper and incessant cracks of the 18-pounders and 4.7s that were closer to the line. There seemed to be a continual stream of shells rumbling and whining overhead on their way to the enemy positions, where the succession of explosions added to the general noise. Fifteen-inch howitzer and 9.2 shells were falling in Gommecourt Wood, whole trees were uprooted and flung into the air, and eventually the wood was in flames. The landscape seemed to be blotted out by drifting smoke; but as part of our scheme was to set up a smoke screen we commenced throwing out smoke bombs.

Under the weight of this cannonade, the Germans crouched invisible in their dug-outs, waiting for the moment it should

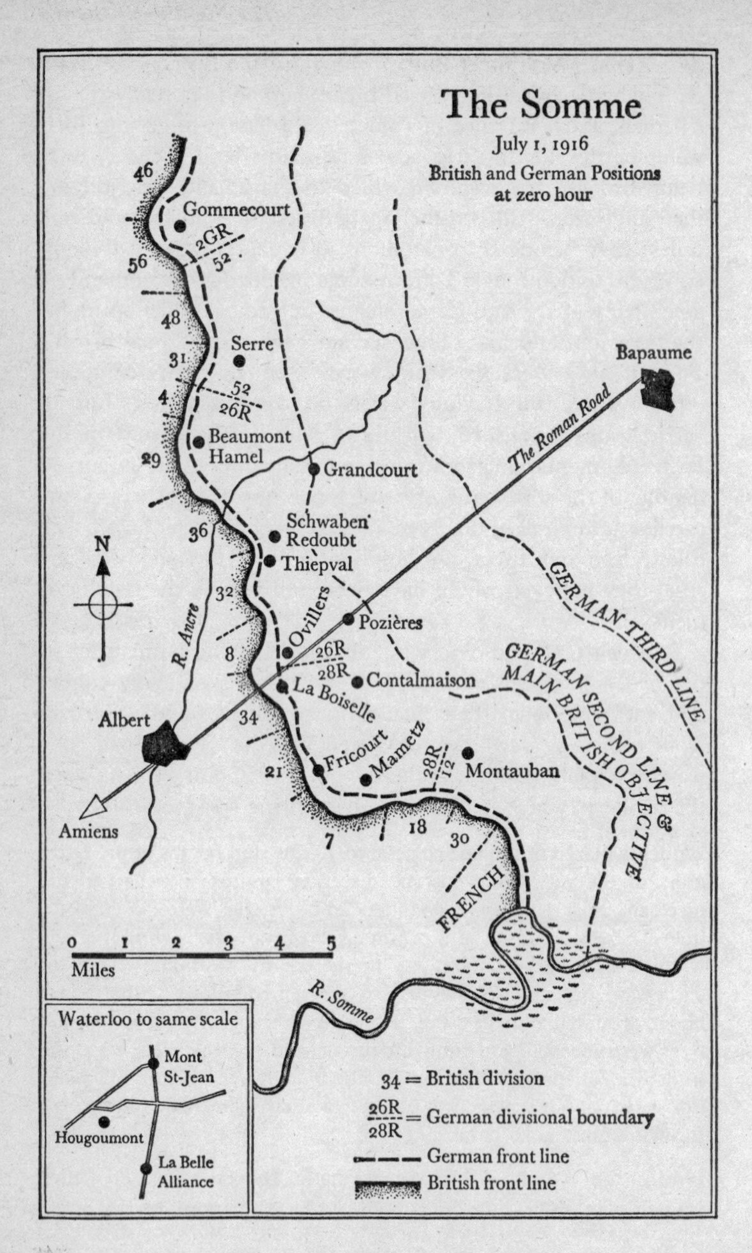
The Somme
July 1, 1916
British and German Positions
at zero hour
46
Gommecourt
2GR
52
56
48
Serre
31
52
26R
4
Beaumont
Hamel
29
Grandcourt
The Roman Road
Bapaume
36
Schwaben
Redoubt
N
Thiepval
32
R. Ancre
Ovillers
Pozières
GERMAN THIRD LINE
8
26R
28R
Contalmaison
GERMAN SECOND LINE &
MAIN BRITISH OBJECTIVE
La Boiselle
Albert
34
Fricourt
Mametz
28R
12
Montauban
21
Amiens
7
18
30
FRENCH
0 1 2 3 4 5
Miles
R. Somme
Waterloo to same scale
Mont
St-Jean
Hougoumont
La Belle
Alliance
34 = British division
26R/28R = German divisional boundary
German front line
British front line

lift as the signal to race up exit shafts. Meanwhile the soldiers of the 1st Somerset Light Infantry sat on the parapet of their trench opposite, laughing and cheering at the sight of the detonations.

Some other soldiers were also already out of their trenches, where the last thing almost everyone had received was a strong tot of rum – Navy rum, and extremely alcoholic. In the 11th Suffolks, two men who had got the teetotallers' share drank themselves insensible and could not be got on to their feet again; and J. F. C. Fuller, investigating a confusion in the Sherwood Forester Brigade, was told that the whole of the leading wave was drunk. He thought the story an exaggeration – which it almost certainly was – but, knowing that 'in many cases men deliberately avoided eating before a battle, for fear of being shot through a full stomach' and discovering that 'through some error' the first line got the rum ration intended for the second as well as their own, he concluded that 'many of the men in the front line must have been drunk well before zero hour.' A strong tot of rum, whatever its functional effect, must have been particularly comforting to the men in those divisions whose commanders had decided to take them out of the trench to lie down in no-man's-land before zero – the 8th, 36th, 46th and 56th and part of the 32nd.

The signal for these men to stand up and advance, for those in the trenches to climb their scaling ladders and leap over the parapet into no-man's-land, was to be the shrill of the platoon officers' whistles, blown when their synchronized watches showed 7.30 a.m. In four places, however – Mametz on the 7th Division's front, Fricourt (21st Division), La Boiselle (34th Division), and Beaumont Hamel (29th Division) – the signal, about ten minutes ahead of zero, was to be the detonation of eight enormous mines which had been tunnelled under the German trenches and filled with dynamite.

The Battle

Despite the immense growth of complexity of the machinery and business of war which had taken place in Western armies since 1815, the Battle of the Somme was to be in many ways a simpler event than Waterloo – not, indeed, in terms of the strains of management it threw on commanders and their staffs, but in the range and nature of the encounters between different categories of armed groups which took place on the ground. At Waterloo we counted seven different sorts of encounters: artillery versus artillery, infantry and cavalry, cavalry versus cavalry and infantry; infantry versus infantry; and single combat. Several of these could or did not occur on the Somme. The horse, for example, had disappeared from the battlefield, though to the regret of almost every soldier – even infantry officers speak lovingly of their horses – and temporary work in the transport section of infantry regiments was eagerly sought after by the men, who seemed to find in caring for animals an outlet for the gentler emotions to which they could give no expression among their fellows. Haig had had three cavalry divisions brought up to the Somme front, but they were neither expected to, nor did they, play any part on 1 July, or any other day in 1916. Single combat, too, had ceased to be an option, for soldiers on a bullet-lashed battlefield could neither assume the posture nor risk the exposure-time necessary for the exchange of blows, even if haphazard encounter brought them together. The nearest thing to single combat in trench warfare ('him or me' bayonet thrusting excepted), was perhaps the game of 'bombing up the traverses', of which the most striking feature, so characteristic of the First World War, was that one did not see one's enemy. Thus there were only three sorts of encounter possible on the field of the Somme: artillery versus artillery; artillery versus infantry; and infantry versus infantry – though, if we treat machine-gunners as a separate category, we also get infantry versus machine-gunners and artillery versus machine-gunners.

We have already seen how much or little success artillery had had in attacking infantry and machine-gunners, and in attacking other artillery. Many of the German batteries had had guns disabled and crews killed during the preliminary bombardment. But enough remained to put down spoiling bombardments on the British front trenches at the moment of the attack – a company of the Queen Victoria's Rifles, crowded into their assembly trenches, were struck by a sudden stream of shells at about 7 a.m. and had forty men killed or wounded in a few minutes – and to fire standing barrages into no-man's-land the instant they got the signal that the British had left their trenches. These sorts of encounter apart, how did the infantry fare, on both sides, once the British had left their trenches to advance to the assault?

Infantry versus Machine-gunners

Several survivors have left accounts of the first moments of the attack. Queen Victoria's Rifles, a leading battalion of 56th London Division in VII Corps' diversionary attack on the northern flank of Fourth Army, had about 500 yards of no-man's-land to cross; Royal Engineer companies laid smoke from dischargers to cover their advance. Some time after seven o'clock, the Germans 'began spraying our parapets with machine-gun bullets, but sharp to the minute of zero' (7.25 a.m. for this division) 'we erected our ladders and climbed out into the open. Shells were bursting everywhere and through the drifting smoke in front of us we could see the enemy's first line from which grey figures emerged ... We moved forward in long lines, stumbling through the mass of shell-holes, wire and wreckage, and behind us more waves appeared.' Towards the centre of the Fourth Army's front, the 9th Royal Inniskilling Fusiliers, of the 36th Ulster Division, were in the leading wave. Their commanding officer, Ricardo,

> stood on the parapet between the two centre exits to wish them luck ... They got going without delay, no fuss, no running, no shouting,

everything solid and thorough – just like the men themselves [these were farming people from County Tyrone]. Here and there a boy would wave his hand to me as I shouted good luck ... through my megaphone. And all had a cheery face. Most were carrying loads. Fancy advancing against heavy fire with a big roll of barbed wire on your shoulder!

Describing a second wave attack, in an account which holds good for the first, Gilbert Hall of the 1st Barnsley Pals (13th York and Lancasters, 31st Division), heard his officer blow his whistle 'and C Company climbed over the parapet and moved forward to be confronted with ... a long grassy slope rising gently to a series of low crests about six hundred yards in front. The German trenches were clearly visible, three lines of fortifications with sand-bagged parapets, enabled by the slope of the ground to fire over each other into the advancing British infantry. In front of the enemy lines lay thick belts of uncut wire, breached by a few narrow gaps.' Towards that wire the Barnsley Pals set off, as up and down the line at zero did 60,000 other infantrymen. In some battalions, the men were able to walk upright, with arms sloped or ported, as they had been expecting. In others they were soon bent forward, like men walking into a strong wind and rain, their bayonets fixed and their rifles horizontal. 'Troops always, in my experience,' wrote Lord Chandos, whose observation this is, 'unconsciously assume this crouching position when advancing against heavy fire.'

Most soldiers were encountering heavy fire within seconds of leaving their trenches. The 10th West Yorks, attacking towards the ruined village of Fricourt in the little valley of the River Ancre, had its two follow-up companies caught in the open by German machine-gunners who emerged from their dug-outs after the leading waves had passed over the top and onward. They were 'practically annihilated and lay shot down in their waves'. In the neighbouring 34th Division, the 15th and 16th Royal Scots, two Edinburgh Pals' Battalions containing a high proportion of Mancunians, were caught in flank by machine-guns firing from the ruins of La Boiselle and lost several hundred men in a few minutes, though the survivors marched on to enter the German lines. Their neighbouring battalions, the 10th Lincolns and 11th Suffolks (the Grimsby Chums and the Cambridge Battalion) were caught by the same flanking fire; of those who pressed on to the German trenches, some, to quote the official history 'were burnt to death by flame throwers as [they] reached the [German] parapet'; others were caught again by machine-gun fire as they entered the German position. An artillery officer who walked across later came on 'line after line of dead men lying where they had fallen'. Behind the Edinburghs, the four Tyneside Irish battalions of the 103rd Brigade underwent a bizarre and pointless massacre. The 34th Division's commander had decided to move all twelve of his battalions simultaneously towards the German front, the 101st and 102nd Brigades from the front trench, the 103rd from the support line (called the Tara-Usna Line, in a little re-entrant known to the brigade as the Avoca Valley – all three names allusions to Irish beauty spots celebrated by Yeats and the Irish literary nationalists). This decision gave the last brigade a mile of open ground to cover before it reached its own front line, a safe enough passage if the enemy's machine-guns had been extinguished, otherwise a funeral march. A sergeant of the 3rd Tyneside Irish (26th Northumberland Fusiliers) describes how it was: 'I could see, away to my left and right, long lines of men. Then I heard the "patter, patter" of machine-guns in the distance. By the time I'd gone another ten yards there seemed to be only a few men left around me; by the time I had gone twenty yards, I seemed to be on my own. Then I was hit myself.' Not all went down so soon. A few heroic souls pressed on to the British front line, crossed no-man's-land and entered the German trenches. But the brigade was destroyed; one of its battalions had lost over 600 men killed or wounded, another, 500; the brigadier and two battalion commanders had been hit, a third lay dead.* Militarily, the advance had achieved nothing. Most of the bodies lay on territory British before the battle had begun.

In the neighbouring 32nd Division, the 16th Northumberland

*In the Tyneside Scottish Brigade, all four battalion commanders were killed on 1 July.

Fusiliers (Newcastle Commercials) and the 15th Lancashire Fusiliers (1st Salford Pals) were also hit by machine-gun fire from Thiepval as they got out of their trenches, the Newcastle Commercials following a football kicked by a well-known north country player. Several waves were cut down at once and the commanding officers ordered the untouched companies to stay in their trenches. In the swampy valley of the Ancre, several battalions of the Ulster Division were enfiladed by German machine-guns as the men tried to cross no-man's-land, there 400 yards wide. Casualties were worst in the 9th Royal Irish Fusiliers (the Armagh, Monaghan and Cavan battalion of the U.V.F.), 532 officers and men going down as rush after rush towards the wire – orthodox tactics learnt on Irish hillsides in make-believe battles four years before – was stopped by bursts of bullets; losses in two others were almost as heavy. In the regular 29th Division on the Ulstermen's left, several battalions suffered the worst of First World War experiences: to advance across no-man's-land under heavy fire only to find the enemy's wire uncut (it was uncut at many places elsewhere also) and to be machine-gunned down while searching for a way through. Among them were the 1st Royal Inniskilling Fusiliers, the same Inniskillings who had stood in square to be cannonaded throughout the afternoon near the crossroads at Waterloo 101 years before. Opposite Beaumont Hamel, fired on by German machine-gunners who had emerged from the recesses of Y Ravine, into which the Division's amateur gunners had tried but failed to drop a shell during the bombardment, 568 Inniskillings became casualties in a few minutes, of whom 246 died. Shortly afterwards the only battalion from the Empire to take part in the Somme attack, the 1st Newfoundland Regiment, raised exclusively from native-born Newfoundlanders, tried to find a way where the Inniskillings had found none and in the attempt lost more men killed, wounded or missing – 710, including all the officers – than any other battalion was to do on 1 July (though the 10th West Yorks had just lost exactly the same number opposite Fricourt, on the 21st Division's front). Finally, on the 46th (North Midland) Division's front, at the extreme northern edge of the battlefield, the 1/6th North Staffs and the 1/6th

South Staffords, Territorial battalions from Wolverhampton and Hanley, each had their leading companies caught by fire opposite uncut wire, on which most who arrived there were shot or bombed by the defenders.

Infantry versus Infantry

Perhaps twenty battalions of the attacking force, out of sixty committed to the first wave, had thus been disabled in no-man's-land by machine-gun fire, to which they had been unable to reply and whose source they had generally been unable to identify. A number had also suffered casualties from German barrage fire – true barrage fire, in that it took the form of a continuous fall of shells along a predetermined line in no-man's-land – laid by guns which had either escaped destruction by the British batteries or had remained 'masked' (present but silent, and so undetectable), or else had arrived on the Somme front towards the conclusion of the bombardment – an event which had given the German high command all the notice to reinforce any general could require. The barrage had been particularly heavy on VIII Corps' front, perhaps because its gunners were notably less well-trained even than the rest in that underskilled army; it was there that there was most uncut wire, additional proof of incompetent gunnery.

But for the battalions which had got through, the worst, in a collective sense, was now over, for entry into the German positions meant that the German gunners could no longer put down on them a barrage, their own troops being mixed up with the attackers, while the attackers themselves could take advantage of the Germans' own trenches to shelter from the German infantry's bombs and bullets. In practice, things were less simple and far more dangerous than this thumbnail analysis suggests, for the British could not remain in the German trenches they had reached, having objectives to reach which lay much deeper within German trenches, yet *had* to remain to fight for a while if they were not to be attacked in the rear when they pushed on. The Germans, moreover, had enough of their telephone cable network intact sometimes to be able to inform their batteries which trenches were in British hands, and so to be able to call down fire on them. The British had no such link with their artillery, the telephone lines they had trailed across no-man's-land having almost without exception, and to no one's surprise, been cut. As a result, and to complicate the pattern of activity in those battalions which had entered the German front line, the attackers were under an obligation, both to consolidate – clear the captured trench of any resisting Germans – and to 'follow the barrage' – to walk off whither the next curtain of shellfire played on the second or third line of German trenches or on an intervening 'shellhole position'; some divisional artilleries had allowed for as many as six of these 'lifts'.

Following a shrapnel barrage was, for all the tumult produced, not in itself a dangerous thing to do, given accurate gunnery, for the cast of shrapnel is forward, only the occasional base-plate whining back to inflict injury on the infantry behind. By late 1917 British infantrymen had learnt, and were glad, to walk as close as twenty-five yards in the rear of a boiling, roaring cloud of explosive and dust, accepting that it was safer to court death from the barrage than to hang back and perhaps be killed by a German whom the shells had spared and one's own tardiness allowed the time to pop up from his dug-out. In July 1916, however, few gunners knew how to make a barrage 'creep' at a regular walking pace across a piece of enemy-held territory and, prudently, few infantrymen would risk approaching too close to a barrage line until they saw it lift and move to the next target. The consequence was that the advance, even when it worked to plan, took the form of a series of discontinuous and quite literally breathless jerks forward, the lift of the barrage to the next objective being the signal for the waiting infantry to leave their positions of shelter and race the intervening two or three hundred yards to regain its protection. For the 18th and 30th Divisions, flanking the French at the extreme southern end of the battlefield, this programme worked very well, and both gained all their objectives within the limit of time set, though each at the loss of about 3,000 casualties. They benefited, however, from their proximity to the French, whose gunnery, after two years of war, was much superior to the Royal Artillery's, and whose infantry, here belonging to the XX Corps, were among the best soldiers on the western front. The British III Corps' advance was, therefore, in the language of the period, to some extent a 'sympathetic' one.

Where, farther north, the British had to make their way alone, most battalions, even if they managed to get into the German trenches, sooner or later 'lost' the barrage, which they had to watch grinding noisily and remorselessly away from them, according to a pre-arranged timetable, and could only very rarely and with the greatest difficulty recall to work over the objective on which they were stuck. The reasons for this 'loss' of the barrage were everywhere the same: the infantry arrived in the German front trench either too disorganized by losses to be able to push on at the time required, or else were held there so long by the resistance of the German defenders that the barrage left without them, or else, exhausted by physical effort and nervous strain more quickly than the staff had allowed for, stopped to rest.

Exhaustion was what stopped the advance on the northern sector of the 21st Division's front, after a very brave and quick rush had carried the two leading battalions of the 64th Brigade deep into the German position. Despite heavy losses from machine-gun fire in no-man's-land, the 9th and 10th King's Own Yorkshire Light Infantry crossed it, found good gaps in the German wire, shot or bombed the defenders who opposed them and took the front trench. Joined there by their support battalions, the 15th Durham Light Infantry and the 1st East Yorkshire, they moved on behind the barrage to the next German trench, from which they extracted and sent back about 200 German prisoners. All this had taken about ten minutes, during which half of the soldiers and most of the officers had been killed or wounded. The barrage now moved forward, and the battalions were still able to follow, but they now came across more and more Germans, hidden in shell-holes or bits of trench, who threw bombs or opened fire with their rifles. At the end of half an hour from zero, by which time they had covered a mile of ground from their own front line, the Yorkshiremen reached an old sunken road. Here the majority stopped – 'a general halt was called' is the euphemism employed by the official historian – though some parties from all four battalions went on a little farther, to Crucifix Trench, the brigade's first objective. Five hundred yards ahead, Germans had begun to fire machine-guns from positions hidden in woodland and now 'the barrage had passed on'. The 9th and 10th K.O.Y.L.I.'s advance was over for the day, the men too worn out by fatigue and fear to go any farther themselves, the officers who might have led them dead or wounded. After a long delay, news of the circumstances reached those 'L.O.O.B.' – Left Out of Battle to form a cadre if losses were very heavy – and one of the officers, a young captain, Basil Liddell Hart, went forward through German machine-gun fire to take command. He remained with survivors throughout the afternoon, while men were killed and wounded in a succession of British attacks and German counter-attacks. At nightfall they were withdrawn.

On the 8th Division's front, it was disorganization that caused the British battalions to lose the barrage. The German artillery opposite was unsubdued and put down a heavy fire into no-man's-land as soon as the infantry reached to within eighty yards of the wire. They had already suffered from long-range machine-gun fire during the process of scaling the British parapet, filing out through the gaps in their entanglements, forming up into waves (each battalion was in four waves at fifty paces distance), and marching across no-man's-land, here, in places, 800 yards wide. Under the lash of the German barrage, the British infantry broke formation and rushed the wire. At several spots the survivors got into the German front trench and even beyond it. But most of their officers had gone down and the groups of attackers were small and separated from each other. Guessing at their circumstances, one of the commanding officers, Lieutenant-Colonel Bastard of the 2nd Lincolns, crossed no-man's-land alone from the British side, collected his scattered soldiers and those of a neighbouring battalion and organized a stretch of the captured trench for defence. The British barrage, which might have helped him in this, was now far ahead, playing on positions which the British had no hope of reaching, let alone taking, and it was not until 9.15 a.m., nearly two hours after zero, that the matter of recalling it was even discussed between the divisional commander and his brigadiers. They then told him that there could be no question of renewing the attack with their broken brigades and that to put down a barrage again on the German front line might kill more friends than enemies. So there things were left. Bastard and his handful eventually withdrew.

The most frustrating circumstances in which to lose the barrage were those experienced by battalions which got more or less intact into the German lines but there encountered such resolute resistance that they were unable to leave on schedule for the next objective. Typical of such an experience, though such experiences were rare on 1 July, was that of the London Scottish. Part of the 56th London Division, which had been given the diversionary role of attacking the Gommecourt salient to the north of Fourth Army's front, the London Scottish was one of the most famous Territorial and best of all battalions in the B.E.F. First of the Territorial infantry to land in France, it had fought at the First Battle of Ypres in 1914, Givenchy, Festubert and Loos in 1915 and had been in line opposite the Gommecourt salient since early May. Gommecourt was an exceptionally strong sector of the trench-line, where the terrain favoured the defenders, who had done much over the years to improve on it, and was garrisoned by an excellent German division, the 2nd Guard Reserve. German batteries were numerous, the fire of those of the attacked divisions being supplemented by that of batteries still farther to the north, beyond the limit of the British offensive. Yet the London Scottish, together with the other London Territorials, Queen Victoria's Rifles, the Queen's Westminster Rifles, the Rangers, the London Rifle Brigade, were confident of their ability to get into the German position and secure it.

Their confidence was justified. The barrage plan allotted them half an hour to clear their objectives and join hands with the 46th North Midland Division's leaders, who were to attack concentrically. Leaving their trenches behind a smoke screen at 7.30 a.m., the London Scottish were at once caught by a severe standing barrage in no-man's-land, but pressed on and, though several officers lost direction in the smoke and led their men to the wrong sector of trench, got almost everywhere into the German front line. In several places the Germans had abandoned it, probably deliberately, at the Scots' approach, so there and elsewhere they were able to press on quickly to the German second trench. On the right, one of the companies was able to reach and overpass its objective, retiring when it discovered its mistake. On the left, D Company had met heavier resistance all along and could not reach its final objective, a fourth trench line beyond the three it had already captured. The time was now 8 a.m. and the divisional artillery shifted its fire, as the timetable required, to the expected point of junction with the 46th Division, far away on the London Scots' left. That left them to their own devices. Worse, they were now physically isolated for the German standing barrage did not move, continuing to rain down a curtain of shells in no-man's-land all morning and afternoon, preventing the movement of essential supplies from the British to the captured German trenches.

A fresh supply of ammunition and bombs, particularly bombs, were what quickly became essential to the Scots. For in their isolation they were subjected to a series of counter-attacks by the Germans, who now surrounded them on three sides. These counter-attacks provide an excellent example of what trench fighting in the First World War was like and why it took the form it did.

The weight of fire overhead, from both field-guns and machine-guns, kept attackers and defenders alike in the trenches. The trenches, nevertheless, gave them access to each other, being part of a continuous system or grid, fire trenches running in one direction, communication trenches athwart them in another. The two sides, however, would rarely see each other, because both fire and communication trenches were 'traversed' – dug in angular kinks to deny an attacker the chance of firing down the whole length of the trench and to localize the blast of any shell which fell into it. In these circumstances, friend and foe could approach very close without being able, though aware of each other's presence, to do each other much harm in the conventional way. An impasse could result – to be resolved sometimes by an individual or group on one side or the other deciding to 'go out over the top' or 'go above ground'. Sergeant Gurney of D Company was killed doing that during the initial advance, jumping up the side of the trench to get at some Germans who were holding up the attack round a corner. The normal method of resolving the impasse, however, was by 'bombing', the throwing of a hand-grenade over the top of the traverse, and running round to arrive just after it exploded. If played seriously, it was an extremely dangerous game, for one could run into the explosion of one's own grenade, or into the fire of an unwounded enemy soldier, or into the grenade of someone bombing from the next traverse up. Equally, it could be nearly a sham combat with the two sides sticking prudently to their own traverses and the grenades falling harmlessly in the bay between them. Here in the trenches which the London Scottish had captured it became something else: a static attritional affair, the Scots having blown in several sections of trench around them, using explosives brought by accompanying Royal Engineers, and so having enclosed themselves in an earthwork stockade. Inside it they ought to have been secure and could have expected eventually to have the section of trench they had captured incorporated in the British system on the other side of no-man's-land by new digging. Several circumstances militated against this outcome: they were overlooked from three sides; the trenches had been so knocked about by the British bombardment that the occupants were exposed to fire in many places; the Germans had artillery available to bombard them from close range, and fresh infantry to deliver counter-attacks; the barrage in no-man's-land prevented either supplies or reinforcements from crossing. On the far side, the London Scots' commanding officer, who had been Left Out of Battle, became aware of his companies' plight, and got together a relief party. It set off in three groups between 9 and 10 a.m., each burdened with bandoliers of ammunition and boxes of bombs. Only three of the fifty-nine who started got through, and though this does not mean that all the rest were hit, some presumably deciding to take shelter in shell-holes, the figures do testify to the weight of fire which the Germans were laying on and over the London Scots' position. Two of the four company commanders were by then out of action, and, soon after 2 p.m., a third was killed. The burden of managing the defence now fell wholly upon the fourth, Captain Sparks. 'The better to direct the fighting, he was often seen standing and moving on the unbroken ground between the trenches' – conduct which would have attracted admiration at Waterloo and, when displayed on a First World War battlefield, beggars powers of eulogy. His men, though less exposed, were steadily being wounded or killed by bomb-blast or sniping shots, and though his garrison was occasionally reinforced by refugees from even harder pressed battalions on his left flank – the Rangers, Kensingtons and Queen Victoria's – it was dwindling in strength. The men who were left were running out of ammunition fast and, like defenders of some imperial fortlet on the veldt or the Frontier, kept their rifles going throughout the hot afternoon with rounds extracted from the pouches of the casualties. By 4 p.m. Captain Sparks recognized that his tiny force, now under attack by thirteen German infantry companies from three different regiments, was about to go under. He sent the following message back across no-man's-land: 'I am faced with this position. I have collected all bombs and [cartridges] from casualties. Every one has been used. I am faced with three alternatives: (a) to stay here with such of my men as are alive and be killed. (b) to surrender to the enemy (c) to withdraw such of my men as I can. Either of these first two alternatives is distasteful to me. I propose to adopt the latter.' Using discarded German rifles and ammunition, he and four N.C.O.s made a final stand in the German front trench while the other survivors escaped into no-man's-land. There most of them, including Sparks, hid until darkness fell and allowed them to regain the British lines. During the day, the London Scottish, which had numbered 856 at dawn, had been reduced by death or wounds to 266.

Such was one result, duplicated at thirty of forty other points up and down the Somme battlefront on the evening of 1 July, of trusting in the power of contemporary artillery to destroy an enemy position and 'shoot the infantry through' its ruins. The four forms of failure just examined do not exhaust the list of mishaps consequent on such an undertaking. The Ulster Division failed to carry its final objective, after a very rapid advance to its first, because the British barrage actually held it up, so allowing the Germans time to man with reinforcements brought from the rear positions which the Ulstermen would otherwise have found empty. Understandably, therefore, the Ulster Division counted 1 July a victory and the date, which also happens to coincide with the anniversary of the Battle of the Boyne (Old Style), is observed by the Protestants of the province as one of their holy days. Again, some battalions' attacks failed because their supports could not or would not follow the trail they had blazed into the German positions, so leaving them cut off deep within the enemy lines. This seems to have been the fate of some of the 12th York and Lancasters, whose graves were found on 13 November, at the very end of the battle, when the village of Serre, one of the uncaptured first-day objectives, at last fell into British hands. Many battalions of the second wave failed to make their attacks work because, arriving in the German trenches, they became so intermingled with the survivors of a battalion which had gone over before them that they lost order and cohesion themselves. But some battalions, one should not forget, succeeded. The 7th Division took some, the 18th most, the 30th all its first objectives, and battalions of the 21st and 34th secured sizable sections of the German trenches opposite their own. The French, better-trained, more experienced, and with much more heavy artillery, had taken all their first-day objectives, and would have gone on if the plan had provided for unexpected success. The first day of the Somme had not been a complete military failure.

But it had been a human tragedy. The Germans, with about sixty battalions on the British Somme front, though about forty in the line, say about 35,000 soldiers, had had killed or wounded about 6,000. Bad enough; but it was in the enormous disparity between their losses and the British that the weight of the tragedy lies: the German 180th Regiment lost 280 men on 1 July out of about 3,000; attacking it, the British had lost 5,121 out of 12,000. In all the British had lost about 60,000, of whom 21,000 had been killed, most in the first hour of the attack, perhaps the first minutes. 'The trenches,' wrote Robert Kee fifty years later, 'were the concentration camps of the First World War'; and though the analogy is what an academic reviewer would call unhistorical, there *is* something Treblinka-like about almost all accounts of 1 July, about those long docile lines of young men, shoddily uniformed, heavily burdened, numbered about their necks, plodding forward across a featureless landscape to their own extermination inside the barbed wire. Accounts of the Somme produce in readers and audiences much the same range of emotions as do descriptions of the running of Auschwitz – guilty fascination, incredulity, horror, disgust, pity and anger – and not only from the pacific and tender-hearted; not only from the military historian, on whom, as he recounts the extinction of this brave effort or that, falls an awful lethargy, his typewriter keys tapping leadenly on the paper to drive the lines of print, like the waves of a Kitchener battalion failing to take its objective, more and more slowly towards the foot of the page; but also from professional soldiers. Anger is the response which the story of the Somme most commonly evokes among professionals. Why did the commanders not do something about it? Why did they let the attack go on? Why did they not stop one battalion following in the wake of another to join it in death?

Some battalions were stopped. On the northern face of the Gommecourt salient, where the 46th North Midland Division's attack had failed completely with heavy loss in the morning, one of the brigade commanders, Brigadier-General H. B. Williams, who had seen the 1/6th North Staffordshire and 1/6th South Staffordshire massacred shortly after zero, declined to send forward their sister battalions, the 1/5th North Staffords and the 1/5th South Staffords, later in the afternoon. The whole of the 10th and 12th Brigades, in 4th Division, were held back from a pointless renewal of the attack north of Beaumont Hamel about the same time, and in the evening General de Lisle, commanding the 29th Division, countermanded orders for the 1/4th and 1/5th King's Own Yorkshire Light Infantry, which had been brought forward from X Corps' reserve, to make a further attempt on the corpse-strewn slopes of Thiepval. There were other reprieves, but the majority of battalions scheduled to attack did so, no matter what had happened to those which had preceded them. There are a number of ways of explaining why this should have been so. Normal military sense of commitment to a plan was one reason, the spirit of contemporary generalship, schooled to believe in the inevitability of heavy casualties, another, the mood of self-sacrifice which had the Kitchener armies in its grip a third. But most important of all was the simple ignorance of what was happening which prevailed almost everywhere on the British side of no-man's-land throughout most of the day.

Even sixty years later, it is very difficult to discover much that is precise, detailed and human about the fate of a great number of the battalions of the Fourth Army on 1 July. Many of the London Territorial regiments, with a strong and long-established sense of identity, a middle-class character, and personal connections with metropolitan journalism and publishing, produced after the war excellent regimental histories in which the official chronicle is supplemented and illuminated by a great deal of personal reminiscence from literate and articulate survivors. The regular battalions of the Guards and the regiments of the line added copiously to their existing histories. But, as we have seen, the Somme was predominantly a battle of humbler and more transient groups than these, over which the regular army had temporarily cast the cloak of its identity, but which at the peace vanished from public memory almost as quickly as they had been conjured into existence. It was not a deliberate act of obscuration. The regular regiments which had raised the greatest number of 'Service' battalions were often the least affluent (the rough rule of thumb in calculating the social status of an English regiment is that the farther from London its depot, the less fashionable it will be, and the less monied its officers) and the least able therefore to stand the expense of printing a really exhaustive history. There was moreover a difficulty about sources. The principal source of a unit's history is the War Diary, which the intelligence officer is supposed to write up daily. When he is an amateur its contents tend always to be sketchy, and when action is intense and casualties heavy it may run for days in arrears, later to be written up for form's sake from a single, sometimes second-hand, memory of events. All these caveats apply to the War Diaries of 1 July. Consequent uncertainty about the experience of particular Kitchener battalions has added an extra poignancy to their collective story. The uncertainties might have remained forever undispersed. At the very last moment, however, a Lincolnshire farmer, Martin Middlebrook, in whom a chance visit to the war cemeteries of the Somme in the late nineteen-sixties aroused an obsessive curiosity about the nature and fate of the Kitchener armies, embarked on a quest to discover survivors of 1 July, and in a truly heroic effort of historical fieldwork, found and interviewed 546 of them, by then, with the exception of a few enlisted under age, men of seventy or over.

The book which he made from his interviews is a remarkable achievement, comparable with Siborne's history of Waterloo, constructed on the same basis, and certainly fit to stand beside it, as well as being a great deal more readable.* But whereas Siborne addressed his inquiries to all surviving officers, and from their replies was able to piece together a meaningful account of the battle, Middlebrook's answers came, of course, only from the junior ranks whose view was a very local one and which collectively depict almost indecipherable chaos. We have already seen what Siborne's correspondents had to say about the limitations which the circumstances of battle imposed on the grasp of the passage of events. What Middlebrook's evidence emphasizes is the extent to which 100 years of technological change had further reduced the range of effective vision on the

***The First Day on the Somme* by Martin Middlebrook.

battlefield, particularly in those not familiar with the realities of war. 'On my left,' wrote a private of the 1/8th Royal Warwicks, of the scene at zero hour, 'I could see large shell bursts as the West Yorks advanced and saw many men falling forward. I thought at first they were looking for nose-caps (a favourite souvenir) and it was some time before I realized they were hit.' On the far side of no-man's-land, 'I found the German wire well cut,' wrote a private of the 4th Tyneside Scottish, 'but only three of our company got past there. There was my lieutenant, a sergeant and myself. The rest seemed to have been hit in no-man's-land ... the officer said, "God, God, where's the rest of the boys?"' Private Tomlinson, of the 1/7th Sherwood Foresters, accompanied his commanding officer across no-man's-land, who had gone to find out for himself what was happening to his battalion. 'When we got to the German wire I was absolutely amazed to see it intact, after what we had been told. The colonel and I took cover behind a small bank but after a bit the colonel raised himself on his hands and knees to see better. Immediately he was hit on the forehead by a single bullet.'

With a view of events so hard and dangerous to come by at close quarters with the enemy, it is to be taken for granted that in the British lines a composite picture of the battle was even more difficult to piece together. Rowland Feilding, a Coldstream Guardsman who had come up to observe the battle from a point opposite Mametz, wrote to his wife, 'the sight was inspiring and magnificent. From right to left, but particularly opposite the French ... the whole horizon seemed to be on fire, the bursting shells blending with smoke from the burning villages ... this is a district of long views. Never was there a field better suited for watching military operations.' But Feilding was there as a sight-seer, seeking sensation, not precise information. A sight-seer with a more professionally inquisitive motive, J. F. C. Fuller, found on arrival 'an intense bombardment ... in full swing, and so much dust and smoke [covering] the Gommecourt salient that it was difficult to see anything clearly. At five minutes to zero a somewhat scattered smoke barrage was put down, then came the attack across no-man's-land. I cannot say that I saw it. All I can vouch for is that a little later on through my glasses I did see several groups of men, presumably of the 139 Brigade, moving towards Pigeon Wood.' A commanding officer on the Gommecourt sector, almost under Fuller's eyes, Colonel Dickens, of Queen Victoria's Rifles, saw even less than he did: 'For two hours after zero, no news whatsoever was received from the front' (which was only about 1,000 yards distant) 'all communications, visual and telephonic having failed. Beyond answering appeals from the Brigade' (next headquarters upward) 'for information, we had leisure to observe what was going on.' But he learnt nothing until, after nine o'clock, he was visited by 'two plucky runners who [had] returned to our line through the barrage'.

Why should he have had to depend on runners? The reason is simple to explain. The communication system in Fourth Army, resembling in essentials that installed up and down the Western Front and on both sides of no-man's-land, was a comprehensive one. It was based on the telephone and the telegraph, the latter replacing the former where amplification was difficult to ensure, and ran through an extremely elaborate network of 'land line' and 'air line'. Air lines from the major headquarters – G.H.Q. at Montreuil and Fourth Army H.Q. at Querrieux, fifteen miles from the front – to Corps, and Division, with as much lateral branching as was necessary to make communication to a flank possible. Forward of Division, to Brigade and Battalion, the lines left their poles to descend earthwards, becoming 'land lines', by this stage of the war no longer strung vulnerably along the walls of the communication trenches, but buried under the duckboards on the floor. The nearer it approached the front trench, the deeper was it buried, until in the forward zone it reached a depth of six feet. The installation of this 'six-foot bury' had been one of the most time-consuming preparations for the offensive, but was justified by the security of communication it provided even under the heaviest enemy shellfire. It had, however, one disabling shortcoming: it stopped at the edge of no-man's-land. Once the troops left their trenches, as at 7.30 a.m. on 1 July, they passed beyond the carry of their signals system into the unknown. The army had provided them with some makeshifts to indicate their position: rockets, tin triangles sewn to the backs of their packs as air recognition symbols, lamps and flags, and some one-way signalling expedients, Morse shutters, semaphore flags and carrier pigeons; but none were to prove of real use on 1 July. Indeed, these items seem only to have further encumbered men already heavily laden, in a fashion more reminiscent of explorers setting off on an expedition than soldiers entering battle. The story of Scott's Last Expedition, news of which had magnetized the English-speaking world on the eve of the war, may have seemed, as it does in retrospect, of special significance to a reflective soldier of the Fourth Army as, bowed under the weight of rations and protective clothing, he prepared to leave base-camp for the dash to the final objective on the evening of 30 June; it has parallels in the fate of the vanished party of the 12th York and Lancasters, whose bodies were discovered five months after the attack in the heart of the German position.

That a party could disappear so completely, not in the Antarctic wastes but at a point almost within visual range of their own lines, seems incomprehensible today, so attuned are we to thinking of wireless providing instant communication across the battlefield. But the cloud of unknowing which descended on a First World War battlefield at zero hour was accepted as one of its hazards by contemporary generals. Since the middle of the nineteenth century, the width of battlefields had been extending so rapidly that no general could hope to be present, as Wellington had made himself, at each successive point of crisis; since the end of the century the range and volume of small-arms fire had been increasing to such an extent that no general could hope to survey, as Wellington had done, the line of battle from the front rank. The main work of the general, it had been accepted, had now to be done in his office, before the battle began; and indeed one of the pieces of military literature most talked of in the British army before the First World War was a short story, *A Sense of Proportion*, by General Sir Edward Swinton, which had as its central character a general – obviously based on the great Moltke – who, having made his dispositions on the eve of battle, spends its hours casting flies for trout, serene in the assurance – which the story's conclusion vindicates – that he had done all he could.

No British general spent 1 July fishing. But the spirit which informs the plans laid by the Fourth Army, whether those of a formation like XIII Corps (equivalent in size to Wellington's Waterloo army) which ran to thirty-one pages (Wellington issued no written plan for Waterloo), or a unit like Queen Victoria's Rifles, a force of under 1,000 men, which ran to twenty numbered paragraphs, is essentially Swintonian. It is a spirit not of providing for eventualities, but rather of attempting to preordain the future; a spirit borne out by the language of the orders: 'infantry and machine-guns will be pushed forward at once ...'; 'the siege and heavy artillery will be advanced ...' 'After the capture of their final objective the 30th Division will be relieved by the 9th Division ...' Man's attempts at preordination are always risky and require as a minimum precondition for success the cooperation of all concerned. Upon that of the Germans the British could not of course count. Consequently, at every point where the future threatened to resist preordination, Haig and Rawlinson had reinsured themselves – by lengthening the duration of the bombardment, adding to the targets to be destroyed, increasing the ratio of troops to space.

The effect of these reinsurances was to complicate the plan. And the complication of a plan which would depend for its success on the smooth interaction of a very large number of mutually dependent elements invited its frustration. Interaction requires articulation, to adopt the language with which J. F. C. Fuller was fond of obscuring military truths; which means that if major operations are to be carried through in the teeth of enemy resistance, commanders must at all times be able to talk to their troops, troops to their supporting artillery and so on. Such conversations were easily arranged while everyone was on the same side of no-man's-land. But once the infantry departed on their journey, conversation stopped, to be carried on, if at all, through the medium of the battalion runners, upon whose messages Colonel Dickens, for example, had to rely for news, two hours old, of the progress of his fighting companies.

Discontinuities of this order in the receipt of information, particularly when the information concerned difficulties or failure, made the management of a battle, in the tactile and instantaneous fashion open to Wellington at Waterloo, impossible. Commanders could not discover where the soldiers were: 'Observation of troops, on account of the smoke and dust, was extremely difficult, and although two contact patrol aeroplanes displayed amazing daring, flying along the front sometimes only fifty feet above the troops under heavy small-arms fire, no definite information could be obtained' (of the 4th Division in mid-morning); hard-pressed battalions could not ask their supporting batteries for fire support: the 14th Brigade, pinned down by heavy machine-gun fire from Thiepval at 8.45 a.m., could not get the promise of a barrage until 12.05 p.m.; carefully rehearsed soldiers failed to cooperate in changes of plan which events made necessary: 'a party (of the 56th Division) told off to carry concertina wire could not for some time be got to understand that they must drop their loads and help to form an attacking wave.' Throughout the morning and afternoon, Rawlinson, at Querrieux, and Haig, in his advanced headquarters at the Château de Beauquesne, ten miles to the north, attempted to follow the battle from scraps of imprecise information several hours old. Neither made real sense of it. Neither, very wisely, ordered any substantive changes of plan. Many of the gunners, whose fire, if properly directed, would have been so effective in saving British lives, also remained, though closer at hand, inactive spectators: 'On the whole,' wrote Neil Fraser Tytler of a Lancashire Territorial Field Brigade, 'we had a very delightful day, with nothing to do except send numerous reports through to Head Quarters and observe the stupendous spectacle before us. There was nothing to do as regards controlling my battery's fire, as the barrage orders had all been prepared beforehand.' Throughout this period, the only group of soldiers with precise information to offer of the whereabouts and circumstances of their units were the battalion runners. It is ironic to reflect that the taunt thrown into the faces of so many highly trained German Great General Staff officers, excluded by official policy from service in the trenches, by Hitler, ex-runner of the 16th

Bavarian Reserve Regiment – that he knew more about the realities of war than they – had after all a coarse grain of truth to it.

The Wounded

Given that the high command spent most of 1 July in ignorance of how the Fourth Army was faring, its passivity is excusable; given its silence, the persistence of lower headquarters, too, in sticking to the plan is quite understandable. Nor were the military consequences of their inertia disastrous, not at least in the short term. For the Germans on the Somme were far too weak, in men, in artillery, in ancillary means, to have thought of snatching profit from the British army's disorganization by a quick counter-offensive. Looked at in these terms, therefore, the continuation of the attack throughout the day did not compound the military error. But it did multiply the scale and prolong the duration of the human suffering which the battle brought, by a factor difficult to quantify but certainly very large indeed. It caused many more men to be wounded than any sort of necessity required and left those wounded early on to agonize in no-man's-land throughout the day.

Waterloo wounds, when not instantly lethal, had had in many cases extremely unpleasant consequences, fatal septicaemia and peritonitis being the most obvious. But Waterloo wounds (cannon wounds apart) had been in general single and simple: penetrations or perforations by lances or low-velocity bullets, cuts by swords. If the bleeding they caused was not too severe, if dirt was not carried into the wound or the intestines opened, the patient's chances of survival were better than we would expect. The wounds suffered by the human body on the Somme were of a far greater variety and degree of severity than any Waterloo surgeon would have seen. Edged-weapon wounds would have almost disappeared, for though the marks of bayonets were found on a number of bodies, it was presumed that they had been inflicted after the victim was dead; the best statistic available is that edged-weapon wounds were a fraction of one per cent of all wounds inflicted in the First World War. Bullet wounds were far more frequent, amounting to about thirty per cent of all new wounds, but probably forming a greater proportion on the first day of the Somme, because of the unusual targets presented to the machine-gunners. Shell and bomb wounds were probably reduced in proportion for the same reason, but usually amounted to about seventy per cent of those inflicted. Shell wounds were the most to be feared, because of the multiple effects shell explosion could produce in the human body. At its worst it could disintegrate a human being, so that nothing recognizable – sometimes apparently nothing at all – remained of him: 'a signaller had just stepped out,' recalled the medical officer of the 2nd Royal Welch Fusiliers, of a later battle, 'when a shell burst on him, leaving not a vestige that could be seen anywhere near.' A little beforehand he had witnessed another equally frightening and fatal consequence of shell fire: 'two men suddenly rose into the air vertically, 15 feet perhaps, amid a spout of soil 150 yards ahead. They rose and fell with the easy, graceful poise of acrobats. A rifle, revolving slowly, rose high above them before, still revolving, it fell. The sight recalled, even in these surroundings, a memory of boyhood: a turn that thrilled me in a travelling circus at St Andrews.' Less spectacular, but sometimes as deadly, shell blast could create over-pressure or vacuums in the body's organs, rupturing the lungs and producing haemorrhages in the brain and spinal cord. It was effects of this sort which killed three Welch Fusiliers 'sitting in a shell-hole ... with no more visible mark on them than some singeing of their clothing'. Much the most common wounding by shell fire, however, was by splinter or shrapnel ball. Such projectiles travelled fairly slowly, and rapidly lost their velocity; to that extent, they were less to be feared than bullet wounds. But they often travelled in clusters, which would inflict several large or many small wounds on the same person. The splinters were irregular in shape, so producing a very rough wound with a great deal of tissue damage, and they frequently carried fragments of clothing or other foreign matter into the body, which made infection almost inevitable. Very large shell fragments could, like Water loo cannon-balls, amputate limbs, decapitate, bisect or otherwise grossly mutilate the human frame.

As a killing agent over long as well as short ranges, however, the bullet was champion. Unlike the musket-ball which, moving at slow speed and without rotating, merely drove a clean path for itself through soft tissue, the high-velocity conical bullet, spinning quickly about its long axis, could produce inside the human body a variety of extremely unpleasant results. At best, it left a neat channel with the 'exit' wound the same size as the 'entry'. Should it be caused to 'tumble' inside the body, however, either on hitting bone or for some ballistic reason, its path beyond the point of tumble became very much enlarged and the 'exit' wound – often mistaken by amateurs for the entry – 'explosive' in appearance. The effects of a tumble produced by striking bone were enhanced by the bone's splintering under the impact, its own fragments then becoming secondary projectiles which produced massive damage to tissues round about. Some bullets also set up hydraulic effects, their passage driving body fluids away from the wound track at pressures which surrounding tissues could not withstand. The lower frequency of bullet wounds, noted in the medical statistics, may thus have been due to the bullet's greater initial lethality, for doctors noted the nature of wounds brought to them for treatment, not those inflicted on the bodies of all soldiers. It is suspicious, for example, that, at one major Casualty Clearing Station, the proportion of bullet wounds of the chest to all other gunshot wounds was, during a year of serious fighting, about three per cent.

Much of the wounding produced by the weapons of trench warfare, to say nothing of its volume, was new to surgeons. All armies, and particularly the British, had made remarkably successful efforts, however, to meet the surgical problems with which the war confronted them. For the Somme, and for some time beforehand, bed space had been provided to receive every wounded soldier evacuated from a battle; anaesthetics, antiseptics, dressings and instruments were freely available (and every soldier carried, sewn to his uniform, a packet containing a sterile 'first field dressing'). Blood, typed by Landsteiner just before the war, was now known to be safely transfusable, and was fairly freely transfused from 1917 onwards. Surgical technique was highly developed and was adapted with admirable flexibility to the types of wounds the war produced; the principle of 'debridement', or the removal of damaged tissue ('debris') from the site of the wound, was established by 1915, and, when applied early on as a general rule, was remarkably effective in preventing the onset of one of the most feared consequences of major muscular damage, gas gangrene (an infection contracted from soil-organisms carried into the wound when the victim fell to the ground). But much surgery had perforce to be radical, decisions for amputation being far commoner than they would be in the Second World War, when bone-grafting and other reconstructive techniques had been brought to a higher level of development. And, in cases of gross abdominal, chest and head wounds, where infection was present from the outset, surgery offered no remedy. None would be found until antibiotics became available towards the end of 1943.

Still, despite the intensification of the hazards with which battle threatened the soldier of the First World War, the care which medicine could provide had been made to keep pace. The infrastructure of the medical service was impressive in scope. Parallel to the complicated networks of supply and communication, emplaced to transport the soldier into battle and provide him with essentials while there – the buried cable system, railheads, roadheads, tramways, Corps and Divisional dumps, workshops, Field Ordnance Parks – was an equally elaborate system to evacuate him from battle if he was hit, treat his wounds and restore him to health (and so to the fighting line). The first point in the system was the Regimental Aid Post, where the battalion medical officer sorted the casualties, separated the dying, and sent those he could not treat (all but the very lightly wounded) off down the line, sedated and bandaged. Their destination was the Casualty Clearing Station, via the Dressing Station and the Collecting Post, from which they would be transported by train to a Stationary or Base Hospital, either in France or in Britain. At the beginning of the war, the practice was to delay operation until the patient had been brought to the comprehensive facilities available at the base. But as it became clear that several categories of wounds 'travelled' badly, notably those of the chest and abdomen, the practice developed of initiating major surgery farther and farther forward. By 1916 the Casualty Clearing Station had become the principal seat of life-saving surgery, but it was also undertaken by special travelling surgical teams in Advanced Dressing Stations, within medium-gun range of the enemy.

There remained, nevertheless, a brutal selectivity about military surgery which the practitioners did their best to hide from the patients but could not disguise from themselves. It was called 'triage', or 'the division into three', from the vocabulary of the French army which had instituted it, and required surgeons, from the press of casualties flowing in during a battle, to send on those who could stand the journey and to choose, from the group remaining, which men were worth subjecting to serious surgery and which must be left to die; the greater the press of casualties, the larger the latter group. A Casualty Clearing Station usually contained a special tent, known tacitly as the moribund ward, in which the unlucky third of the triage group were congregated.

But 'unlucky' is a very relative term to use in such circumstances. The patients of the moribund ward were sedated, washed, fed, given to drink, comforted by female nurses, shielded from the fact of their approaching death. To have reached the moribund ward was to have been spared the worst which battle could inflict on the wounded soldier of that or indeed of any war, and which the wounded of the Somme suffered in their thousands on 1 July: to die alone and in pain on the battlefield. For in the elaborate chain of medical arrangements provided by Fourth Army was a weak or perhaps missing link. To the rear of the Advanced Dressing Station, transport was plentiful; forward of the Regimental Aid Post, transport was very sparse. The evacuation of the wounded in a battalion fell on thirty-two stretcher bearers, capable of carrying sixteen wounded between them, and needing an hour or more for each journey. But in many battalions on 1 July, over 300 men were wounded; or, rather, arrived wounded at the Regimental Aid Post. These included a majority of lightly wounded, who could evacuate themselves, and those of the more seriously wounded found still alive on the British side of no-man's-land. But what of those lying farther away?

Some men wounded close to the British front line, shot through the head, chest, or the major blood vessels, would have died however quickly collected by the stretcher bearers; that went also for men wounded in the same way nearer the German lines. But there was a category of wounds of intermediate gravity – punctures of the lungs ('sucking wounds'), leading to a fatal loss of oxygen in the blood, punctures of the abdomen, causing internal bleeding and threatening peritonitis – which would leave the victim alive for some time and from which he would eventually recover if he could be got swiftly to treatment. Both these and other less serious wounds would be associated with the phenomenon known as 'shock' in which the blood volume falls and can only be restored by outside assistance. It is very difficult to estimate what proportion of the wounded of 1 July would have suffered this type of injury. But the chest and abdomen form about fifty per cent of the surface of the body presented, when upright, to enemy projectiles; skin surface covering the spine and great vessels – the heart and the major arteries – is less than half of that. We may therefore conclude that about a quarter of wounds received which were not immediately fatal were to the chest and abdomen. To this category of critical wounds we should add fractures of the thigh, with which massive shock and heavy bleeding are almost always associated, and penetrations of the skull which, without antibiotics, frequently led to fatal abscesses of the brain if surgery was delayed longer than twelve hours. Perhaps a third, therefore, of the wounds inflicted on 1 July were of the 'critical but not immediately fatal' variety.

Those who suffered such wounds on the wrong side of no-man's-land, or inside the German positions, particularly on the sectors where the British attack failed or was violently resisted, had a very poor expectation of life. In many places, and at a surprisingly early stage of the battle, the Germans offered the British an unofficial and unilateral truce: at about 2 p.m. on the front of the 56th London Division, where 'a German medical officer ... came out with a white flag and said that there was no objection to the removal of wounded on the British side of the wire, so long as no firing took place', and on VIII Corps' front, where the Germans allowed stretcher bearers to move freely about no-man's-land between noon and 4 p.m. But elsewhere the wounded had to lie out until darkness, or until next morning, when the Germans again in several places offered a humanitarian truce, or indeed until even later than that. On the III Corps' front it was not until 3 July that all wounded had been carried off the field, on the X Corps' front not until noon on 4 July, and that despite the conscription of large numbers of engineers and gunners to act as stretcher-bearers. Gerald Brenan, crossing the battlefield to visit his friend Ralph Partridge, three weeks later, stumbled upon the consequence of the delay: 'The wounded, who could not be brought in, had crawled into shell holes, wrapped their waterproof sheets round them, taken out their Bibles, and died like that.'* It may be, therefore, that as many as a third of the 21,000 killed and missing on 1 July died as a result of wounds from which they would have had a chance of recovering if they could have been brought even to the Regimental Aid Post within the first hours of injury.

The Will to Combat

The late-twentieth-century soldier does not expect to be left to die of his wounds on the battlefield. For the reduction of delay in evacuating the wounded soldiers from the point of injury to the point of first aid in the period since 1918 has been striking, and represents easily the most important of the many advances achieved in modern military medicine. In Western armies during the Second World War, the delay between suffering injury and arriving at the aid post was commonly reduced to under an

**A Life of One's Own* by Gerald Brenan (Eyre & Spottiswoode, 1962).

hour; in Vietnam, where 'casualty evacuation' helicopter pilots were trained to land in the fire-zone and under fire, it averaged fifteen minutes – rather less than the victim of a civilian traffic accident might spend in an ambulance. The patient moreover can now be transfused in the helicopter and landed at a hospital offering a complete range of resuscitation facilities and expert surgery in a variety of specialities. Today the victim is unlucky who, escaping death outright, succumbs to the effects of his wounds.

The early-twentieth-century soldier already perceived the hope of life which the novel combination of expert surgery, anaesthetics and asepsis offered if he should be hit; but it was a perception less immediate than that which he had of the dangers, enormously magnified since Waterloo, of the battlefield on which he had to move and of the gravity, equally magnified, of the sort of wound he could suffer on it. What, given these perceptions, was it which impelled him to leave cover, advance and engage in combat in such circumstances?

Not everyone, even at high moments of the war, was prepared to do so. On the eve of the Somme, quintessentially a high moment, a number of soldiers inflicted wounds on themselves to avoid having to 'jump the parapet'.* At low moments of the war, of which the later stages of the Third Battle of Ypres ('Passchendaele') yielded many, some British battalions realistically accepted that some men would not or could not stand the strain of action and excused them from duty: 'to gratify a mawkish humanitarianism,' wrote the excessively tough-minded

*Is the self-inflicted wound ('S.I.W.') a phenomenon produced by the First World War, risked despite the legal penalty attached because of recent advances in medicine? Nineteenth-century Russian serfs had been given to knocking out their front teeth, with which soldiers bit the old musket cartridge, so as to avoid conscription. But instances of self-wounding, before the development of antiseptics, have escaped me. Modern medicine has however made a wound inflicted by the enemy or by genuine accident – the latter at least five per cent of all woundings between 1914 and 1918 – a very desirable passport off the battlefield: 'a comfortable wound,' wrote Maurice Bowra, 'was an act of God much to be welcomed'; 'to them a wound,' wrote Hanbury-Sparrow of his soldiers, 'no matter how slight, terminates all moral obligation to go on'; 'When the whistle blew,' recalled F. W. A. Turner of his jump-off on 1 July, 'the first man up my ladder was an American, Private Martin. As soon as he reached the top he was shot through the wrist. He came straight back. "I've got mine," he said, "I'm off."'

medical officer of the 2nd Royal Welch Fusiliers on 15 October 1917, 'two or three score mean fellows are encouraged to slip away every time there is risk to their skins, so more and more average men learn to shirk with impunity, attacks fail, and losses run into untold thousands, because the most dutiful of our men are not backed up.' And, taking a very broad view of the war, a point was reached in every army at which either a majority or a disabling minority refused to go on. This point was reached by the French army in May 1917, when 'collective indiscipline' occurred in 54 of the 100 divisions on the Western Front; in the Russian Army in July 1917, when it failed to resist the German counter-attack consequent on the collapse of the 'Kerensky Offensive'; in the Italian army in November 1917, when the Second Army disintegrated under German–Austrian attack at Caporetto. In March 1918, the British Fifth Army collapsed, as much morally as physically, and in October the German army in the west signified to its officers its unwillingness to continue fighting. In each case, excepting Germany's, and if we count the battle of Loos in September 1915 as Britain's introduction to heavy losses, the moment of collapse occurred between two years six months and two years eleven months after the outbreak. However, it is probably not the lapse of time which is significant, but the relationship of total casualties to the number of fighting troops engaged; a rough calculation, and anything better than a rough calculation is difficult with such notoriously unreliable statistics as casualty figures, suggests that the break came soon after the total number of deaths suffered equalled the number of fighting infantry in the divisions. Counting the fighting infantry of a division at 10,000, and the number of British, French, Italian and Russian divisions engaged against the central powers as 60, 110, 45 and 120, we get figures of 600,000, 1,100,000, 450,000 and 1,200,000 which are more or less the totals of deaths suffered by each combatant power at the moment its army underwent collapse or crisis. The German army, which certainly suffered a great many more deaths before cracking, escapes from the pattern; but it is important to recall that, almost until the end of the war, it had been fed on a diet of victory: in 1914 Tannenberg, in 1915 Gorlice-Tarnów, in

1916 the defeat of Romania, in 1917 Caporetto and the Russian armistice, in 1918 a succession of breakthroughs of the British and French fronts.

Broad views of this sort tell us nothing, however, about what did (or did not) motivate the soldier to fight in a specific combat situation in trench warfare. We have seen that on the Somme, on 1 July, there were special factors at work which were implicit in the composition and experience – or inexperience – of the Fourth Army. But over and above its cohesion, sense of mission, mood of self-sacrifice, local as well as national patriotism, there were other elements in play. Self-confidence and credulity were certainly present, and powerfully effective at persuading the Pals to jump the parapet. But to emphasize the populist character of the Kitchener armies is to minimize the importance which leadership played in taking it into battle. And arguments can be found to suggest that leadership – conscious, principled, exemplary – was of higher quality and greater military significance in the First World War, at least in the British army, than before or since.

The 'Lost Generation' and the 'Public School Officer' are clichés which seem too well-worn to be worth repeating or re-examining in this context. Certainly by the end of the war, the officers of the British Expeditionary Force had ceased to be a socially exclusive group, indeed were perhaps more broadly recruited than their equivalents in the Second World War. For the British army of 1939–45 put officer selection on to a scientific basis, making all applicants for commissions submit to tests of their intelligence, stability, companionability, leadership potential and the like, considerations which favoured the middle-class over the working-class candidate. The officer casualty rate of the First World War would have made such a system unworkable, even if it had been thought of, and as the campaign on the Western Front dragged on it became the practice to promote direct from the ranks on a commanding officer's recommendation. Thus men whose only qualification for a commission was that they had proved themselves good soldiers, and who would in peacetime have had no chance, or thought, of becoming officers, found themselves suddenly lieutenants, later captains, majors, or even lieutenant-colonels. In the longer term, this wholesale conferment of officer-status (avoided in the German army by recourse to such expedients as the creation of 'deputy officers' and 'sergeant-major-lieutenants') on lower-middle and upper working-class youth was to have a highly beneficial effect on the English class-system. The immediate difficulty remained, nevertheless, and it was one which each new officer had to face, of finding and adopting a personal style to match the rank which chance had thrust upon him.

The demeanour of the regular officer ought to have provided the temporary ('temporary gentleman' was the unkind wartime gibe) with an ideal type. But the regular officers disliked serving in the Kitchener battalions. ('The inspection proceeded. The General found that many of the men came out in August 1914. He was at home with these – he had just come from inspecting the 20th Royal Fusiliers [a Kitchener battalion]. He chatted and chaffed, pinched their arms and ears, asked how many children they had, and if they could be doing with leave to get another. As he passed from one 1914 man to another he dug his elbows into the Commanding Officer's ribs and exclaimed, "You're a lucky fellow." When it was over he said ... "That's been a treat. That's the sort we've known for thirty years."'*) Most Kitchener battalions knew as regular officers only the colonel and the adjutant. What served to ensure, in the remarkable way that happened, that new officers should resemble old was the decision to choose the first temporary officers, as R. C. Sherriff recalls from painful experience, from among public schoolboys, at a time when the regular British officer was a public school rather than a distinctively military type. This threatens to be a tautology, for the critics of the pre-1914 public schools commonly condemned them as militaristic. In a sense they were. But unlike the German military schools, which segregated the future officer from childhood and brought him up in a strictly military regime, the public schools educated the whole English upper-middle class and exposed it to a variety of influences, athletic, scholastic, ethical and religious as well as military. The eighteen-year-old who went on to the Royal

*From *The War the Infantry Knew*, Royal Welch Fusiliers.

Military College was treated when he arrived there as someone already formed in character and attitude and only needing tactical training to take his place in his regiment.

Thus it was that the amateur officers of the New Armies knew from the outset what was expected of them. And they knew too into what sort of institution the embryo regiment they had joined should grow. For the British regiment, with its complex and highly individual accretion of traditions, local affinities, annual rituals, inter-company rivalries, fierce autonomy and distinctive name – King's Shropshire Light Infantry, Loyal North Lancashire, Duke of Wellington's, Royal Fusiliers – was an extension, indeed a creation, of the Victorian public school system. Simply by being themselves, therefore, the first amateur officers provided their untrained soldiers both with an environment and a type of leadership almost identical to those found in a regular, peacetime regiment. They organized games for the men, and took part themselves, because that was the public school recipe for usefully occupying young males in their spare time. They organized competitions between platoons and companies – in cross-country running, rifle shooting, trench digging – because competition was the dynamic of public school life. They saw to the men's food, health, cleanliness, because as seniors they had been taught to do the same for junior boys. They administered automatically the military code of rewards and punishments, because it mirrored the system in which they had been brought up. And they took their men to church because it was there on Sundays that the school went *en masse*.

It is important not to exaggerate the piety of the public school officer of the New Armies. Some, like the most famous of them, Rupert Brooke, had become intellectual agnostics at the university. Some were indifferent or non-committal in a way which would have made them instantly at home in Wellington's army. Graham Greenwell, an eighteen-year-old Wykehamist whose diary is an important corrective to some 'Lost Generation' myths, seems to have lived a healthily pagan life for four years, sharing with Julian Grenfell, a regular officer, the view that, in the latter's words, 'War . . . is like a big picnic without the objectlessness of a picnic,' before which he had 'never been so well or so happy'. Grenfell was unlike Greenwell, however, in seeming actually to enjoy killing Germans, for which he had had a special sniping rifle made. Many of the amateur officers, and some of the regulars, in the early stages at least, shrank from killing. Greenwell describes the look-out he kept for the rare sight of a German and how, when at last he saw one within sniping range, he handed his rifle to a sergeant to do the deed. Hanbury-Sparrow, a regular officer, is more specific about his conscientious objection: 'You neither want to be killed nor to kill anybody. Officers, you feel, shouldn't engage in the rough-and-tumble – that's for the men; [yours] is the thinking part.' His objection is not therefore strictly ethical; but he did take an elevated view of what qualities an officer should possess:

Were they or were they not braver? That was your criterion ... For the act of being brave compelled the utilization of the whole reserve of moral force that lay in a man ... every battalion had its own little core of officers around which the battalion clung. Wounds or sickness might get them but sure enough they'd return ... Revolving around this nucleus was an endless changeover of officers. Death claimed many, but of the survivors only the good gravitated towards the centre. The rest ... couldn't stick it, and amongst them almost invariably were the hard drinkers and persistent womanizers – the very men, in fact, whose conduct showed their lack of inner discipline. Here in the trenches your sins found you out.

This equation of courage with morality, a sort of heroic Puritanism, is distinctive of the public school approach to the First World War at least in the early stages – though Sparrow was not an orthodox Christian but a disciple of Rudolf Steiner. A Christian who articulated the approach in words which appear to express the feelings of many New Army officers was Donald Hankey, killed on the Somme in October 1916. The essays in which he had spoken for his generation were written for the *Spectator* over the signature 'A Student in Arms'. Perhaps the most significant of them for the modern reader is 'The Beloved Captain', in which he characterizes the ideal leader.

He came in the early days ... tall, erect, smiling ... For a few days he just watched. Then he started work. He picked out some of the most awkward ones and ... marched them away by themselves ...

His confidence was infectious ... His simplicity could not fail to be understood ... very soon the awkward squad found themselves awkward no longer ... The fact was that he had won his way into our affections. We loved him ... If anyone had a sore foot he would kneel down ... and look at it ... If a blister had to be lanced, he would very likely lance it himself ... There was something almost religious about this care for our feet. It seemed to have a touch of the Christ about it.

The point to which Hankey leads the reader, via a catalogue of the complete officer's virtues but with more art than these extracts convey, is the revelation of the beloved captain as Christ himself. 'We knew that we should lose him ... But how was the company to get on without him? To see him was to forget our personal anxieties and only to think of ... the regiment and honour.' He is killed. 'But he lives ... And we who knew him do not forget. [And] I think that those who went West have seen him. When they got to the other side I think they were met ... And as they knelt before that gracious pierced Figure, I reckon they saw nearby the Captain's smile. Anyway, in that faith let me die, if death should come to me.'

It seems unlikely that many officers would have admitted to sharing in its entirety Hankey's view of what an officer should be. Nevertheless, it appears to indicate the direction, at the beginning of the war, of their aspirations and, if that is so, it eases our understanding of why the thousands of the New Armies climbed so readily into no-man's-land on 1 July and trudged off behind their platoon leaders. But why, once there, they continued to advance and to contest ground with the enemy demands a different explanation, for, amid the conditions of the First World War battlefield, leadership of the close-order variety exercised at Waterloo was not possible. The men were deployed in 'extended order', so that many were too far from the officer to be under his physical control – no question of pushing or thumping the ranks into line on the Somme – while the noise level, higher still than at Waterloo, drowned the human voice at a few feet.

The hope of plunder as a motive may be discounted. Soldiers of 1914–18 could leave their money as credits with the paymaster, while trinkets had declined in relative value, so that there was little for which life was worth risking to be found across no-man's-land. Compulsion, on the other hand, was as important an agent in impelling men into the fight as ever. Crozier describes dealing with some runaways of his battalion on the afternoon of the Ulster Division's fight for the Schwaben Redoubt:

> a strong rabble of tired, hungry and thirsty stragglers approach me from the east ... 'Where are you going?' I ask. One says this, one another. They are ... given a drink and hunted back to the fight. Another more formidable party cut across ... They are damned if they are going to stay ... A young sprinting subaltern heads them off. They push by him. He draws his revolver ... They take no notice. He fires. Down drops a British soldier at his feet. The effect is instantaneous. They turn back ...

Moreover each battalion told off men to act as 'battle police' – in Queen Victoria's Rifles their duties were 'to see that no stragglers are left in the trenches' and to send 'any so found ... up to their companies' – and the topography of the First World War battlefield made the task of the battle police comparatively simple. For men anxious to avoid death would naturally seek to remain in the trench system which, like school corridors, were easily patrolled for truants.

Compulsion, however, is not the whole, nor even part, of the answer. That is best sought, perhaps, by looking at what it was that impelled the defenders, in this case the Germans, to fight. On their motivation the topography of the battlefield would also have exerted a powerful influence. For, as the British plan emphasized, and the Germans knew without being told, possession of the parapet of their front-line trench was to be decisive in determining who lived and who died (not that the plan put it like that). Should the British reach it first, they could kill the defenders, at no risk to themselves, by throwing grenades down the shafts of the dug-outs. The Germans might, of course, dissuade them by offering their surrender. But, questions of honour and fighting spirit apart, surrendering was a ticklish business in trench warfare. Prisoners had no inherent value, so that soldiers did not seek specially to take them. The onus fell rather on the would-be prisoner to get his surrender accepted, something difficult to do when friend and enemy met so rarely face-to-face, when face-to-face encounters tended to provoke hair-trigger reactions, and when a pacific shout from a dark dug-out in a foreign language might be misinterpreted. Even if taken prisoner, the captive's safety was not assured. For prudence required that he be sent instantly across no-man's-land, where he risked stumbling into his own artillery's barrage, or being mistaken for an attacker by infantry of the second wave waiting in the opposite front line. Both these fates overtook German prisoners on 1 July. About eighty out of 300 unwounded Germans being sent back to the trenches of the 56th London Division opposite Gommecourt were killed at 9.30 a.m. by German shelling; later Crozier, commanding the 9th Royal Irish Rifles, realized that his reserves in the front line were firing at prisoners whom some of his wounded were escorting from the far side. At his command, they stopped, but reluctantly. '"After all," I heard a youngster say, "they are only Germans."'

It does not take very much more illustration than these instances provide to explain why it was that the Germans raked so ferociously the advancing British lines at zero hour on 1 July. There was, in a sense, nothing personal about it. To surrender was dishonourable and might be dangerous. To run away was impossible (for the Germans, of course, had their own battle police farther down the trenches). To kill the British was, therefore, a necessity – though the majority would have called it a duty and, to the British on the wrong side of the wire, it may have seemed that they found it a pleasure. Certainly, as we have seen, easy killing does seem to generate in human beings symptoms of pleasure, which the zoologist Hans Kruuk has tried to relate to the compulsive behaviour of certain predatory animals when they come upon groups of their prey which are unable to escape from them. There was, however, probably no vindictiveness in the shots which Germans, in many places, later aimed at the British wounded, lying outside their trenches, whenever they moved. The Germans, recently released from the imprisonment of a terrible bombardment, outnumbered, and just reprieved from the sentence of execution by grenade-blast below ground, were tensed to shoot at anyone who by as much as a gesture threatened to renew the attack on their positions.

If this helps to explain the German 'will to combat', it helps to explain too what motivated the British to dispute with them possession of the front trench. But does it explain what prompted the infantry who got in to leave its shelter and press on to their next objective? Excitement, sense of duty, knowledge of the plan, previous rehearsal of their task would all have been a spur; so too would have been the re-imposition of leadership which possession of the trench made possible; many accounts describe how the officers moved along a captured trench to speak to their men, sort out disorganization, encourage, congratulate, exhort. But again, as with the Germans, mechanical and topographical factors were at work. An enemy front trench was a dangerous place; its defences were the wrong way round and it had no-man's-land, an area of intense hazard, behind it. To be counter-attacked in the front trench was to risk expulsion into the danger-zone one had just escaped – the enemy barrage-line or the killing area of his machine-guns. Safer, in many ways, to press on, particularly if one's own barrage, that explosive safety curtain, was still within reach and offered one safe passage to the next enemy trench. To reach that was to provide oneself with room to manoeuvre to one's rear, if events should subsequently force one to retreat; it was also to vacate a space for the supporting waves to occupy, which would come to one's assistance if the enemy should counter-attack. For how long such a comparatively complex set of perceptions would impel infantry to move forward into enemy territory is very difficult to estimate. These are really officers' perceptions. But the precariousness of life in the enemy's front trench would have been evident to most soldiers; once persuaded to move forward from it, it seems to have been possible to keep them on the go until enemy resistance or the onset of exhaustion forced them to ground. What happened then would have been determined by the rules of trench warfare examined earlier.

Commemoration

1 July was not the end of the Battle of the Somme. The attack was to be renewed several times during the summer, autumn and early winter, and only officially to be closed down on 18 November. The official history of the war names eight 'phases' of the battle after 1 July; the first phase was officially designated after the war, by the Battle Nomenclature Committee, the Battle of Albert (from a little town behind the front), a name now used by no one.

By the time the battle ended, 419,654 British soldiers had become casualties on the Somme, and nearly 200,000 French. The exact number of German casualties has been debated ever since, the official historians seeking to show that they exceeded those of the Allies, their opponents the contrary.

But the Battle of the Somme has, in a sense, not ended yet. The villages and towns of the battlefield had all been rebuilt, in an unpleasing red brick, by the middle 1920s. By then, too, the majority of the bodies of Allied soldiers killed on the Somme had been disinterred from their hasty graves and reburied, the British in fifty of the beautiful garden cemeteries which the Imperial War Graves Commission has created in every country in which Britain has had soldiers killed since 1914. But the principal memorial which the Somme left to the British nation is not one of headstones and inscriptions. It is intellectual and literary, and it turns on the revelation, from which the British had hitherto been shielded by their navy, that war could threaten with death the young manhood of a whole nation. This realization was to have important political after-effects during the Second World War: 'On one occasion when [the American] General Marshall was in England, pouring forth the most cogent and logical arguments in favour of a prompt invasion of the Continent ... Lord Cherwell remarked to him, "It's no use – you are arguing against the casualties on the Somme."' The same realization was to colour British strategic thinking, both official and academic, about what sort of wars she should fight.

For the cause of the defeat on the first day of the Somme needed no analysis: anyone could understand that when a near majority of the soldiers committed to an attack are killed or wounded in its opening moments, the remainder will be too shocked and disorientated to continue.

These impressions might, nevertheless, have faded, had it not been that the experience of the Western Front, of which the Somme marked the opening of a crucial phase, called forth from the generation which underwent it a literature of immense imaginative sweep and power. Much of it was poetic in form, and it was the poetry which was published soonest, a great deal of it while the war was in progress. Precisely because this earliest outpouring of protest was poetry, a literary form which hid from its audience its documentary value, its effect was transitory; or rather, it was to require for a prolongation of its effect some form of verification in prose, some confirmation in ordinary language that what Sassoon, Graves, Blunden had had to say was not private and subjective but an expression of the feelings of a whole generation. 'A whole generation' of course goes too far, including many without the educational of emotional equipment to see or feel as the poets saw and felt, and including too a considerable number, as well-educated and sentient as they, who either tolerated the war or actually enjoyed it. But a silent majority of the war-generation probably perceived in their verse a truthfulness to which they could assent. And ten years after the war, in the period 1928–32, the endorsement in prose, of the poets' cry from the trenches, suddenly found expression and an audience. Remarque's *All Quiet on the Western Front*, Blunden's *Undertones of War* and Graves's *Goodbye to All That* were published in 1928, Aldington's *Death of a Hero* and Hemingway's *Farewell to Arms* in 1929, Sassoon's *Memoirs of an Infantry Officer* in 1930 – autobiographical or fictional-documentary accounts of the war on the German, British and Italian fronts, four by writers who had already acquired reputations as leading 'war poets'.

Writers are as quick as anyone, quicker sometimes than publishers, to detect the drift of the literary market. One should be cautious, therefore, in interpreting the sudden outpouring of war

literature at the end of the nineteen-twenties in the language of psychology – the 'lifting of a collective amnesia' or the 'dissipation of a mass repression'. All great wars of modern times have evoked a literary response, but always at a certain remove from the termination of hostilities themselves. Bloem, whose *Vormarsch* is one of the disregarded classics of the First World War, had made his literary reputation in Germany before 1914 with a series of novels about the Franco-Prussian War in which he had been too young to take part; Crane, in America, had done the same for the Civil War with *The Red Badge of Courage*. Both had recognized a new public's readiness to listen to what they had to say; in capturing its attention, they attracted a crowd of imitators to the same theme.

But if the outpouring of war books in 1928–32 had been a mere exercise in literary share-pushing, one would not have expected much, if any, of what was published to sustain a readership; one good book and a cluster of mediocre imitations would be all that a modern critic might realistically look to turn up on re-examination. That, of course, is not the case. Remarque's work is a little cloying to present-day taste, Aldington's has a bitterness which seems exaggerated. But Blunden's, Graves's, Hemingway's and Sassoon's have not only stood up well to the passage of time; everything about them suggests that they will continue to be read, not as background material for an understanding of the Great War, or as documentary evidence, but as moving and enduring expressions of truth about how man confronts the inevitability of death.

It is the eternal quality contained in the best literature of the First World War (the quality of much of the secondary authors' writing is also admirable) that invests the experience of the Somme with the importance it continues to hold. Nothing which the Second World War evokes stands comparison with it (though a few unfairly overlooked novels of remarkable depth have been published: in England David Holbrook's *Flesh Wounds*, in America James Jones's *The Thin Red Line*). Indeed, the only important category of book which the Second World War established in England was the prisoner-of-war story. Its extraordinary vogue prompts one to speculate about what channels non-combatant perceptions of war move through. If one takes it as axiomatic that the public's interest is in what happened at the point of maximum danger, then the success of the literature of trench life is instantly understandable. Does the public's obsession with what went on behind the barbed wire of *Stalagluft III* or Colditz imply some unconscious recognition that it was to be in a camp – concentration camp, extermination camp, labour camp, prisoner-of-war camp; the differentiation blurs easily – to have been the enemy's chattel, not his opponent, that was really dangerous in the Second World War, and that to have been a fighting soldier was to have lived in relative safety? Or was it rather that the public had recognized that from the literature of the First World War, from the story of the Somme, it had learnt as much as it ever would about what modern wars could do to men and perceived that some limit of what human beings could and could not stand on the battlefield had at last been reached; that none of the refinements of military technique or perfection of weapons achieved by science since 1918 had effectively worsened the predicament of the individual who found himself in the killing zone; and that the voice from the trenches spoke for every soldier of the industrial age? If so, Sassoon's, Graves's, Blunden's readership had perceived an important slice of reality.*

*This book was completed before the publication of Paul Fussell's *The Great War and Modern Memory* (Oxford University Press, 1975), which is wholly and illuminatingly devoted to 'the British experience on the Western Front ... and some of the literary means by which it has been remembered, conventionalized and mythologized'.

5 The Future of Battle

The Moving Battlefield

The Somme brings the story of the development of battle as a human experience and human ordeal into our own times – those of industrial economies, mass electorates and conscript armies. This is true even though the Somme may seem, to a late-twentieth-century way of thinking, an old-fashioned battle – more of a part with, say, Gettysburg than with Kursk or Alamein or the Ardennes or Sinai – and this chiefly because of the absence from the field of any 'fighting vehicle' and from the skies above it of ground attack aircraft. But to dwell on the significance of these missing ingredients is to adopt a narrowly Western point of view. For though it is certainly true that the great battles of the Second World War in France and the Desert were characterized by the employment of tanks and aircraft in abundance, and in high proportion to the number of accompanying infantrymen – those 'naked soldiers' of the Somme and Waterloo whom we have seen standing and dying on the open battlefield – the fighting elsewhere was, for the great majority of combatants and for much of the time, as earth-bound, snail-paced and soft-skinned a business as it had been for the 200 preceding years. The battles in the Pacific Islands and Burma were fought almost wholly without benefit of armour – beneath the jungle canopy almost without intervention by aircraft; the long campaign of Italy was fought by the Germans without air-cover and with few tanks; the great opening battles in Russia were conducted, except in the centre of the front, by vast infantry armies; and, despite interruptions like Kursk, the campaign remained until the last year, when the Russians had assembled their great tank armies, a war of shoe leather and horseflesh – to be eaten when times were hard (men boiled their belts at Stalingrad), otherwise to be flogged across the endless acres of the steppe, now eastward, now westward as the fortune of war directed. Stalingrad, in a sense *the* battle of the war, wryly nicknamed 'Verdun on the Volga', was almost exclusively a battle between infantrymen (pinned beneath the ruins of the city by their competing artilleries), for the tanks which had carried the German advance thither proved useless within Stalingrad itself, while the tank columns with which the Russians eventually encircled their German attackers were unleashed far beyond the city's outskirts.

It is startling, moreover, when one dissects any of the great tank battles themselves, to discover how little of the fighting took the form of the tank versus tank combat commonly thought typical of that particular sort of event. The critical phase of the battle of Kursk, 11–13 July, did see enormous armadas of tanks locked in close-range combat within a comparatively confined arena, which was almost devoid of supporting infantry. In the final stage of the Goodwood offensive east of Caen, the British tanks which arrived at the foot of Borguebus Ridge, there to be destroyed by the heavier guns of the I SS Panzer Corps, had far outstripped their accompanying infantry by the speed of their advance; and time and again in the Desert the Germans forced the British to throw their fragile Crusaders and Stuarts, unsupported by infantry escorts, on to the muzzles of their 88mm anti-tank guns (using their own not very superior Panzer Mark IIIs to bait the trap). Indeed, examples of heavy, important, even decisive tank versus tank engagements can be multiplied: the Avranches counter-attack, the Moscow winter offensive of 1942, the fighting in the Kiev salient in the winter of 1943, and so on down a very long list. But because of the composition of the forces engaged, most of the fighting between armoured divisions was, in practice, fighting not between tanks and tanks but between infantry and infantry; and the longer the war endured, the more was this the case.

This sounds paradoxical; but the paradox is simply resolved by a glance at the types of sub-units going to make up an armoured division. Some of these sub-units were, of course, tank regiments; but others were artillery and engineer regiments, and some were always infantry battalions. It was a wartime fashion to call these battalions 'armoured infantry' (*Panzergrenadiere* in the German army), appropriately enough when they were provided, as were the first *Panzergrenadiere*, with lightly-armoured half-tracks in which to move around the battlefield. But few armies – the Russians never, the British rarely – found themselves able to fit out the infantry battalions of their armoured divisions with such expensive and specialized equipment; all, moreover, as the war progressed, and the vulnerability of tanks to infantry anti-tank weapons emphasized itself, felt obliged to increase the proportion of infantry to tanks within their tank formations. Thus the Germans, whose early panzer divisions contained four tank regiments – and over 500 tanks – as against three infantry battalions, had by the end of the war reversed the proportions, so that four infantry battalions supported two tank regiments (with less than 150 tanks). This reversal was in part forced on them by their inability to produce tanks at the same rate as their enemies – the same failure which prevented them from providing more than one in four of their *Panzergrenadier* battalions with armoured transport – but, though exaggerated in their case, the same trend was detectable in the organizations of the more affluent armies over the same period. The British, who had begun the war with an armoured division containing six tank regiments to a single infantry battalion, ended it with the same division having five infantry battalions to four tank regiments; the Americans, though they kept the number of infantry battalions in their armoured divisions at three throughout the war, progressively reduced the number of tank regiments in it from eight to six and finally to three also.

A Second World War armoured division in action, therefore, little resembled the fast-moving fleet of ironclads, wheeling and shooting in unison, of which the visionaries of *blitzkrieg* had dreamed. Occasionally, of course, as during Rommel's dash from the Meuse to Arras in 1940 or in the British XXX Corps' advance from the Seine to Brussels in 1944, a tank offensive could take on the character of a sea-chase. But, whenever tanks encountered tanks in large numbers, their speed slowed inevitably to a walking pace, punctuated by long spells of immobility, and their operations, losing the simplicity of ship-to-ship actions, became heavily intermingled with confused infantry combats of a kind little different from those which soldiers of the First World War had experienced in many of the great offensives. For 'armoured infantry', *Panzergrenadiere*, motor riflemen, as the Americans, Germans and British respectively entitled the infantry of their tank formations, though they might drive in their vehicles to wherever it was that their commanders wished them to fight, always – if only out of simple prudence, for their carriers made prime targets – dismounted at a safe distance from the enemy before moving to close with him. Once on the ground, they became as vulnerable to fire of every sort as any other infantryman at any other time or place. General von Mellenthin's description of a Russian tank formation's attack on the positions of the 19th Panzer Division on the Dnieper in October 1943 very clearly brings this out:

[During] the artillery bombardment ... no movement was possible, for 290 guns of all calibres were pounding a thousand yards of front ... [They] reached as far back as divisional battle headquarters, and the two divisions holding the corps front were shelled with such intensity that it was impossible to gauge the *Schwerpunkt* ... After two hours' bombardment our trench system looked like a freshly ploughed field, and, in spite of being carefully dug in, many of our heavy weapons and anti-tank guns had been knocked out.

Suddenly Russian infantry in solid serried ranks attacked behind a barrage on a narrow front, with tanks in support, and one wave following the other. Numerous low-flying planes attacked those strong-points which were still firing. A Russian infantry attack is an awe-inspiring spectacle; the long grey waves come pounding on, uttering fierce cries, and the defending troops require nerves of steel.* In dealing with such attacks fire-discipline is of vital importance.

The Russian onslaught made some headway but during the afternoon the armoured assault troops, whom we were keeping in reserve, were able to wipe out those Russians who had penetrated the defence system. We only lost a mile or so of ground.

*'Sometimes,' writes General von Mellenthin elsewhere, 'the Russians supplied vodka to their storm battalions, and the night before the attack we could hear them roaring like devils.'

The interest of this passage lies at several levels. Like the work of many professional soldiers, it is written, whether deliberately or unconsciously is difficult to judge, in a sort of secret language: 'fire-discipline is of vital importance,' the key sentence, means, though the meaning might be lost on the average civilian reader, that the German infantry had to hold the fire of their weapons until the Russians were very – terrifyingly – close to the holes in which they were lying, and then fire in sustained, simultaneous and well-aimed bursts, to kill as many of the enemy as was possible before, they, too, found the shelter of earth and caught their breath and their courage. To fire too late – not in any case the natural reaction, indeed its very opposite – would have been to be killed themselves; to fire too soon would have been to drive the Russians too quickly to cover, sparing many and leaving them with their will to advance unshaken, while calling down on the German positions a further helping of the terrible preparatory bombardment which they had undergone in the two preceding hours.

But at another level the interest is historical and literary: it lies in the similarity between the action portrayed and that so familiar to soldiers of the trench-garrisons of the First World War. Indeed, in almost every respect, this is a First World War battle that von Mellenthin is describing, the presence of the tanks and planes being almost irrelevant and the titles of the units – panzer regiments, motor rifle divisions – completely so. The predicament of the soldiers, whether attacking or defending, is exactly that of their predecessors of 1914–18 in similar circumstances, and their fate identical. 'We lost only a mile or so of ground,' a great deal of territory to have surrendered, admittedly, on the Western Front in 1916, say, but quite within the bounds of concurrent normal experience on the Eastern, is a sentence which buries a lot of soldiers in a narrative of either war.

But if many battles of the Second World War resemble, at the level of human experience, those of the First, what then was the function and achievement of all those thousands of tanks – about a quarter of a million were built in the Second World War as against less than ten thousand in the First –

which ranged the battlefields of 1939–45: Shermans, T-34s, Churchills, Tigers, Panthers, Mark IV Panzers, Cromwells, Matildas, Valentines, Stuarts, Grants, J.S.IIIs, growling and clanking and rumbling across desert, steppe, pasture, tarmac, snowfield, floodplain? This is a complex question. It is best tackled by recognizing that 'function' and 'achievement' can, in this context, be quite different things, and indeed in practice were so. Most of the tanks catalogued above had a narrowly specialized function. The Churchill, for example, like the Matilda and the Valentine, was an 'infantry' tank, descending directly from the trench-crossing, wire-crushing Mother of the First World War, and designed like it to destroy by fire or intimidation the resistance of enemy infantry in strong points. The Tiger, on the other hand, was, in the last resort, an anti-tank tank, the Super-Dreadnought of the armoured battlefield, able to outgun any opponent, and to absorb or deflect its riposte. But, in either case, the achievements of tanks of such specialized function could be only limited and local. The Churchill could 'fight infantry through' a thick belt of wire and pill-boxes, thus forcing forward an advance which without its assistance would have been stopped or repulsed. The Tiger, if at hand when the Churchill appeared, could destroy it and 'restore the front'. But neither could do either of these things at much faster than at walking pace or over any distance, their enormous weight robbing them of speed and causing them so rapidly to wear out their tracks, that like modern pieces of earth-moving machinery, they had to be carried from place to place on specially constructed and quite unwarlike transporters.

The Sherman, however, or the T-34 or the Panzer Mark III, though none of them a match in gun-power or armour for the specialized heavies, could, at rare moments of opportunity, transform the character of a whole campaign. They could not do it often, nor could they do it to order, for it required the concurrence of conditions and circumstances beyond the mere concentration of a superiority of armour. But when this transformation occurred, the focus of fighting could be shifted 100 miles in a week – as it was in France in May 1940, or in Poland in June 1944 – and the routines of the contending armies turned topsy-turvy. *How* this transformation was achieved was not a function of the tanks' speed, nor of their capacity to overcome the resistance of the enemy lying in their path. For tanks which merely break through the enemy lines and motor off into the distance have a short life-expectancy. Mechanical breakdowns, to which tanks are preternaturally prone, will quickly thin their numbers, and exhaustion of fuel, of which they carry enough for only a few hours' driving, will shortly thereafter bring the remainder to a halt. The 'armoured break-through', about which all commanders have, since September 1939, dreamt – or had nightmares – requires therefore considerable preparation.

A great deal will be purely administrative: the concentration of troops, weapons and supplies at the point chosen for the attempt at break-through. Such administrative preparation is essential; some strategic commentators regard it indeed as the be-all and end-all of generalship. But preparation has, in the military context, another and more important sense, as in 'preparatory bombardment'. Here it means something different: to inflict such damage on the enemy as will prepare him – 'set him up' – for the blow, designed to do him the real injury, which is to follow. This is the sort of preparation which is crucial to an armoured break-through. Very occasionally it can be avoided or dispensed with, either because the antecedents of the attack are so cloaked in surprise – as they were before the Ardennes offensive of December 1944 – or because the army which is to receive it is so unfamiliar with the potentiality of armoured forces – as was the French on the Meuse in May 1940 – that the victim's powers of resistance are numbed by the shock of the main attack itself. But circumstances like these occur very rarely, normally only at the beginning of a war or else on a front long 'quiet'. Very much more often, the defender's powers of resistance must be worn down by a protracted process of combat – 'attrition' is the word we would use today – before a general will judge it safe to release his armour for the break-through.

Attrition, however, is too painful a process for an enemy to submit to it willingly. Sometimes, like the Wehrmacht in Normandy in 1944, he must submit willy-nilly; and then the battle he fights will follow the prescribed stages of 'fixing', 'attrition' and 'break-through'. But if he is given warning or has space at his disposal, he will behave differently: given warning, he will dig, wire and mine himself in – as did the Russians at Kursk – so securely that the attacker's effort at attrition will exhaust only the attacker's strength; given space, he will – like Manstein in his Kharkov counterstroke of February 1943 – break contact at the first sign of an attack and withdraw, so 'opening up' the battle and making it 'fluid', forcing the enemy to fight on ground unfamiliar to him – but well-known to the defender – and on terms and to a timetable which the defender, not he, dictates. If the attacker is to achieve his break-through, therefore, the enemy must be made to 'stand': to fight resolutely, that is, on the ground on which he is attacked, replacing the troops progressively consumed in its defence with others from his reserve until he has no more to feed forward. If then the attacker, by better husbandry, still retains a surplus, and if that surplus contains a sizable armoured element, he is in a position to achieve armoured break-through.

Yet break-through will not follow of its own accord, nor even by the tanks making ground on the far side of any gap they open in the enemy's lines. More is necessary than that: the tanks must get the army to follow them. But an army's readiness to advance when the opportunity offers is by no means automatic and spontaneous. There is, indeed, a very powerful resistance to movement in all modern armies, which is partly material and partly psychological in character, and so strong that it may even be compared in its effect to that offered by the enemy. The psychological resistance is perhaps the easier to understand. For though shelter, warmth, recreation, variety of diet are things taken for granted at one's fireside, we know that they are hard to come by on campaign, and may guess that, when found, will be valued all the more highly for that reason and surrendered reluctantly. A good billet, a quiet area, a better hole are, indeed, for all but the most exceptional spirit, what every veteran soldier seeks and his readiness to make of them a temporary substitute for home is something against which a thrusting commander must struggle hard if he is to keep his campaign alive. Even a half-good billet, a downright awful hole, will tempt the soldier to bide. How strong the temptation is was brought home to me when, while studying a large-scale trench map of the Western Front, I asked my father if his battery of 6-inch guns had not been positioned in the area it covered. He agreed that it was and at once began to point out on it, with that faculty of total topographical recall – undimmed by fifty years' absence – apparently possessed by all survivors of the First World War, its salient features. Here was the orchard where the battery had had its fighting post and there ran the gun lines; but, and clearly more important to him, here was the lock on the canal in which they had swum on fine afternoons, here was the field in which they had played football, that was the farmhouse where the family Courvisier had cooked them omelettes (they had a son away at the war and a soft spot for soldiers, particularly if they spoke a little French), there was the Calvary under which he had waited in the evenings for his elder brother to walk over from a neighbouring battery with scraps of news from home, listening in the dangerous darkness for the ring of his spurs along the *pavé*. It was obvious that this square mile of Picardy, for all its devastation and terror, had come to have for him in the few months he spent there something of the familiarity, even the security, of the rural Staffordshire in which he had grown up. And it was not only he who found his Staffordshire in France; all over the *Zone des Armées* – this indeed is the point of Mottram's *Spanish Farm* – there were individuals, groups of friends, whole units shutting out what they could of the war, making their little temporary worlds, resisting change, putting down roots.

Yet the strongest roots which the British or any modern army puts down were and are material. Its face was towards the enemy which stood in its path. But holding it down on to, even dragging it back along that path, was a densely woven net of what staff officers call rear links – to divisional dumps, water-points, telephone exchanges, railheads, ammunition parks, ordnance depots – whose function was to extend the reach of the army but whose effect, centripetal rather than centrifugal, was to attract it backwards towards it own base. These links were in theory elastic, but they were to prove notably rigid whenever the strain of an advance was thrown upon them, while the points to which they were anchored – corps and army bases, field parks, headquarters, forage and shell dumps, hospitals – were virtually immovable; it would take months of peace, in 1918–19, to prise them loose from the subsoil. The armies of the Second World War, which were organized for movement, proved less immobile in an emergency; but, despite the fleet-footed appearance which their flotillas of trucks lent them, they were effectively equipped only for short-range journeying. Asked to advance any distance at speed, they would demand the use of railways, which the enemy or friendly partisans or their own air forces would assuredly have just finished destroying.

The free use of railways guarantees, in any case, no certain rapidity of movement, as the snail-like pace of the French army's advance from its railheads into Lorraine in August 1914 demonstrates. Something much more than mere means of transport – though transport is even more vital than a burning belief in the power of the offensive (and that was very strong in the French army of 1914) – is necessary if an army is to be impelled into rapid forward motion. The army needs a vision, a dream, a nightmare, or some mixture of the three if it is to be electrified into a headlong advance. In 1914 the German army, footing itself twenty miles southward day after day, was possessed by a vision – total victory in six weeks, the overthrow of the French army, entry into *la ville lumière*, the triumphal defile down the Champs Élysées. But visions like these present themselves rarely and, however hard a general may try to conjure them into being, usually defy his artifices. Or, to put it more accurately in the past tense, usually *defied*; for it is possible to argue that while the mechanization of armies *has* produced a revolution in warfare, the real consequence, its effective potential for change, is not material but psychological; that tanks, in short, should be thought of not so much as weapons but as theatrical devices, *dei ex machina*, by the manoeuvring of which a general is enabled so to manipulate the emotions, so to stimulate the responses of his army that its resistance to movement is overcome, its tendency to self-protection transcended and its normal rhythm of campaigning shattered by the imposition of a higher object than that of holding one's ground, driving the enemy off one's front or even registering an incontestable victory. That higher object is the rescue of comrades in danger.

It is an object which the use of parachute troops allows a general to impose in an even more imperative form. In the Arnhem operation, for example, Montgomery was able to use the predicament of the 6th Airborne Division, which had been air-landed deep within enemy-held territory, as a spur to the advance of the Guards Armoured Division, and the exposure of the Guards' tanks on the *via dolorosa* northward from the Allies' lines as a prod to the rest of XXX Corps' infantry trailing behind. The French, in operations like *Lorraine* in northern Indo-China, where they parachuted battalions into the heart of the Vietminh fastness and challenged their road-bound columns to reach them, elevated this technique to the level of a strategic principle. But it is too risky a technique, as the outcome of Arnhem established, to be employed often. The armoured thrust, on the other hand, offers a general the chance both to titillate his soldiers' sense of solidarity with comrades at risk and to control the degree of risk to which they are exposed. It is possible to miscalculate, of course, as did Rommel in the Crusader Battle in November 1941, and then the armoured thrust must withdraw if it is not to wither where it comes to rest. But if its reach is calculated right, as it was by Hitler in May 1940, or by Hoth and Guderian in the Russian summer of 1941, the infantry, haunted by the nightmare of leaving the tank crews to die alone, will struggle forward somehow across the chasm that yawns between their line of departure and the tanks' foremost point of advance, a chasm which in other circumstances they would rightly think unbridgeable, and by a week or a fortnight of unreasonable effort, as much moral as physical in the demands it makes on them, transform by their advance the course of a whole campaign.

There is, then, as much psychological trickery to the consummation of a break-through as there is material preparation and rational control. Certainly without the working of this moral confidence trick, which plays on the soldiers' sense of unity with isolated comrades and feeds on the exhilaration cumulatively generated by the dash to their rescue, no break-through could be engineered. But there is an anterior and yet more important psychological trick to be played before a break-through can occur – one which, as we have seen, has to be pulled off in *both* armies, the attacking and defending: that of getting their soldiers to stand. For unless soldiers have stood, squared up to each other, exchanged blow for blow and felt the heavier tell, a break-through will indeed have no more lasting effect than any other stroke of trickery. Easy victories, between equals, almost never stick. The defeated lick their wounds, nurse their grievances and wait for the odds to even out again. The easiness of Germany's victories in 1870 goes far to explain the bitterness which the French harboured against her for forty years and the magnitude of the price they exacted in revenge on the battlefields of 1914–18. Hitler's easy victories of June–August 1941 bought him the agonies of Stalingrad, in the same way and for the same reason that Pearl Harbor cost Japan the defeat of Leyte Gulf. And in our own decade we have seen the Arab armies, adamant in their refusal to accept Israel's lightning victories of 1967 as a fair test of their relative worths, return to the struggle and insist on repeating the trial. It is for this reason that it is possible to say that the tank, though it has transformed the pace and appearance of modern campaigning, has not changed the nature of battle. The focus of fighting may be shifted twenty miles in a single day by an armoured thrust but wherever it comes to rest there must take place exactly the same sort of struggle between man and man which battlefields have seen since armies came into being.

Battle, therefore – and this is not an idea which must be

pushed to extremes, as it was by Foch and the 'offensive school' of French strategists before the First World War – is essentially a moral conflict. It requires, if it is to take place, a mutual and sustained act of will by two contending parties, and if it is to result in a decision, the moral collapse of one of them. How protracted that act of will must be, and how complete that moral collapse, are not things about which one can be specific. In an 'ideal' battle the act would be sustained long enough for the collapse to be total; and, in practice, that ideal situation was almost perfectly realized at Waterloo. But although one would like to say that 'a battle is something which happens between two armies leading to the moral and then physical disintegration of one or the other of them' – and this is as near to a working definition of what a battle is that one is likely to get – few battles see both armies making so sustained and complete a moral commitment or either coming to so final an end. Armies may indeed commit themselves fervently to the cause of bringing about the other's disintegration and utterly fail to achieve it, despite appalling human loss – as happened on the Somme. Armies again may step quite light-heartedly on to the battlefield and suffer there a shattering moral catastrophe – as overtook the French at Agincourt. But the result – negative or 'wrong' in such less than 'ideal' cases – does not mean that these encounters escape from the definition of what a battle is or, contrarily, vitiate it. Some of the moral effects which a stalemate or a misfired battle have on some of its survivors will be identical to those felt by victors or vanquished in a battle which meets the definition exactly. They, having found, from whatever source (and that is a very complicated matter), the moral resolution to stand, will inwardly have enjoyed their reward or paid the penalty.

When Sir Herbert Butterfield proposes in *Man on His Past*, therefore, that 'every battle in world history may be different from every other battle, but they must have something in common if we can group them under the term "battle" at all,' without mooting what that thing in common may be, we are now in a position to submit a suggestion. It is not something 'strategic', nor 'tactical', nor material, nor technical. It is not something any quantity of coloured maps will reveal, or any collection of comparative statistics of strengths and casualties, or even any set of parallel readings from the military classics, though the classics brilliantly illuminate our understanding of battle once we have arrived at it. What battles have in common is human: the behaviour of men struggling to reconcile their instinct for self-preservation, their sense of honour and the achievement of some aim over which other men are ready to kill them. The study of battle is therefore always a study of fear and usually of courage; always of leadership, usually of obedience; always of compulsion, sometimes of insubordination; always of anxiety, sometimes of elation or catharsis; always of uncertainty and doubt, misinformation and misapprehension, usually also of faith and sometimes of vision; always of violence, sometimes also of cruelty, self-sacrifice, compassion; above all, it is always a study of solidarity and usually also of disintegration – for it is towards the disintegration of human groups that battle is directed. It is necessarily a social and psychological study. But it is not a study only for the sociologist or the psychologist, and indeed ought not perhaps to be properly a study for either. For the human group in battle, and the quality and source of the stress it undergoes, are drained of life and meaning by the laboratory approach which social scientists practise. Battles belong to finite moments in history, to the societies which raise the armies which fight them, to the economies and technologies which those societies sustain. Battle is a historical subject, whose nature and trend of development can only be understood down a long historical perspective.

The Trend of Battle

What is the trend of battle's development? This is too large a question to be tackled without some refinement and, even though I began with the idea of looking only at battles fought inside the same climatic and geographical zone – North-West

Europe – between ethnic in-groups – white Europeans* – and within a framework of the same value-system – Western Christianity – I am not sure that such limitations refine it as much as it needs to be. So many other factors besides climate, terrain and ethos intrude, most obvious among which are those of technology and economics. Luckily, however, if we are looking at battle as a situation which encompasses the individual and his group, within a given timespan and a circumscribed locality, most may be excluded. For though the rise of industry has enormously enhanced the power which states can deploy against each other in war, and the improvement in weapons has almost infinitely extended the range of a general's reach, the predicament of the individual on the battlefield has, at whatever moment we choose to examine, still to be measured on one quite short scale: that of the physical and mental endurance of himself and his group. Men can stand only so much of anything (and dead men are dead whether killed by arrow or high-explosive), so that what needs to be established for our purposes is not the factor by which the mechanization of battle has multiplied the cost of waging war to the states involved but the degree to which it has increased the strain thrown on the human participants.

How do we mark off the degrees on the scale we want to draw? The world of mountaineering offers us a useful analogy. Mountains, like battlefields, are places inherently dangerous for the individual to inhabit. It is less easy to get killed, of course, on a mountain, if one takes sensible precautions, than on a battlefield; yet the risk of death always stalks the climber, just as it attracts him to the mountain in the first place, and numbers of climbers are killed on every major range every year. But the degree of danger to which the climber is exposed varies between quite wide limits, determined by the height of the summit to which he aspires, the steepness and inaccessibility of

*What evidence we have, drawn from studies done in the Pacific during the Second World War, suggests that fighting between out-groups is more ferocious than between in-groups. Of American soldiers who had seen Japanese as prisoners, a near majority stated they felt, as a result, 'all the more like killing them'; of Americans who had seen German prisoners, more than half felt 'it's too bad we have to be fighting them, they are men just like us.'

the face ('exposure' is the technical term) up which he chooses to make his ascent, the severity and predictability of weather conditions the face attracts, and the stability of the material of which it is composed. The higher the summit, at least in Alpine climbing, the colder the ascent, and the longer, too, which adds fatigue to the dangers; the more unpredictable the weather, the higher the risk of being trapped on the face; the sharper its gradient and the more unstable its composition, the greater the 'objective dangers' – avalanche, ice-splinters, falling stones. Falling stones, sometimes – significantly – called 'mountain artillery' by German climbers, are perhaps the most lethal of all mountaineering's hazards, because they materialize with the least warning and are caused by factors least subject to the climber's control.

At the beginning of this century, when climbers began to travel widely in search of new climbs, an attempt was made to collate the difficulties each offered so that a stranger would know whether or not it was within his capabilities. And though the British, the French, the Swiss and the Italians each produced a different system of classification for their own mountains, the systems roughly agreed in recognizing six grades of severity from 'easy' to 'extremely difficult'. Warning that a face was at the upper end of the scale was usually enough to deter beginners from tackling it while most climbers were content to confine themselves to those in the middle band.

The systems thus achieved their purpose. But just before the outbreak of the Second World War a new spirit took hold of top-class European climbers which made the classification of the most spectacular climbs thenceforth attempted and achieved more and more difficult. The spirit was that of 'extreme' climbing – climbing to 'the limits of what is physically and psychologically possible' – by 'artificial' methods: the use of metal pegs, hammered into the rock, on faces where no 'natural' hand- or foot-holds can be found. These methods stimulated a violent hostility among the traditional Alpinists who had developed a Romanticist philosophy of mountaineering which laid stress on its spiritual value to man through the harmony it engendered between him and nature, leading them to describe the extremists' feats as 'perversions', 'degradations' and 'evil demonstrations'. And the outcome of the best-publicized of the early essays in extreme technique, the 1935 and 1936 attempts on the 'unclimbable' North Face of the Eiger which killed all six of those who set foot on it, lent force to their disapproval by suggesting that there were indeed affronts which the spirit of the mountains was not prepared to tolerate.

In 1938, however, the North Face was conquered by extreme technique and since the war has been climbed again and again. By the nineteen-sixties the mere ascent was commonplace. Additional hazards were sought to add spice and sensation to that climb and to others: climbing in the depth of winter, or climbing 'direct' ('*superdirettisima*') up the line which 'a drop of water would follow if it fell directly from the summit', finally climbing both 'direct' and in winter, at first on lesser peaks like the Cima Grande in the Dolomites, ultimately on the Eiger itself. But by the time this stage had been reached the classical grading systems had lost most of their meaning. Much of the climbing was of standard five or six; but the technical difficulties paled beside the objective dangers – the volleys of stones travelling at killing speeds down the face, the showers of ice-splinters, the avalanches, the lightning strikes, to which the 'extreme' climber, hung about with the ironware of his fad, actually acted as a point of attraction – while the 'objective dangers' were themselves overshadowed by what we may call – though it is not a term mountaineers use – the 'subjective dangers'. For several days on the big faces, and several days was what *superdirettisima* demanded, drove men to the end of their physical resources, and with their strength went their will and their courage, upon which everything else in extreme climbing depends. Climbing, always a test of nerve and of physical skill, had been transformed by the mania of the extremists, who were now using electric drills, expanding bolts and what look to the ignorant suspiciously like pieces of scaffolding in their search for more and more 'direct' lines, into a battle of attrition in which will-power and endurance were paramount. And the casualties which they suffered bore comparison with those inflicted in attritional warfare: of the first seventy climbers who attempted the *Eigerwand* between 1935 and 1958, seventeen were killed on it, either by falling or from exposure. These figures provide material for an arresting comparison. Two of these, Hinterstoisser, after whom one of the most difficult traverses on the face is called, and Kurz, whose heroism in death has become one of the legends of Alpine climbing, were, as it happens, both taking leave from the German army to tackle the climb. Their regiment, the 100th *Gebirgsjäger*, was that subsequently chosen, during the airborne invasion of Crete in May 1941, to crash-land on to the runway at Maleme airport under the guns of the defending New Zealanders – perhaps the single most reckless operation of the war, though the one which turned the battle from a disaster to a victory for the Germans – and in doing so suffered about 150 casualties out of a strength of 800 – an 18 per cent ratio, contrasting with 24 per cent for the first thirteen Eiger attempts. Thus an operation of war of the most 'extreme' kind was actually proved slightly less dangerous to the unit involved than the chosen diversion of its bravest spirits.

If, then, we are asking the question, 'Has mountaineering got more dangerous over the past century?' the answer is, 'As practised at the top of the league, yes.' What had begun as a one-day event, a scramble up the easiest route to the top of any mountain which took the fancy of a group of friends, either because of its prominence or its promise of a prospect, a day to be enjoyed for the pleasure it brought in exercising one's agility, testing one's nerve, practising team spirit and enjoying God's great outdoors, has become in our own time a sort of military operation, in which sport imitates war, and war of the dreariest, deadliest, most long drawn-out sort. Indeed the hard men of the 'Winter Eiger Direct', crouched shivering day after day in their tiny, filthy, smelly snow holes, hacked with infinite labour out of the face, depressed by the death of comrades, short of food, and expecting from moment to moment to be swept out of existence by the explosion of an avalanche, recall none so vividly as the soldiers of Paulus's Sixth Army, freezing to death in identical snow-holes among the ruins of Stalingrad.

But the question we want to ask, of course, is whether we can put Agincourt, Waterloo and the Somme on a comparable scale of severity and say with confidence 'two', 'four' and 'six'? Can we grade them for 'technical difficulty', 'exposure', 'length' and 'objective dangers'? Even if we decide we can, will that lead us to the conclusion that the risk to the individual front-line soldier on the battlefield has been rising throughout the period under review?

LENGTH

One statement can be safely made – is, indeed, a commonplace: battles have been getting longer. Agincourt could have been timed in hours and minutes. And Waterloo, though part of a three-day ordeal, as we have seen, for several regiments, was for others a one-day affair; for that reason it was rated less severe by Wellington than Talavera, which had lasted two days and a night. But fifty years later, Gettysburg, bloodiest of the battles of the American Civil War, endured three days, from mid-morning on the first to late afternoon on the third. And by the beginning of the twentieth century battles between large armies, like that of Liao-yang between the Russians and Japanese in Manchuria, could occupy a fortnight. By the middle of the First World War their span had reached several months: the Somme had an official duration of four and a half months (1 July–18 November), Passchendaele of just over three (31 July–10 November 1917), Verdun of ten (21 February–20 December 1916). Indeed were it not for the instance of Stalingrad (23 August 1942–31 January 1943) and Normandy (6 June–25 August 1944) it might even be argued that battles have been getting shorter since 1918. But that argument would probably straggle off into a discussion of what the word 'battle' means, it being a perfectly tenable view that much of the fighting of the First and Second World Wars was not 'battle' as that concept has generally been understood, but 'siege', something much more limited and concrete in its aims and almost always much more protracted in its conduct. If indeed we compare the battles of the First World War with the sieges of Petersburg in the American Civil War, which lasted ten months, or of Sebastopol in the

Crimea, which lasted a year, their comparative prolongation looks more apparent than real.

But just as no soldier fought through the entire period of the Somme, neither did his counterpart at Petersburg or Sebastopol spend every day under fire in the trenches; sieges, as Roger Fenton's or Mathew Brady's photographs remind us, have their quiet moments. Even the Somme had its quiet moments; and, more important than that, it was run as a battle on a strict system of turn and turn about. The regiments which had made the great attack were almost all withdrawn the following day; most of the German regiments which repelled it were relieved within the week. And so to demonstrate the lengthening of battles is not necessarily to prove a heightening of risk to the individual. Nevertheless, big, protracted battles make insatiable demands on a general's stock of regiments, often requiring him to use them time and again on the same piece of ground. Sometimes the very intensity of the conflict precludes relief. And some armies are organized in a way that commits the individual to long spells in the line, either without interruption or at best with only the shortest intervals of relief.

The Red Army – and this must be counted one of the most insidious cruelties which the Second World War inflicted on the Russian people – granted no home leave to its soldiers from beginning to end; men remained with their units until killed or disabled, and while they lived many, having mentally abandoned their families, took 'field wives' from among their women comrades. Curiously the American army also so ran itself during that war that a man, once assigned to a fighting unit – which it was American policy to keep continuously in the line for long periods, making up losses by individual replacement – could look forward to a release from danger only through death or wounds. A sensation of 'endlessness' and 'hopelessness' resulted, so depressing and widespread in its effects that it eventually prompted the high command to institute fixed terms of combat duty, of which the controversial 'Vietnam year' is the best-known consequence.

Even during the most reluctant soldier's year in Vietnam, however, he could find himself in combat, as at the siege of

Khe San, for day after day. The character of the battle itself precluded his relief. Even more so was this the case, say, during the battle of Normandy, whose intensity required that British regiments, which would normally have expected regular breaks in their spells in the line, be left wherever they had been put after landing for week upon week; 3 Commando, a raiding force intended and equipped for the briefest exposure to the enemy, spent two months in the Bois de Bavent on the extreme left flank of the Normandy bridgehead, losing meanwhile most of its officers and over half its men. And in the First World War regiments not uncommonly underwent the worst possible combination of ordeals: that of being relieved from a battle after intense engagement and then sent back until losses had exceeded 100 per cent of the original strength. In this way, at Verdun, the German 3rd *Jäger* and 87th Infantry each lost, in a few weeks, more than the original number of soldiers with which they had entered the battle.

OBJECTIVE DANGERS

Thus the prolongation of battles, while it may not mean that the modern soldier has to submit to any single spell of combat longer than that which one of Grant's or Wellington's men underwent (though in practice it generally does), has certainly heightened the risk to the individual by multiplying the occasions on which battle – the same battle – may summon him to its service. What of the risks he runs on the battlefield when he reaches it, whether for the first, second or third time? This is a complex matter to unravel, because it is entangled with the dimensions of the battlefield, and with the sort of weapons deployed and the degree of protection which the soldier can find on it. Clearly it has always been very dangerous to be in the 'killing zone', whether that be 200 yards wide, as it was at Agincourt, half a mile as at Waterloo, or upwards of five miles as on the Somme. But the widening of the zone, besides enlarging the number of people in hazard, has probably also intensified the danger to its occupants, particularly at the front of the zone

(Forward Edge of the Battle Area, FEBA, as Staff College students are taught to call it). We can be fairly certain about this because, though the percentage of casualties suffered in battles as far apart in time as Waterloo and the Somme are of the same order of comparison, the rate at which they were inflicted in time sharply diverged. In the two battles the 1st Battalion, Inniskilling Fusiliers suffered 427 and 568 casualties, out of 698 and 801 soldiers engaged: casualty rates of 61 and 70 per cent respectively. But at Waterloo, as we have seen, the infliction of casualties was spread out over three hours; on the Somme, the losses were probably suffered in the first thirty minutes. The battalions on both occasions were ruined; but the process of ruination occupied only one-sixth of the time on the Somme as it had at Waterloo.

Moreover, despite improvements in medical care of the wounded, and even allowing that arrangements for the collection of the wounded broke down on the Somme, it is significant that the proportion of fatal to non-fatal casualties suffered by the battalion in the two battles differed sharply; at Waterloo, 117 soldiers were killed or succumbed to their wounds, on the Somme, 245 – a fatal casualty rate of 27 and 43 per cent respectively. This, of course, is far too tiny a sample on which to erect any sort of an argument; but it is offered not as evidence in a dubious case but as illustration of something that does not really need demonstration: that the killing power of weapons and the volume of munitions available to feed them has been rising throughout the last two centuries, with predictable results. The longbows of the Agincourt archers, the muskets of the Waterloo infantrymen were very effective agents for the temporary transformation of an airspace of modest dimensions into an atmosphere of high lethality. But 'modest' and 'temporary' are the important qualifications. By the beginning of the First World War, soldiers possessed the means to maintain a lethal environment over wide areas for sustained periods. Hence the titles of some of the war's most deeply felt novels, *Le Feu* (*Under Fire*) by Henri Barbusse, *A Man Could Stand Up* by Ford Madox Ford and *In Stahlgewittern* (*Storm of Steel*) by Ernst Jünger, through which each of these soldier-authors sought to convey in a phrase to their readers what it was about the new warfare which made it different from all other warfare men had hitherto experienced: that it marooned them, as it were, on an undiscovered continent, where one layer of the air on which they depended for life was charged with lethal metallic particles, where man in consequence was forced to adopt a subterranean dwelling and an abject posture, where the use of day and night were reversed and here, by a bizarre modification of Erewhonian logic, good health was regarded as a burden, but wounds as a benefaction to be sought and enjoyed. It was as if the arms-manufacturers had succeeded in introducing a new element into the atmosphere, compounded of fire and steel, whose presence rendered battlefields uninhabitable (giving them that eerily empty look which, to an experienced twentieth-century soldier, is a prime indicator that danger lies all about). The introduction of poison gas into warfare (first employed by the Germans at Ypres in 1915) did of course actually bring about a chemical change in the atmosphere of the battlefield; and for a time, which persisted into the nineteen-forties, its deadliness was thought unsurpassable. But subsequent advances in metallurgy and projectile design, accompanied by reductions in the cost and availability of high-explosive, have allowed mechanical killing-agents to overhaul it in lethality once more. Today, a ready abundance of anti-personnel mines (first widely used in the Second World War), claymore mines (giant static shotgun charges), high-fragmentation grenades and shells* and sub-calibre ammunition for the automatic weapons now universally carried by infantrymen, provide even quite small units with the means to so deluge their fronts with airborne metal as to make them virtually unapproachable by anyone lacking armoured protection. So abundant have these killing agents become (to say nothing of those fired from larger, more distant weapons or launched from the air) that the underlying aim of weapon-training has now in many armies been changed: for the traditional object, that of teaching the soldier to hit a selected target,

*Yielding metallic segments sometimes so small that the fatal wounds they inflict are almost undetectable. Keith Douglas, the Oxford poet, killed by a mortar fragment in Normandy in 1944, suffered a wound of this sort.

has been substituted that of teaching a group to create an impenetrable zone – akin in character to one of those meteorite belts which it is supposed will offer such hazards to travellers when and if men venture into deep space. The soldiers of French infantry platoons are taught to 'fire out' only to 200 metres, marksmanship, where necessary, being left to a pair of *tireurs d'élite*; the Italian infantry platoon is equipped almost exclusively with sub-machine guns, effective only for spraying the immediate neighbourhood with bullets, and requiring no greater skill to use than a housewife needs to spray her kitchen with insecticide from an aerosol can; and, lest these instances be thought a bit by-the-by, the Russian, German and American infantry companies are each armed with automatic weapons only, firing the modern lightened ammunition, of which the infantrymen can carry twice or three times the supply provided to his counterpart of the Second World War. 'Wasting ammunition', for decades the cardinal military sin, has in consequence become a military virtue; 'hitting the target', for centuries the principal military skill, is henceforth to be left to the law of averages. Perhaps only in the British army, traditionally a guild of sharpshooters, and in Northern Ireland in the nineteen-seventies embroiled in a campaign which requires its soldiers to fire back at terrorist gunmen without touching the bystanders whom the gunmen use as cover, is marksmanship still lauded and taught.

EXPOSURE

Danger buried beneath the soil of the battlefield, wafted by its breezes, suffusing in solid form its air space – mines, gas, projectiles – these 'objective dangers', some new, some as old as warfare, have, through a superabundance of supply, made the killing zone, even at its foremost edge, a yet more dangerous place for the soldier to inhabit in the twentieth century than it has ever been before. Indeed the new superfluity of killing agents has brought about a situation of which none of the classical strategists glimpsed even a prefiguration: the transformation of the very environment of the battlefield into one almost wholly

– and indiscriminately – hostile to man. Moreover, and this is a development, at least from the standpoint of the individual, of perhaps even greater significance, the size of the area which this hostile environment encompasses grows constantly bigger, yet within boundaries of progressively greater rigidity. Of what this portends for the individual, mountaineering again offers a view. For the modern fashion of combining 'extreme' technique with very long ascents has increased the degree of 'exposure' (danger of falling, risk of stone-falls) to almost intolerable limits, while making retreat from exposed situations more and more difficult. The fate of Sedlmayer and Mehringer, the first Alpinists to attempt the *Eigerwand*, illustrates the hazards of the trend. At the end of five days and four nights on the face, for most of which they had been lost from sight of the watchers below, they reappeared, still moving upwards. Tourists expressed optimism. The mountain guides and experienced climbers kept silent. They realized that the pair's line of retreat had been cut off by avalanches and stone-falls on the lower slopes and that their 'only hope now was to fight a way to the top'.

Before they could fight their way out, the cold of the North Face killed them both, at a point on its centre too far for any rescue party to reach from the summit or flanks of the peak. Thus it was the very size of the North Face and its unrelentingly hostile character which, as much as anything, did for them. And in the same way it is the very size of modern battlefields which, given the 'objective dangers' present, invests them with such peril for the individual soldier. For it is now almost impossible to run away from a battle. 'A rational army *would* run away,' thought Montesquieu, implying that in his time soldiers had the choice. In practice it was a choice which their leaders devoted a great deal of effort to prevent them from exercising ('the common soldier must fear his officer more than the enemy': Frederick the Great) but, when the lesser fear overcame the greater, run men could and did. The first moments of flight, as du Picq convincingly demonstrates, were probably the most dangerous of any a soldier could spend on the battlefield, because it was then that he was most exposed to the enemy's blows, but if he could clear the killing zone without being shot in the back or sabred by a pursuing cavalryman he had a good chance of getting off unscathed. At Agincourt, as we saw, a large number of the French cavalry force, savaged by the arrow cloud, veered off into the neighbouring wood which, within a few seconds' riding, offered them perfect safety; and at Waterloo, a considerable body of Belgian troops, having taken refuge in equally convenient woods, waited there around their cooking fires until the decrescendo of the evening persuaded them danger had passed. This 'right to flight' is naturally not one which generals are willing to concede. But its availability is one of the things which in the past have made battle bearable, by allowing the soldier to believe that his presence on the battlefield was ultimately voluntary, and it has been frequently exercised by armies of all nations, not always with results fatal either to individuals or the greater cause: First Bull Run, the Second Battle of the Somme, and Kasserine provide the most obvious modern verifications of the half-truth that he who fights and runs away lives to fight another day.

Indeed for an army to run away can be to inflict a very serious frustration on its enemy's plans. Schlieffen's fear that the Russians would refuse to stand their ground was the principal factor in persuading him to frame his notorious design for a lightning victory against France, which the French retreat to the Marne brought to naught in 1914. And the rapidity with which an intact French army recoiled towards the Marne before the *blitzkrieg* in 1940 prompted Hitler, his mind awash with memories of 1914, to spare the British at Dunkirk, lest he lose in battle with them the tanks he would need to rewrite that page of history. It is not, however, the frustration of the enemy's plans but the preservation of his own person which a soldier wishes when he turns tail on the battlefield. And the evidence very strongly suggests that flight will less and less well serve his purpose. For, to the soldier on foot, the dimensions of modern battlefields, perhaps 100 miles wide by twenty deep, perhaps even more, and certainly thirty to fifty times as large as those of the eighteenth century, put their boundaries almost beyond his reach. Even if reached, they are likely to prove impenetrable, moreover, for the modern battlefield, unlike that of the past, is more crowded at its rear than at its forward edge. Fighting soldiers are now in a minority in armies (a fact over which staff officers chronically agonize), and the fighting soldier who has decided to fight no more will find his passage rearwards impeded by a thickening host of administrative soldiers, not to say by military policemen whose duty it precisely is to prevent fugitives from making good their escape. All the more will this congestion make things difficult for the soldier who tries to leave the battlefield by vehicle, for control of roads and bridges is a principal task of rear-area troops; while enemy aircraft, though they may ignore, and perhaps not even see the foot soldier on the ground, are magnetized by vehicles in motion.

The chances are, however, that the errant foot soldier will be as visible from the air as he is from ground-level. For modern battlefields, if fought over at all long, quickly wear threadbare. Trees and bushes disappear, buildings are levelled, even the contours of the ground disturbed. Movement on the surface becomes impossible by day, and because of artificial illumination and, now, infra-red surveillance, hazardous by night. The nocturnal and subterranean pattern of living which these phenomena impose is commonly thought characteristic only of the First World War, in which the art – and strategic advantages – of defoliation were discovered by accident. But it was a pattern also dominant on many Second World War battlefields, very generally in the Korean war and at a variety of places in Vietnam and Israel. The would-be fugitive, trapped on the naked face of the modern battlefield may therefore find, like Sedlmayer and Mehringer, no alternative but to fight on, hoping to gain through the defeat of the enemy the release he knows he cannot win by retreat.

Yet there is one alternative, familiar to students of siege warfare and christened by them 'internal desertion'. Impracticable in a well-organized fortress, it flourished, particularly among civilians, whenever a commandant failed to concentrate all food in his own stores. Modern battlefields, because so difficult to escape from, encourage among soldiers a siege mentality; but, because of the prodigality of modern military supply, are often littered with preserved food. A soldier who has decided to soldier no more, who prefers not to desert to the enemy and who can find somewhere to hide may, therefore, sometimes manage to sit out the fighting, if it remains static, for considerable periods. The wastes of the 'old Somme battlefield', pitted with dug-outs and trenches over many square miles, were, during 1917, colonized by a freebooting gang of Australians, who lived by raiding military dumps and eluded the search of the military police for many months, some say until the end of the war. More significant was the desertion of a large number of the non-French garrison at Dien Bien Phu, who burrowed themselves holes in the banks of the little river which traversed the *enceinte* of the fortress and pilfered what they needed to live from the loads parachuted inside the perimeter each night, sometimes fighting the combatants for shares. At the end of the siege they are believed to have outnumbered the active garrison.

ACCIDENT

An addition to the other benefits of internal desertion (largely theoretical, of course, for the activity remains very unusual, perhaps because the necessary conditions arise so infrequently) is its elimination of the accidental dangers attendant on soldiering. Accident has always caused a proportion of battle's deaths and wounds, though exactly what that proportion is, for earlier battles, is difficult to estimate. The men-at-arms suffocated beneath the press of bodies at Agincourt and the Frenchmen who were certainly injured in the 'return cavalry charge' must have been numerous; but it was lethal intent which, to an overwhelming degree, killed in the age of edged weapons. With the appearance of firearms, accidents became much more common; I have described several which occurred at Waterloo, mostly as a result of what the army calls 'accidental discharges' – guns going off unexpectedly. And it was not only small arms which were dangerous to users or their friends; great guns could also kill. Mercer describes how one of his gunners stumbled beside the mouth of the cannon he was serving at the moment of firing: 'As a man naturally does when falling, he threw out both his arms before him, and they were blown off at the elbows' (probably by the stream of explosive gas rather than by the ball itself); Mercer later heard that he had bled to death on the way to the surgeon. As armies have accumulated more and heavier machinery, and more volatile and more powerful explosives, the toll of accidents has risen still further. Tanks are notoriously dangerous to the infantry who accompany them into action, their drivers' visibility being very limited, and armoured cars are dangerous to their own crews, being easily overturned when driven fast over rough going: the 2nd Household Cavalry Regiment actually suffered more casualties in training accidents during the Second World War than at the hands of the enemy. Modern artillery is also a double-edged weapon of support, the practice of firing 'indirect', or from map references rather than at visible targets, resulting in its shells falling sometimes (British infantrymen affect to believe always) among friendly instead of enemy soldiers. And the guns themselves, even more than was the case in Mercer's day, are a peril to their servants: 'prematures' – the explosion of the shell in the barrel instead of beyond it – are, though rare, certain to kill the crew. Mine-laying and, even more so, mine-lifting are procedures which kill sappers, who are also very much at risk when arranging demolitions, and there are a variety of other ways in which military engineering can harm its practitioners: the officer who detonated the great mine under Spanbroekmoelen in June 1917 was electrocuted by a shock from the triggering mechanism.

But it is probably the mechanization of armies which has done most to increase the accident figures: young men are regarded by insurance companies as the worst class of risk, and wars put thousands of young men in charge of powerful vehicles on unsupervised roads fraught with hazards. Collisions, skids, petrol fires, ditchings, overturnings are a commonplace on manoeuvres. In real warfare they are yet more frequent, so much so that, during quiet weeks in the Vietnam campaign, traffic accidents often killed more American soldiers than did the Viet Cong.

Some attempts have been made to calculate the proportion of accidental deaths to all death in battle. Attempts they remain, but the evidence is unarguably demonstrative of a very high level of accidental death in warfare (running in the British army at about one-fifth of battle deaths in the Crimean War, one-seventh in the Boer War) and of a considerable and rising proportion of such accidents being suffered in and as a result of battle itself.

TECHNICAL DIFFICULTY

The mechanization of war which underlies the rising accident rate might be thought to have had, as another direct result, a marked complication of the soldier's role. If we are pursuing our mountaineering analogy, developments in that field would also lead us to the same conclusion; for the extreme climber must be master not only of traditional ropework and balance and grip holds but of cramponing, piton and bolt placement and recovery, ice-screwing, prusiking, and the hanging of étriers. At that point the analogy must fail, however, for while the mountaineer necessarily remains an all-rounder, the modern soldier is increasingly a specialist. Indeed, to flatter its humblest members, the American army has largely replaced the title 'Private' by that of 'Specialist'. Yet it is a name with little substance for, though it imputes to its holder a delicate expertise, he very often possesses no more than is necessary to perform the very simple function which the continuing division of labour within armies has left him; feeding a belt of ammunition into a machine-gun, turning the dials of a wireless set, pulling the trigger of an automatic weapon. It would be perverse to suggest that the modern front-line soldier is less skilled than the musketeer or cannoneer at Waterloo, for it was the purpose of the drill each of them was taught to make him an automaton, and that the modern soldier is not. And it is certainly the case that the man behind the 'Specialist' – the armourer, radio mechanic, gunnery computer operator, helicopter pilot – practises skills of an order of difficulty beyond the comprehension of most soldiers outside this century. Nevertheless, it can be argued, and argued forcibly that the archer at Agincourt exercised a greater range and depth of skills than the modern rifleman, and the mounted man-at-arms even more so. Archery, *épée* and horsemanship are athletic feats, demanding poise, timing, and judgement which few modern military functions require and which correspondingly few soldiers, stronger and healthier though the majority certainly are than the soldiers of the age of edged weapons, can emulate.

The Inhuman Face of War

Warfare in the age of edged weapons required yet another vanished military quality, perhaps even more crucial to skill-at-arms than agility or good reflexes: a sort of empathy with one's adversary, lending the ability to anticipate his actions and forestall his blows, combined with a physical brazenness which would allow a man to look a stranger in the face and strike to fell him without provocation and compunction. Prizefighters, of course, possess this quality, whether learned or inherited, and by reason of that fact alone have for the common man an intense, almost zoological, fascination. For direct, face-to-face, knock-down and drag-out violence is something which modern, middle-class Western man encounters rarely if at all in his everyday life. Yet, despite popular enthusiasm for prizefighting, and fashionable encomiums of the 'social value of violence' to the contrary, it may be doubted whether its disappearance has left an aching void in Western man's pattern of desires. Killing people, *qua* killing and *qua* people, is not an activity which seems to carry widespread approval. It is not only in India that public executioners form a despised and outcast tribe. Even in pre-Revolutionary France the profession had become narrowly hereditary – the family Sanson practised it for seven generations – and executioners who lacked a family refuge 'were lodged in abominable hovels, did not dare to enter the towns except to do their work and even then had to be given an escort for safety's sake'; in twentieth-century England, the appointment was, until its abolition, also monopolized by a single family, the Pierrepoints – by the account of one of them against strong competition, though, by his own admission, competition from the lowest sort of person.

Killing on the scaffold and killing on the battlefield are, of course, markedly dissimilar activities. Yet, for all the elaborate explanations used by civilized societies to exculpate the soldier who kills in battle from taint of personal guilt or social disapproval – that he undergoes the same risk of death as his opponent, that he kills in order to overcome a greater evil than killing – it is worthy of note that the one sort of front-line soldier who has some choice over whether he will kill or not – the officer – has, throughout the period at which we have been looking, consistently and steadily withdrawn himself from the act itself. This withdrawal is symbolized, in a way we have already seen, by the increasingly emblematic weapons which officers have carried; at the beginning of the eighteenth century, when the pike was losing its battlefield utility, a sort of miniature pike; at the beginning of the nineteenth century, when the sword was going out of use, an ornamental sword; at the end of the nineteenth century, when the machine-gun had asserted its dominance, a pistol, usually kept holstered; during the First World War, often no lethal weapon at all, just a walking-stick. And this impression of a distancing of the officer from the infliction of death is reinforced by reading the citations which are written to explain and endorse the award of high decorations for bravery: those written for soldiers lay stress on their success at killing – 'Lance-Corporal — courageously worked his way round the flank of the machine-gun which was holding up the advance and then charged it, firing his carbine from the hip, so accounting for six of the enemy' (citation writers, flinching from 'kill', deal largely in 'account for', 'dispatch', 'dispose of'); on the other hand, those written for officers minimize their direct responsibility for killing and emphasize their powers of inspiration and organization when all about are losing their heads (in the metaphorical sense; nothing so nasty as decapitation ever creeps into a citation) – 'Captain —, taking command at a difficult moment of the battle, quickly rallied his men and, without regard for his own safety, led them back over the open to the position they had earlier been forced to leave ...' To be fair to the citation writers, however, their subject matter is to a certain extent determined for them, since soldiers on the whole are given medals for killing and officers for doing other things. But that merely shifts responsibility for recognizing approved conduct back one step, from the writer of the citation to the one who awards the medal. We could, no doubt, push this regression some way further back again. It would ultimately come to rest against the immovable obstacle of the military value system, of which one major tenet would seem to be 'Officers do not kill' or 'killing is not gentlemanly'.

Killing, none the less, was once a highly officer-like activity, when practised between equals and with a strict regard for the rules. Gronow, the Guardsman with such illuminating memories of Waterloo, was a notable duellist; Wellington himself duelled and one party to the last major duel fought on British soil, in 1852, was a Colonel Romilly. Indeed, professional extinction could follow a refusal to duel when honour required it, quite far into the nineteenth century. To return, moreover, to a much earlier moment in military time, that of Agincourt, is to encounter a world in which killing, or if not killing then certainly fighting, was the *only* gentlemanly activity. How and why has come about this progressive deprecation of the central act of warfare by its directing class, throughout a period when, as we have seen, the amount of killing attempted and achieved on the battlefield has increased from century to century?

The answer is almost certainly comprehended in the question. For killing to be gentlemanly, it must take place between gentlemen; the rules of duelling were, indeed, specific on that point, and the laws of chivalry, though less exigent and exclusive, were equally insistent that the only feats of arms worth the name were those conducted between men of gentle birth, either one to one or in nearly (ideally in exactly) matched numbers. But every trend in warfare since the end of the Middle Ages has been to make personal encounters on the battlefield between men of equal social status more and more difficult to arrange –

drill, the most important military innovation of the sixteenth century, requiring that a man stay where put instead of wandering about looking for a worthy adversary, and smoke, the most obtrusive side-effect of musketry, making such a search improbably successful – and such encounters, even if possible of arrangement, less and less representable as 'fair fight'. For 'fair fight' requires equality of skill. But firearms reduced skill-at-arms to an irrelevance – it was for this that the knight principally condemned them. The sword stroke, practised a thousand times, polished and refined and measured to pass unerringly beneath an opponent's parry, was beaten flat by a musket-shot. The musketeer, militarily speaking, was as good or better than the man-at-arms and, when drilled and mustered and properly led, the superior of any number of horsemen. Given that that was so, the gently born began during the fifteenth and sixteenth centuries, reluctantly and with many a backward glance (though less reluctantly in southern than in northern Europe) to abandon the excitements of single combat for the routines of drill and the duties of leadership, realizing that, if they did not, they risked surrendering their social along with their military station to the mercenary chieftains whose mastery of the new warfare was so irritatingly complete.

Yet in doing so, they accelerated, of course, the trend of which they were so resentful. Knightly warfare was probably already nearly a century out of date by the time of Agincourt, Crécy prefiguring its course and results in almost every respect. The passion for single combat had kept it alive, none the less, and in so doing had held in check many of those military innovations and inventions which were, when unleashed, to make Renaissance and post-Renaissance battles yet more costly than those of chivalry had been. Morally, therefore, the late-medieval resistance of the gently born to military change had exerted a beneficial restraint. The echoes of the rearguard action they fought can be heard sounding through the din of the battles of the gunpowder age. And they reverberate still.

But the distaste for mere killing which those echoes have communicated to the officer-class of Western Europe – and a class officers remain in several real senses – is of course a great deal less important for the rest of us than the facts before which the well-bred professional recoils. Battle, always unpleasant for a minority of the participants, has increasingly become an intolerable experience for the majority. What has been happening is perhaps best described as an exaggerated social and cultural divergence from normality. Battle is always an abnormality. But in violent and technically primitive societies, the facts of battle come as less of a shock to those who first face them, and leave presumably less of a scar, than they do in ordered, technically developed states. This is not to say that a medieval soldier would adapt better to life on a modern battlefield than would a contemporary. To say that would be nonsense – if only because the noise level, for which nothing in his experience would have prepared him, would of itself probably suffice to disorientate and disarm him. But it is to say that, reared in a rural world where disputes between neighbours, humble and great, were common and commonly settled by violence; practised, if not in the use of weapons themselves, then certainly in that of everyday tools which closely resembled them; accustomed to the company of horses and knowing their virtues and vices intimately, he would not have found in a battle of his own day, not at least until the killing began, and unless in the remarkable display of colour and dress in which the chroniclers took such pleasure, anything greatly to shock or surprise him. There was, in short, considerable congruence between the civil and military facts of medieval life and a minimum – admittedly a very substantial minimum – of divergence between them on the battlefield.

Today, in the late twentieth century, there exists also a considerable congruence between the technology of civilian and military life. Armoured vehicles have their counterparts in agricultural and earth-moving machinery, trucks are trucks, whether bringing detergent to the supermarket or taking troops to the front, wireless keeps one *au courant* from minute to minute whether in the bath or a slit trench, civil aircraft are as noisy as military, the quality, though not the volume, of battlefield noise is made familiar by the showing of war films – and this is to mention only artefacts, or their side-effects, with which the general population has an everyday acquaintance. Men and women employed in continuous-process industries are made indirectly familiar with many more modern battlefield phenomena: they are to a considerable degree inured to very high constant noise levels and to emissions of intense light, they work in proximity to dangerous machinery and chemicals, including poison gases, and they are involved in high-speed automatic processes – stamping, turning, reaming, cutting, moulding, the pouring of molten metals and plastics – which require perfectly timed human cooperation and imitate in many respects the actions of modern weapons systems, such as automated artillery pieces, self-loading tank guns, machine-guns, flame throwers, rocket dischargers and the like.

Modern industry, moreover, teaches its work people – though the same lessons are learnt by almost all citizens, first in school and later as the *administrés* of the states' bureaucracy – habits of order, obedience and uniform behaviour which the embryo armies of the sixteenth century could not expect to find in any of their doltish recruits, though they rightly recognized their possession to be essential to the new warfare and devoted a lengthy and brutal effort to their inculcation. If to this pre-conditioning for battle we add the undoubted power which nationalist and ideological feeling exerts in opposition to the human instincts for self-preservation, we ought to conclude that twentieth-century man is potentially a better soldier than he of any other age.

Yet that seems to me an improbable conclusion. In the first place, the climate of family, school and cultural life, for all the respect we accord to the military virtues (without so naming them), has in the aftermath of two world wars become suffused with a deep antipathy to violence and to conflict. The abolition of capital punishment in almost all Western countries is but the most striking example of this distaste; with it belongs the gradual elimination of corporal punishment from education, the right to conscientious objection now conceded even by those states, like France, which have always castigated it as unsocial, the pursuit of economic and political cooperation between nations at the expense even of a partial surrender of sovereignty and a spreading belief in the attainability of a social millennium without passage through the fires of class warfare. It is important, of course, not to make too much of this climate. Moods are contrapuntal, so that the quietism of the drop-out is matched by the insurrectionary beliefs of the parlour revolutionary. Moods are also fickle, and the very absurdity of much of the propaganda of social pacifism is calculated to hurry forward a turn of the tide. We ought, therefore, to be prepared for a dialectical swing away from fraternalism back towards the doctrines of self-reliance and self-defence *coûte que coûte* (of which the Israelis and the Palestinians are each purveying a highly exportable version). Yet were such a swing to complete its travel, I very much doubt whether the thereby changed outlook of Western youth would fit them for service on the battlefield of the future. For, despite the congruence of civilian and military technology which is such an arresting feature of the modern world, where motor cars mimic missiles and machine tools machine-guns in a realization of a Futurist fantasy, the divergence between the facts of everyday and of battlefield existence is not only greater than ever before but is widening year by year.

What are the indices of this divergence? First among them is what one must call – it is not an agreeable word – the impersonalization of battle. Its progress is something we can chart without too much difficulty. Medieval soldiers not only saw their opponents at very close hand (the high-born among them indeed were very often acquainted with one another) but fought them face to face. The rhythm of the fighting and its duration were in consequence dictated by human limitations: a man gained ground on his opponent, scored a hit, felt his sword arm tire, knew he must win in the next five minutes or be done for; and *pari passu* the same rhythms imposed themselves on his opponent. Because medieval armies were small, and battles were often fought without either side holding men in reserve, these rhythms determined the length of combat. And because the power of weapons was not very much greater than the muscle power of those who wielded them, the wounds inflicted were little different from the wounds of everyday life, those suffered in the field or workshop, to be judged at a glance trifling, disabling or fatal. In brief, the terror and brutality of battles could yet be comprehended on a human time scale and in a human way. This close relationship to everyday life was emphasized, as at Agincourt, by the opportunity offered for local civilians to take part; there it was an *ad hoc* raid on the baggage park by the peasantry of the district, under the leadership of their lord, which provoked Henry's orders to massacre the prisoners. And his soldiers' reluctance to carry out those orders, whatever their mixture of motives, is further evidence of the mediated inhumanity of medieval battle: the preservation of the lives of prisoners, even if of wealthy knights rather than poor archers, was an important diminution of the frightfulness of war, and one tending to set a rule for the general good.

The lot of the prisoner on the battlefield of the gunpowder age benefited from the generalization of the principle of ransom. Once armies had become properly regularized, and the care and exchange of prisoners legally regulated, soldiers of any rank, as we saw at Waterloo, could safely offer their surrender in the expectation that it would be accepted without their suffering hurt or indignity; though it was probably safer, as we also saw, to offer surrender to soldiers of one's own sort – infantry to infantry, for example, 'inter-specific' surrender – infantry to cavalry – seeming occasionally to provoke inter-specific acts of cruelty. The face-to-faceness of combat, still one of its marked characteristics despite the increasing range of weapons, could also work to mediate its violence; and it is interesting that, harden their hearts though they might at the spectacle of suffering among their fellow men, soldiers were much touched by the sufferings of the horses, which they were reluctant to kill even to put out of their misery.

Yet gunpowder battles were already to a marked degree more impersonal than those of the age of edged weapons. The wearing of uniforms, however variegated, however splendid, diminished the individual identity of the combatants, which it has been one of the principal functions of medieval panoply to emphasize. So too did the imposition of a rigid chain of command, which robbed subordinates of that independence by which the headstrong nobleman had always set such store, while the new insistence on drill reduced the individual soldier's status to that of a mechanical unit in the order of battle. Battle itself, because of these inbuilt mechanisms, steadily acquired throughout the passage of the gunpowder age a mechanical dynamic of its own, the action of the artillery, firing systematically over several hundred yards at blocs of human beings whom the gunners perceived only as differently coloured masses, being sufficient of itself to keep a battle in progress whether or not the armies were in intimate confrontation.

For all that, gunpowder battles, fought during the daylight hours of a single series of twenty-four, at short ranges, over the span of a few fields, whose farmers might watch the carnage from the safety of a neighbouring hilltop or wood (the Forest of Soignes was crammed with peasants from the Waterloo district during the battle), calculating the cost of the damage to their crops against the income which the windfalls of war would leave them, were events which belonged demonstrably to the world of men. Of the battles of the twentieth century that is something of which it was increasingly difficult to say, if, that is, the sensations of the combatants are accorded the weight they deserve. For what almost all the soldiers of the First World War and many of the Second, even from the victor armies, testify to is their sense of littleness, almost of nothingness, of their abandonment in a physical wilderness, dominated by vast impersonal forces, from which even such normalities as the passage of time had been eliminated. The dimensions of the battlefield, completely depopulated of civilians* and extending far beyond the boundaries of the individual's perception, the events supervening upon it – endless artillery bombardments, sudden and shatteringly powerful aerial bombings, mass irruptions of armoured vehicles – reduced his subjective role, objectively vital though it was, to that of a mere victim. And a victim too was what he risked becoming, even if he took or had forced upon him the decision to stop fighting and give himself up as a prisoner. For men,

*Also of animals: the great, protracted battles on the eastern frontier of France in the autumn and winter of 1944 provoked a westward migration of much of its major fauna. Wild boar, for example, not seen in the Seine valley since the nineteenth century, had become comparatively plentiful again in the nineteen-fifties.

rarely coming face to face, seen by each other, if at all, only as indistinguishable figures in shapeless and monotone uniforms, generally lacked the means to communicate such intentions to each other. A shout of surrender from the darkness of a dug-out was too often an invitation to receive a grenade, the wave of an arm from the hatch of a disabled vehicle the signal to unleash a burst of automatic fire; killing the crews of stopped or burning tanks as they bailed out was normal practice among Second World War infantrymen. It must be counted one of the particular cruelties of modern warfare that, by inducing even in the fit and willing soldier a sense of his unimportance, it encouraged his treating the lives of disarmed or demoralized opponents as equally unimportant.

At another level, the fostering and infliction of deliberate cruelty marks a second major divergence between the facts of everyday and battlefield existence in the twentieth century. Weapons have never been kind to human flesh, but the directing principle behind their design has usually not been that of maximizing the pain and damage they can cause. Before the invention of explosives, the limits of muscle power in itself constrained their hurtfulness; but even for some time thereafter moral inhibitions, fuelled by a sense of the unfairness of adding mechanical and chemical increments to man's power to hurt his brother, served to restrain deliberate barbarities of design. Some of these inhibitions – against the use of poison gas and explosive bullets – were codified and given international force by the Hague Convention of 1899; but the rise of 'thing-killing' as opposed to man-killing weapons – heavy artillery is an example – which by their side-effects inflicted gross suffering and disfigurement, invalidated these restraints. As a result restraints were cast to the winds, and it is now a desired effect of many man-killing weapons that they inflict wounds as terrible and terrifying as possible. The claymore mine, for instance, is filled with metal cubes (how far have we come from Thomas Puckle's famous gun, firing round bullets against Christians and square against infidels!), the cluster bomb with jagged metal fragments, in both cases because that shape of projectile tears and fractures more extensively than a smooth-bodied one. The HEAT and HESH rounds fired by anti-tank guns are designed to fill the interior of armoured vehicles with showers of metal splinters or streams of molten metal, so disabling the tank by killing its crew. And napalm, disliked for ethical reasons even by many tough minded professional soldiers, contains an ingredient which increases the adhesion of the burning petrol to human skin surfaces. Military surgeons, so successful over the past century in resuscitating wounded soldiers and repairing wounds of growing severity, have thus now to meet the challenge of wounding-agents deliberately conceived to defeat their skills.

These intentional inhumanities seem worthy of notice because the societies which sanction them are dedicated, in their treatment of human beings away from the battlefield, to standards of consideration, compassion even, higher than those adopted by any others of which we have knowledge. The modern Western state accepts the responsibility not merely to protect the individual's life and property, traditionally the legal minima, but to educate and heal him, support him in old age and when unemployed, and increasingly to guarantee his prosperity. Might the modern conscript not well think, at first acquaintance with the weapons the state foists on him, that its humanitarian code is evidence either of a nauseating hypocrisy or of a psychotic inability to connect actions with their results?

The third, and in its fashion perhaps most disturbing, divergence between life on and off the battlefield is seen in the role coercion plays in keeping men in the killing zone. Coercion is a word to which the vocabulary of democracy gives grudging house room. The liberal state likes to believe that it works by consent and persuasion, that compulsion is a method of dealing with citizens to which only the lower forms of polity have resort. The truth is, of course, that all armies, whether of democracies or dictatorships, depend on the coercive principle (most armies have a code of law and punishment separate from that administered by the civil courts), that it is a vital element in making battles work, and that it is one which the character of modern warfare invests with more not less force. Remembering the extent of the direct coercion applied at Waterloo – the positioning of cavalry behind the rear rank of unwilling infantry battalions so that they should not be able to break and run, the flogging forward of soldiers by their officers, the firing at 'friendly' cavalry by infantry disgusted at their cowardice – that latter point might seem a difficult one to sustain. But the fact that coercion was indeed direct and personal on the gunpowder battlefield, that the officer who flogged too hard risked a bullet should he turn his back (and sometimes got it, as did Colonel Breyman from one of his grenadiers whom he had hit with his cane during the battle of Saratoga), that bullying cavalry who crowded infantry too close might feel their bayonets, set limits to its scope. It is a function of the impersonality of modern war that the soldier is coerced, certainly at times by people whom he can identify, but more frequently, more continuously and more harshly by vast, unlocalized forces against which he may rail, but at which he cannot strike back and to which he must ultimately submit: the fire which nails him to the ground or drives him beneath it, the great distance which yawns between him and safety, the onward progression of a vehicular advance or retreat which carries him with it willy-nilly. The dynamic of modern battle impels more effectively than any system of discipline of which Frederick the Great could have dreamt.

The Abolition of Battle

Impersonality, coercion, deliberate cruelty, all deployed on a rising scale, make the fitness of modern man to sustain the stress of battle increasingly doubtful. And this seems to me true even though 'modern man' is too vague a figure around whom to frame so general a statement. We must take account of the undoubted willingness of some men at all times to risk, even apparently to enjoy, extreme danger and arbitrary cruelty. Though the life required an oath of submission to be 'burnt with fire, shackled with chains, whipped with rods and killed with steel (*uri, uinciri, uerberari, ferroque necari*)' there were volunteers as well as slaves in the ranks of the gladiators. And not all of them social refugees: Mark Antony's brother Lucius fought as a gladiator in Asia Minor. In our own times, almost all professional soldiers can recall acquaintance with men for whom the terrors of battle seemed to have little meaning. 'Corporal Lofty King,' Brigadier Durnford-Slater wrote of one of his commandos, 'was very tall and very tough. He was a hard fellow in many ways and very hard with his men; he didn't give a damn if he knocked a man down. Sometimes I told him he was being too rough. Lofty would say, "It's good for them, Colonel, it won't do them any harm." He would mean it and believe it. He genuinely enjoyed fighting and looked happiest, indeed inspired, in battle. In the field he was kinder to his men, as if the fighting were a kind of release for him.' Lofty King is a significant figure whose outlines can be discerned in the thick of the fighting on many battlefields (Legros, *l'enfonceur* of the great gate of Hougoumont, belongs to the type) and whose power to impose his superior will on his comrades lends support to one's suspicion that, after all, battle *is* to the strong; that without the presence of the Lofty Kings and the Legros most battlefields would empty of soldiers at the firing of the first salvoes; and that one of the subtlest forms of coercion practised in armies is the patronage by the grandees in the upper-ranks of their bully-boy opposites in the lower. Battle is also to the young. Its physical ordeals – discomfort, loss of sleep, hunger, thirst, burdens – are not only better borne by men under thirty; so too are its terrors, its anxieties, its separations, its bereavements. And young men are also moved more deeply than older men by the moral consolations with which battle compensates the soldier – it would be foolish to deny that there are compensations – for its cruelties: the thrill of comradeship, the excitements of the chase, the exhilarations of surprise, deception and the *ruse de guerre*, the exaltations of success, the sheer fun of prankish irresponsibility. Lord Robbins, the eminent economist, describes in his autobiography how, during the few days of mobile warfare he experienced as a young gunner officer on the Western Front, he was brought wholly unexpectedly to realize by his release from the grim routines of the trenches what an absorbing and enjoyable activity battle could be, and why in times past it had fulfilled the energies and imagination of the European upper class to the exclusion of almost all else.

Yet the prospect of battle, excepting perhaps the first battle of a war or a green unit's first blooding, seems always to alarm men's anxieties, however young and vigorous they be, rather than excite their anticipation. Hence the drinking which seems an inseparable part both of preparation for battle and of combat itself. Alcohol, as we know, depresses the self-protective reflexes, and so induces the appearance and feeling of courage. Other drugs reproduce this effect, notably marijuana; the American army's widespread addiction to it in Vietnam, deeply troubling though it was to the conscience of the nation, may therefore be seen if not as a natural, certainly as a time-honoured response to the uncertainties with which battle racks the soldiers. The choice of that particular army, moreover, had local precedents: the pirates of the South China Sea traditionally dosed themselves with marijuana before attacking European ships.

Hence too, it would seem, the stirring or rekindling of a desire for spiritual fortification before battle. In primitive warfare, the enactment of tribal rites is often an absolutely vital preliminary to any planned encounter with the enemy; and in the Christian armies of the high Middle Ages, like Henry's at Agincourt, the saying of mass and the hearings of confessions (Henry heard mass three times in succession) seems to have been regarded in much the same light, though these sacraments were of course offered by the priests strictly as means to a personal renewal of grace, not of corporate inspiration. Indeed, wherever the light of religion has not died out from armies, men seem to hunger for its consolations on the eve of action; in the Kitchener armies waiting for 1 July to dawn, it was not enough to have written home, made one's will and shaken hands with friends; to have been to church was for many a necessary capstone of the preliminaries. Rum was welcomed to stifle the flutters of panic as the seconds ticked down to zero, but was not accepted, as it seems to have been by Wellington's impious majority, as provision enough against the imminence of combat. Whether or not, however, it is with religious observance that the men preface their entry to battle, or with some solemn military ritual, like Napoleon's grand review of his troops on the morning of Waterloo or the proclamation of an order of the day or some other ceremony, it does seem that something – a pause, a moment of recollection, a summoning of force, a dedicatory act, a prayer of intercession – must be added to the purely material and administrative dispositions made by an army if its men are to commit themselves to battle with the stoutest hearts they can find. That is perhaps why a battle which begins with one army surprising another does not always yield the success it theoretically ought; for unless an army has inwardly hardened itself for the shock, it will not stand to be beaten.

Whatever the process of inward hardening, the shock nevertheless will shake some men's resolution to breaking. It is unfortunately impossible to represent this pattern of breakdown in any comparative style, for it is only since the beginning of this century that armies have been taught to accept that 'courage and cowardice are [not] alternative free choices that come to every man, overriding all emotional stress, that a man [cannot] simply choose which he prefers and . . . be courageous if he is told he must.'* Running away, refusing to fight, getting the shakes or going inert were all stigmatized, less than seventy years ago, as displays of cowardice; and it was only with the greatest difficulty that even an army so comparatively humane in spirit as the British was led to think differently. Men whose symptoms we can now recognize as those of true psychiatric breakdown were shot for desertion during the first two years of the First World War, and the fear of the death penalty yielded a multitude of 'hysterical conversion symptoms' (by which men lose the use of limbs, speech or sight rather than demonstrate straightforward displays of anxiety). The army eventually reconciled itself to the inescapable fact of the breakdown of so many of its soldiers by inventing the notion of 'shellshock' which suggested for it a single physical cause; and treated the soldiers so affected in what were called N.Y.D.N. (Not Yet Diagnosed, Nervous) hospitals. But any statistics of the proportion of psychiatric casualties to all battle casualties for 1914–18 remain hidden. In

*From *Men Against Fire* by S. L. A. Marshall.

the Second World War, however, the psychiatrists of the British and, to an even greater extent the American, medical corps were able to insist on a proper recognition and treatment of psychiatric cases, their hand being much strengthened by their success in teaching the armies how to identify among recruits those particularly suited for the specialist military functions it was increasingly necessary to fill and those most likely to make no sort of soldier at all. As a result, we now have some reliable statistical material: and it reveals that, despite the system of rejection the psychiatrists instituted, psychiatric casualties at every stage of the war formed a significant percentage of all battle casualties, diagnosed as 'exhaustion' cases in their simplest form and as 'neuro-psychiatric' in their more aggravated. 'Depending on the type of battle,' wrote one of the British army's senior psychiatrists, '2% to 30% of all casualties may be psychiatric.' His evidence revealed that, of all battle casualties, ten to fifteen per cent were psychiatric during the 'active' phase of the Battle of France in 1940, ten to twenty per cent during the first ten days of the Normandy battle and twenty per cent during the two latter months, seven to ten per cent in the Middle East in the middle of 1942 and eleven per cent in the first two months of the Italian campaign. Many of these, perhaps as many as ninety per cent, were eventually returned to some form of duty, more or less demanding, but even among those judged fit to be returned quickly to their fighting unit '(figures varied from 70% to 56%) ... some 5% of these broke down again in the same battle.' Moreover, as time dragged on, almost all soldiers exposed to continuous or semi-continuous combat broke down. As the authors of the American official report *Combat Exhaustion* explain:

> There is no such thing as 'getting used to combat' ... Each moment of combat imposes a strain so great that men will break down in direct relation to the intensity and duration of their exposure ... psychiatric casualties are as inevitable as gunshot and shrapnel wounds in warfare ... Most men were ineffective after 180 or even 140 days. The general consensus was that a man reached his peak of effectiveness in the first 90 days of combat, that after that his efficiency began to fall off, and that he became steadily less valuable thereafter until he was

completely useless ... The number of men on duty after 200 to 240 days of combat was small and their value to their units negligible.

The fighting of the Second World War, in short, led to an infantryman's breakdown in a little under a year. The indirect effects of this consequence of battle were many; but some of the most interesting were those felt and betrayed by the leaders of armies. Generals, since the end of the First World War, had become markedly sensitive to the disparity between the risks suffered by the men who framed the plans and those who carried them out. On earlier battlefields, that disparity had been small, if at all apparent. Wellington, indeed, was arguably at greater risk on the field of Waterloo than many of his subordinates, and at Agincourt Henry (though 'generalship' and 'planning' are concepts one can doubtfully apply to medieval warfare, where the setting of an example was all) deliberately courted risk throughout the battle. Hindenburg, Haig, Joffre, on the other hand, never smelt powder; Haig, for motives which he was adept at rationalizing, would not even visit his wounded. Their château-generalship (a style which, to be fair, they inherited rather than created) caused deep if unexpressed offence to the generation of officers who, subalterns in 1914, were senior commanders by 1940, and stimulated in them a new risk-sharing style of leadership, publicly justified for the closer control of the battle it permitted (at a moment when the proliferation of wireless sets made command from a château sensible), privately desired, one suspects, because it quelled a vicarious, anticipatory sense of guilt.

Rommel's variant of the style was to command from the leading tank, Guderian's to roam his battlefields in an armoured wireless truck, Montgomery's to inhabit a 'tactical headquarters' within earshot of the fighting. Generals began also, in a reversal of that long-established trend for officers to distance themselves from killing, to carry weapons: Patton habitually sported a pair of pearl-handled revolvers, Ridgway a pair of grenades, Bock a revolver, Wingate a rifle; and more and more of them, as Grant, eccentric as it was thought, had done, began to dress as private soldiers. Montgomery, Bradley, Stilwell are scarcely distinguishable in their uniforms from the humblest soldiers under their command. Perhaps because of these efforts to identify with their men, however, many generals seemed unable to reproduce that necessary resistance to stress which so noticeably stamped the characters of an older generation of chiefs. Sorrow and anxiety spare only the rarest even among leaders; Wellington wept copiously after Waterloo, Frederick the Great had his surgeons bleed him during his battles to lower the tension he felt, and poor Henry VI keened an endless discordant song throughout all the battles which his courtiers obliged him to attend. But the military code traditionally required composure even at moments of personal agony; and it evoked it: Castelnau and Foch each continued to direct operations after receiving news of the deaths of their sons in the Battle of the Frontiers in 1914, Ludendorff to command despite the loss of both his cherished stepsons at the height of the First World War. During the Second World War the code seemed unable to sustain its votaries. Incompetent generals always become casualties: that war broke competent generals also. Rommel, for all his derring-do, experienced agonies from a nervous stomach, which twice took him away from the front at moments of crisis, Guderian was invalided from Russia with heart-failure, Reichenau suffered a stroke during the campaign, Ridgway had a severe blackout in September 1945 and was advised to retire. Mere hardness of character of the sort demonstrated by Zhukov or Model, rather than any particular strategic or tactical flair, increasingly became the principal military virtue as the Second World War dragged on. Other commanders who appeared to stand the strain did so only by cultivating a curious detachment from the conduct of the battles themselves. The three most admired generals of the British, American and German armies – Alexander, Eisenhower and Rundstedt – were each, in their different ways, not really generals at all, non-generals, almost anti-generals. Alexander, hell-raiser though he had been as a young officer, insisted on leaving control to his subordinates and confined himself to fostering good relations within his multi-national army. So to an even more marked degree did Eisenhower, whose aura became eventually papal rather than military. Rundstedt, revered throughout the German regular officer corps as its last archetypal Prussian, refused to deal with detail or to look at small-scale maps, as if the fighting itself were distasteful to him, but spent his days reading detective stories and thrice resigned his command.

But perhaps the most interesting, even if the best-known and most overworked example of a general's reaction to stress of modern battle is Patton's chastisement of the 'psychoneurotic' soldier in a Sicilian hospital. The publicized incident was in fact the second of two; in each case he had expressed his anger that a 'coward' should be treated in the same way and same place as honourably – physically – wounded soldiers. May we not understand this bepistolled, risk-taking general's outburst as a transmuted expression of his disgust that he, who shared his soldiers' lot, sought them out in the front-line to praise their courage, stood ready to sacrifice his life as readily as any, should have been repaid for identifying himself with them by behaviour which questioned his sincerity – which silently accused him of not knowing the ultimate reality of the ordeal he asked his soldiers to undergo and so made a mockery of his impersonation of the hero? Something of a concern not to be caught out in such a falsity, a refusal to frame orders whose detailed consequences he would not directly suffer, appears to have underlain Alexander's curious aloofness from the mechanics of command.

The chance to intervene directly at the very forward edge of the battlefield, at the height of the fighting, in almost instant response to summons, explains the contemporary general's enthusiasm for the helicopter. In a fashion Patton would genuinely have envied, the helicopter does carry the general back to the stance from which Wellington commanded, and returns to him the power to observe, to manage, to exhort, to manoeuvre, to look battle in the face. In the helicopter, the general has the impression of controlling the battle and shaping it to his ends, of remaking battle as a useful and decisive exercise of power. But is this a valid impression or only an illusion? For the helicopter does not only bring the general to battle. It brings also the 'air-mobile' soldier, whose experience is one of an extraordinary divergence between the normality of comfortable barrack life and the terrors of the battlefield to which, in under half an hour, he can be smoothly transported; the experience resembles, though in a heightened form, that of combat aircrew, in which it produces strains so uniform and intense that the number of operational missions they may fly has to be limited if they are not inevitably to crack. The 'air-mobile' soldier, in his turn, is an element in a new sort of army, one mechanized and tracked and armoured to a degree unmatched in any of the armoured formations of the Second World War. Everything in all fighting units on both sides of the border on the Central Front in Germany is mechanized, including supply, maintenance and bridging equipment, and most of it, including the artillery, is armoured and tracked. Armoured and tracked infantry; the infantry section of ten men cocooned inside its armoured personnel carrier; there lies the revolutionary difference between the armies of the 1970s and the 1940s. Armies on the move, whether in attack or defence, are trained to manoeuvre and expected to operate at thirty miles an hour, moving in dense waves across country, stopping only if so ordered or opposed, and seeking to overwhelm opposition by the weight of fire from their guns and their infantrymen's weapons.

We lack a detailed picture of what an encounter between two such armies would be like in reality; and fortunately so, for it would be the preliminary to a firing of nuclear weapons. The Yom Kippur war between the Syrians, Egyptians and Israelis nevertheless provides a few clues. The battle, to begin with, would be as noisy as any experienced in the First or Second World Wars, there being added to the constant crash of the projectiles of indirect fire weapons – field artillery and rocket projectors – the explosion of mines (with which the future battlefield is to be liberally sown) and the cascading explosions of cluster bombs dropped from the air, a very great deal of mechanical clatter and whine and the unmistakable, periodic clang of high-energy rounds hitting tanks. The noise a solid block of tungsten makes on striking armour is highly distinctive, a high-pitched ringing clang, and although that note would not predominate over the future battlefield's cacophony, it would be the keynote for which the occupants of the armoured vehicles listened, tolling as it would the disablement of a vehicle and extinction of its crew. Not all of its crew perhaps but, being dependent almost exclusively on their sense of hearing for knowledge of events outside their shell, the occupants of tanks and armoured personnel carriers might be inclined to imagine so.

The state of mind of these occupants exercises the leadership of modern armies very considerably. It has already been grasped that to enclose men in a confined and windowless armoured box for long periods is to risk, among other effects, seriously disorientating them. It is therefore intended, when the next type of armoured personnel carrier is built, to provide a quartz peephole for every passenger, so that he shall be able to maintain some picture, however fragmentary, of where he is being taken. It is also understood that soldiers cannot be cramped and congested for long periods without losing their efficiency, and the interior of the infantry carriers are, as tanks are already, to be padded and air-conditioned, provided with means to heat food and cool drinks. Yet one wonders whether all these measures will realize that fighting efficiency they are designed to assure? For what can they be but minor alleviations of a further impersonalization of warfare, a greater alienation of the soldier from anything recognizably human or natural on the field of battle, a steeper reduction of his status to that of a mere adjunct to machinery, the software in the system? And while it is undoubtedly possible for picked men to sustain for short periods conditions of the sort which shut-down armoured warfare will impose – tank and air crews have consistently done so over the last fifty years, naval turret crews for more than a century – it is important to remember that depictions of future battle suppose all fighting soldiers, picked and unpicked, will be able to tolerate something analogous to the aircrew environment for periods not of hours but of many days and nights. The concept of 'continuous operations' which it is proposed to conduct in an armoured battle in Europe, and for which the most elaborate electronic night-fighting equipment – infra-red searchlights, image-intensifiers, ground-surveillance radars and movement sensors – is provided to the armies, requires soldiers to remain continuously in action for periods of 100 or 150 hours. There is even talk of attempting to keep them awake for eighty hours at a stretch, using if necessary doses of one of the amphetamines as the agent; ironic if official condemnation of the private use of hallucinogens and tranquillizers in battle is to partner an official administration of stimulants. In practiee, the Israelis and the Arabs, on whom night-fighting equipment had been lavished, found themselves so exhausted at the end of the day-long battles of October 1973, that they relapsed gratefully into sleep as soon as darkness fell. But the NATO powers cannot count, as can all parties to the Palestinian problem, on having their wars stopped by outside intervention whenever a defeat looms. Their armies therefore must train in all seriousness for 'the land battle in Central Europe', must learn to live for days in stifling gas-masks and clammy radiation suits (which would have to be worn as a precaution even during conventional operations), isolated inside their armoured vehicles from sight or smell of the outside world, connected to it only by disembodied voices received through their wireless sets and able to form an impression of the events transpiring beyond their carapace only from whatever fragments of fact higher authority vouchsafed to communicate.

'In all seriousness' requires to be qualified in the light of these circumstances; 'train with high dedication' would certainly be correct; 'train with a firm conviction that the battle they practise is one likely to be fought' seems much more doubtful. For all the initial advantage which the communist armies' superiority in numbers gives them, their soldiers are not physiologically different from those on the other side. And we are faced now with a prospect of battle which through the physical and nervous strain, the 'multiple stress pattern' it will impose on the combatants, threatens to break them down whether or not they come into direct contact with the enemy. Allied military psychiatrists had learnt by the end of the Second World War that the very first hours of combat disable ten per cent of a fighting force. A major intensification of the strains which broke those men (such as that imposed by several days of 'continuous operations') suggests that it might break the majority, and that 'decision' would be brought about not by the direct infliction of death and wounds but by the immersion of an army in a situation which would prove psychologically intolerable.

'Decision', as I began by saying, is a concept which military historians use in an ambiguous fashion. By 'decisive battle' they can mean simply a battle which has a result, which ends in the clear-cut victory of one side over the other; but by it also a battle whose result causes some real shift in the direction of human affairs far away from the battlefield, bringing about the downfall of a heretofore dominant power, setting the term to a hitherto irresistible tide of imperial expansion, toppling a political system, cutting short the career of a conquering hero. By a curious function of his *déformation professionelle*, the military historian's search for results is almost always directed at one or other of these two levels: at the immediate effect of the battle on the strength of the army and the mind of its commander, or else at its impact on the morale and resources of the war-waging power. Yet, as I have tried to argue, the most important, the really 'decisive' effects of a battle are more immediate and personal than those belonging to these other categories. It is armies which fight battles, and armies which contain the men who, in any society, can and will and know how to fight. Battles, or more precisely defeats, are immediately decisive because they kill some of these men and dissuade the rest, for a longer or shorter period, from wanting to fight any more. As to the longer-term consequences: where a preponderance of the fighting men are drawn from the governing stratum, as in a feudal army or a patrician militia, we should look for them first in the rearranged pattern of personalities which death, cowardice or displays of prowess will have brought about, then in the mood and aspirations which the army will carry home with it. Where the warriors form a unique and expensive specialist group, as in the armies of dynastic and post-dynastic Europe, we should look elsewhere; at the economic cost of the state's effort to reform from the urban crowd or the rural peasantry whence the beaten army was drawn, a substitute for it; at its political costs also, in terms of the concessions the tax-paying classes will wring in return for financing the rebuilding, and the demands for a guarantee of their privileges the military classes will present in competition. Where the army is levied directly on the male youth of the country by general conscription, as in the liberal and not-so-liberal states of twentieth-century Europe and America, we should look far more widely and deeply. The very scale of the First and Second World Wars has determined that, look as we may, we cannot yet categorize all those results or chart their dimensions. But one at least denies contradiction: that the experience of violent and sudden death has been brought through battle into many, perhaps a majority of families, that fear of the suffering – arbitrary and accidental as well as deliberate and purposive – battle can cause to human societies is profound and almost universal, and that the usefulness of future battle is widely doubted.

The young have already made their decision. They are increasingly unwilling to serve as conscripts in armies they see as ornamental. The militant young have taken that decision a stage further: they will fight for the causes which they profess not through the mechanisms of the state and its armed power but, where necessary, against them, by clandestine and guerrilla methods. It remains for armies to admit that the battles of the future will be fought in never-never land. While the great armoured hosts face each other across the boundary between east and west, no soldier on either side will concede that he does not believe in the function for which he plans and trains. As long as states put weapons in their hands, they will show each other the iron face of war. But the suspicion grows that battle has already abolished itself.

Bibliography

Historiography

GEOFFREY BARRACLOUGH, *History in a Changing World* (Blackwell, 1955).

ISAIAH BERLIN, *The Hedgehog and the Fox* (Weidenfeld & Nicolson, 1953).

MARC BLOCH, *The Historian's Craft* (Manchester University Press, 1954).

HERBERT BUTTERFIELD, *Man on his Past* (Cambridge University Press, 1955).

P. L. GARDNER (ed.), *Theories of History* (Allen & Unwin, 1960).

PIETER GEYL, *Debates with Historians* (Meridian Books, New York, 1958).

—— *Encounters in History* (Collins, 1963).

G. GOOCH, *History and Historians in the Nineteenth Century* (2nd ed.) (Longman, 1952).

J. R. HEXTER, *Reappraisals in History* (Longman, 1961).

—— *Doing History* (Allen & Unwin, 1971).

ROBERT RHODES JAMES, 'Thoughts on Writing Military History' in *R.U.S.I. Journal*, May 1966.

JAY LUVAAS, *The Military Legacy of the Civil War* (University of Chicago Press, 1959).

—— *The Education of an Army* (Cassell, 1965).

ARNALDO MOMIGLIANO, *Studies in Historiography* (Weidenfeld & Nicolson, 1966).

G. J. RENIER, *History, Its Purpose and Method* (Allen & Unwin, 1950).

BERYL SMALLEY, *Historians in the Middle Ages* (Thames & Hudson, 1974).

STEPHEN USHER, *The Historians of Greece and Rome* (Hamish Hamilton, 1969).

S. WILLIAM (ed.), *Essays in Modern European Historiography* (University of Chicago Press, 1970).

Agincourt

C. T. ALLMAND, *Society at War, The Experience of England and France during the Hundred Years' War* (Oliver & Boyd, 1973). (Contains a very full bibliography for the whole period.)

RICHARD BARBER, *The Knight and Chivalry* (Longman, 1970).

JOHN BARNIE, *War in Mediaeval Society* (Weidenfeld & Nicolson, 1974).

MARC BLOCH, *Feudal Society* (Routledge & Kegan Paul, 1961).

A. H. BURNE, *The Agincourt War* (Eyre & Spottiswoode, 1956).

CHRISTOPHER HIBBERT, *Agincourt* (Batsford, 1964).

M. H. KEEN, *The Laws of War in the Late Middle Ages* (Routledge & Kegan Paul, 1965).

FERDINAND LOT, *L'Art Militaire et les Armées au Moyen Age* (Paris: Payot, 1946).

R. A. NEWHALL, *The English Conquest of Normandy 1416–1424* (Harvard University Press, 1924).

—— *Muster and Review* (Harvard University Press, 1940).

SIR HARRY NICOLAS (ed.), *History of the Battle of Agincourt* (London, 1832).

SIR CHARLES OMAN, *A History of the Art of War in the Middle Ages* (2nd ed.) (Methuen, 1924).

MICHAEL POWICKE, *Military Obligation in Mediaeval England* (O.U.P., 1962).

Waterloo

There is an excellent bibliography in Jac Weller's *Wellington at Waterloo* (Longman, 1967), but it includes few of the British regimental histories of which there are about forty to be read. To identify them, the order of battle should be used as a key to A. S. White's *Bibliography of Regimental Histories of the British Army* (Society for Army Historical Research, 1965).

MARQUESS OF ANGLESEY, *One-Leg, The Life and Letters of Henry William Paget* (Cape, 1961).

LT-COL. NEIL BANNANTYNE, *History of the Thirtieth Regiment* (Liverpool: Littlebury, 1923).

ANTONY BRETT-JAMES (ed.), *The Hundred Days* (Macmillan, 1964).

DAVID G. CHANDLER, *The Campaigns of Napoleon* (Weidenfeld & Nicolson, 1966).

CHARLES DALTON, *The Waterloo Roll Call* (2nd ed.), (Arms and Armour Press, 1971).

LT-COL. C. GREENHILL GARDYNE, *The Life of a Regiment* (The Gordon Highlanders), (2nd ed.) (Medici Society, 1929).
RICHARD GLOVER, *Peninsula Preparations, The Reform of the British Army 1795–1809* (Cambridge University Press, 1963).
HENRI HOUSSAYE, *1815, Waterloo* (Black, 1900).
DAVID HOWARTH, *A Near Run Thing* (Collins, 1968).
LADY DE LANCEY, *A Week at Waterloo* (John Murray, 1906).
REV. WILLIAM LEEKE, *The History of Lord Seaton's Regiment* (53rd Light Infantry) (Hatchard, 1866).
CAVALIÉ MERCER, *Journal of the Waterloo Campaign* (Peter Davies, 1927).
SIR CHARLES OMAN, *Wellington's Army, 1809–1814* (Methuen, 1913).
MAJOR-GENERAL H. T. SIBORNE (ed.), *Waterloo Letters* (Cassell, 1891).
CAPTAIN WILLIAM SIBORNE, *The Waterloo Campaign 1815* (Constable, 1904).
GENERAL SIR EVELYN WOOD, *Cavalry in the Waterloo Campaign* (Pall Mall Press, 1897).

The Somme

A thesis written for fellowship of the Library Association, *The Battle of the Somme 1916: A Bibliography*, by Arthur T. E. Bray (University Microfilms Ltd., 1968) refers to almost all the published and unpublished sources in English and to the most important French and German printed sources.

BERNARD BERGONZI, *Heroes' Twilight* (Constable, 1965).
CHARLES CARRINGTON, *Soldier From the Wars Returning* (Hutchinson, 1965).
GUY CHAPMAN, *A Passionate Prodigality* (McGibbon & Kee, 1965).
F. P. CROZIER, *A Brass Hat in No Man's Land* (Cape, 1930).
SIR JAMES EDMONDS, *Military Operations, France and Belgium 1916*, vol. 1 (Macmillan, 1932).
A. H. FARRAR-HOCKLEY, *The Somme* (Batsford, 1964).
ROWLAND FEILDING, *War Letters to a Wife* (Medici Society, 1919).
V. W. GERMAINS, *The Kitchener Armies* (Peter Davies, 1930).
DONALD HANKEY, *A Student in Arms* (Andrew Melrose, 1916).
ERNST JÜNGER, *Storm of Steel* (Chatto & Windus, 1929).
MAJOR C. A. CUTHBERT KEESON, *The History and Records of Queen Victoria's Rifles* (Constable, 1923).
B. H. LIDDELL HART, *Memoirs*, vol. I (Cassell, 1965).

LT-COL. J. H. LINDSAY, *The London Scottish in the Great War* (privately printed, 1925).

MARTIN MIDDLEBROOK, *The First Day on the Somme* (Allen Lane, 1971).

R. H. MOTTRAM, *Journey to the Western Front* (Bell, 1936).

G. A. PANICHAS (ed.), *Promise of Greatness* (Cassell, 1968).

REICHSARCHIV, *Schlachten des Weltkrieges; Somme-Nord 1* (Stalling: Oldenburg, 1927.)

ROYAL WELCH FUSILIERS, *The War the Infantry Knew* (P. S. King, 1938).

JOHN TERRAINE (ed.), *General Jack's Dairies* (Eyre & Spottiswoode, 1964).

BASIL WILLIAMS, *Raising and Training the New Armies* (Constable, 1917).

CAPTAIN G. C. WYNNE, *If Germany Attacks* (Faber, 1940).

Warfare, Conflict and Society

HANNAH ARENDT, *On Violence* (Allen Lane, 1969).

L. BRANSON and G. W. GOETHALS (ed.), *War, Studies from Psychology, Sociology, Anthropology* (Basic Books, New York, 1964).

W. L. BURN, *The Age of Equipoise* (Unwin University Books, 1964).

NIGEL CALDER (ed.), *Unless Peace Comes* (Allen Lane, 1968).

PETER CALVERT, *Revolution* (Macmillan, 1970).

ELIAS CANETTI, *Crowds and Power* (Gollancz, 1962).

DAVID CAUTE, *The Left in Europe Since 1789* (World University Library, 1969).

I. F. CLARKE, *Voices Prophesying War, 1768–1914* (O.U.P., 1966).

TREVOR CLIFFE, *Military Technology and the European Balance* (Adelphi Paper 89, Institute for Strategic Studies, 1972).

SIR EDWARD CREASY, *The Fifteen Decisive Battles of the World* (36th ed.), (Richard Bentley, 1894).

CHRISTOPHER DUFFY, *The Army of Frederick the Great* (David & Charles, 1974).

CYRIL FALLS, *A Hundred Years of Warfare* (Duckworth, 1953).

LT-COL. A. A. HANBURY-SPARROW, *The Land-Locked Lake* (Arthur Barker, 1932).

LT-GENERAL JOHN H. HAY, *Vietnam Studies, Tactical and Material Innovation* (Washington: Department of the Army, 1974).

E. J. HOBSBAWM, *Revolutionaries* (Weidenfeld & Nicolson, 1972).

MICHAEL HOWARD, *The Franco-Prussian War* (Hart-Davis, 1961).

JAMES JOLL, *The Second International* (Weidenfeld & Nicolson, 1955).

MICHAEL MALLETT, *Mercenaries and their Masters* (Bodley Head, 1974).
MAJOR-GEN. F. W. VON MELLENTHIN, *Panzer Battles* (Cassell, 1955).
J. U. NEF, *War and Human Progress* (Routledge & Kegan Paul, 1950).
GERALD PRIESTLAND, *The Future of Violence* (Hamish Hamilton, 1974).
MICHAEL ROBERTS, *Essays in Swedish History* (Weidenfeld & Nicolson, 1967).
GEORGE RUDÉ, *Paris and London in the 18th Century* (Fontana, 1970).
Sunday Times Insight Team, *The Middle East War* (André Deutsch, 1974).
JAC WELLER, *Weapons and Tactics* (Nicholas Vane, 1966).
QUINCY WRIGHT, *A Study of War* (Chicago University Press, 1952).
PETER YOUNG, *The British Army 1642–1970* (Kimber, 1967).

Battlefield Stress, Combat Motivation and Military Medicine

ROBERT H. AHRENFELDT, *Psychiatry in the British Army in the Second World War* (Routledge & Kegan Paul, 1958).
ROBERT ARDREY, *The Territorial Imperative* (Collins, 1967).
JOHN BAYNES, *Morale, A Study of Men and Courage* (Cassell, 1967).
ALAN BLACKSHAW, *Mountaineering* (Penguin, 1965).
SHALOM ENDLEMANN (ed.), *Violence in the Streets* (Duckworth, 1969).
P. GILMAN and D. HOUSTON, *Eiger Direct* (Collins, 1966).
MICHAEL GRANT, *Gladiators* (Weidenfeld & Nicolson, 1967).
ROY G. GRINKER and JOHN P. SPIEGEL, *Men Under Stress* (McGraw-Hill, 1963).
HEINRICH HARRER, *The White Spider, The History of the Eiger's North Face* (Hart-Davis, 1959).
E. J. HOBSBAWM, *Bandits* (Weidenfeld & Nicolson, 1969).
HANS KRUUK, 'The Urge to Kill', *New Scientist* (29 June 1972).
HUGH L'ÉTANG, *The Pathology of Leadership* (Heinemann, 1969).
KONRAD LORENZ, *On Aggression* (Methuen, 1963).
SIR A. SALUSBURY MCNALTY and W. FRANKLIN MELLOR (ed.), *Medical Services in War* (H.M.S.O., 1968).
SIR W. G. MACPHERSON (ed.), *History of the Great War: Medical Services. Diseases of the War II* (H.M.S.O. 1923); *Surgery of the War II* (H.M.S.O. 1922); *General History III* (H.M.S.O. 1923).
S. L. A. MARSHALL, *Men Against Fire* (William Morrow, 1947).
LORD MORAN, *The Anatomy of Courage* (Constable, 1945).

Bibliography

ARDANT DU PICQ, *Battle Studies* (Macmillan, New York, 1921).
PIERRE-HENRI SIMON, *Portrait d'un Officier* (Éditions du Seuil, 1958).
W. F. STEVENSON, *Wounds in War* (Longman, 1897).
S. A. STOUFFER & OTHERS, *The American Soldier* (Princeton University Press, 1949).
T. D. M. STOUT, *Official History of New Zealand in the Second World War, War Surgery and Medicine* (Wellington: 1954).
LIONEL TERRAY, *Conquistadores of the Useless* (Gollancz, 1963).
LIONEL TIGER, *Men in Groups* (Nelson, 1969).
B. URLANIS, *Wars and Population* (Moscow: Progress Publishers, 1971).
U.S. INFANTRY JOURNAL, *Digest of Army Ground Forces Committee on Battle Casualties Report* (September, 1949).
War Office Committee of Enquiry into 'Shellshock' Report (H.M.S.O., 1922).
MARTHA WOLFENSTEIN, *Disaster* (Routledge & Kegan Paul, 1957).
G. F YOUNG, *Mountain Craft* (Methuen, 1934).

Index

accidents as cause of death, 195–6, 317–19
Adam, General Sir Frederick, 135
Advanced Dressing Stations, 272
aerial torpedoes, 208
Agincourt, 134, 140, 175, 302, 320, 336; the campaign, 78–86; killing of prisoners, 84–5, 108–12, 327; the battle, 86–92; waiting period, 88–9, 139; archers versus infantry and cavalry, 92–3; infantry versus infantry, 97–107; 'wall of dead', 107, 197; the wounded, 112–14; will to combat, 114–16; religious preparation, 114–15, 137, 333; types of combat, 144, 145; length, 308; objective dangers, 310, 311; 'right to flight', 315
Agis, King of Lacedaemonians, 66, 67
'air-mobile' soldier, 338–9
Aisne, river, 213, 214
Albemarle, George, 5th Earl of, at Waterloo, 130, 131, 136, 185, 190–91, 204
Albuera, 20, 35–9, 63
Aldington, Richard, 286, 288
Alençon, Duke of, 105
Alexander, Viscount, 337, 338
All Quiet on the Western Front (Remarque), 286
Amiens, 209
archers, at Agincourt, 83, 88, 89–107, 320
Ardennes, 75
Argives, at Mantinea, 65–7
Argonne, 208, 209, 214
'armoured break-through', preparation for, 296–7
armoured divisions, of Second World War, 292–3
'armoured infantry', 291–2, 293
Arnhem, 300
artillery, on Somme, 208, 216–19, 231–40, 241, 243–5, 251–8, 259
artillery, at Waterloo, 126, 127, 128, 129, 131, 140–44, 239; cavalry versus, 151–3; versus infantry, 160–62
Artois, 214, 215
attrition, 215–16, 296–7
Aubers Ridge, 210
Australian Official History of the Great War, 47
Avranches, 291

Background to Napoleonic Warfare (Quimby), 40
Balaclava, 40, 45, 63
Bannantyne, Lieutenant-Colonel, 135n.
Barbusse, Henri, 311
Barnsley Pals, on Somme, 241, 248
barrage fire, 216–19, 251–4

Bastard, Lieutenant-Colonel, on Somme, 254–5
battle: abolition of, 331–43; definition of, 302; impersonalization of, 326–31; nature of, 301–3; preconditioning for, 323–5; battle, trend of, 303–20, 339–42; accident, 317–19; exposure, 313–17; length, 308–10; objective dangers, 310–13; technical difficulty, 319–20
'battle pieces', 28–9, 35–45, 61–7
battlefield: escape from, 314–17; increased movement on, 290–300; widening of, 310–11, 313–17
battlefield existence, divergence between everyday life and, 326–31
Baudelaire, Charles, 177
Bazaine, Marshal, 54
Beau Quesne, Château de, 267
Beaumont Hamel, 245, 250, 261
Bebel, F. A., 177
Belcher, Ensign (32nd Regiment), at Waterloo, 187
Bell, Sir Charles, 205
'Beloved Captain, The' (Hankey), 280–81
Bergen-op-Zoom, 155, 187
Bethencourt, 81
Béthune, river, 81
Bloem, Walter, 24, 287
Blücher, Marshal, 132, 138, 206
Blunden, Edmund, 286, 288, 289
Bock, Field Marshal Fedor von, 336
Bonaparte, Jérôme, 131
Borguebus Ridge, 291
Borodino, 29, 44
Bowra, Sir Maurice, 275 n.
Brabant, Duke of, at Agincourt, 84, 108
Bradley, General Omar, 336
Braine l'Alleud, 123
Braine le Comte, 135
breakdown, 334–8, 341
Brébant, Clignet de, 94
Brenan, Gerald, 274
Bresle, river, 81
Breyman, Colonel, 331
Brigades (on Somme): 14th, 267; 64th, 253; 101st, 249; 102nd, 249; 103rd, 249
Britain, Battle of, 74–5
British Army 1642-1970, The (Young), 40
British Battles and Sieges (Napier), 39
British Expeditionary Force of 1916: composition of, 219–29; on Somme, 229 *et seq.*
British Official History of the First World War, 29–30
Brown, Lieutenant (4th Regiment), and Waterloo, 129
Bull (gunner officer), at Waterloo, 138, 162, 193
Burne, Lieutenant-Colonel A. H., 33; on Agincourt, 87, 91, 98
Butterfield, Sir Herbert, 302
Bylandt, General, at Waterloo, 161, 171
Byron, Lord, 118, 192

Cadogan, Sir Alexander, 186
Caesar, Julius, and narrative tradition, 62–5, 66–7, 73
Cambridge Battalion, at Somme, 249

Camerone, 20
Camoys, Lord, 88, 98
Campaigns of Napoleon, The (Chandler), 40–41
Canning, Colonel, at Waterloo, 202
Caporetto, 276, 277
Carnot, Lazare, N. M., 175
Castelnau, Vicomte de, 337
casualties, 276; at Waterloo, 128, 161–2, 188, 204–5, 311; at Verdun, 216; on Somme, 249–51, 258, 260, 274, 285, 311; psychiatric, 334–5, 341
Casualty Clearing Station, 271, 272
Cathcart, Sir George, 131, 132
cavalry at Agincourt, 83–4, 87, 88; and infantry, archers versus, 92–3; versus infantry, 94–6
cavalry at Waterloo, 126–7, 128, 141, 144, 145–7; versus artillery, 151–3; versus cavalry, 147–51; versus infantry, 154–60
Champagne, 210, 214, 215; Second Battle of (1915), 107
Chandler, David, 40–41, 42, 44
Chandos, Lord, 248
Chapman, Professor Guy, 48, 49
Chasseurs, at Waterloo, 171
Châtiments, Les (Hugo), 118
Chavasse, Captain N. C., twice awarded Victoria Cross, 30
Chemin des Dames, 208, 210
Cherbourg, 165
Chetwynd-Stapleton, Lieutenant, on Somme, 242
Childe Harold (Byron), 118
Christie, Ensign (44th Regiment), at Waterloo, 187
Churchill, Winston, 74–5, 228
Churchill tanks, 295
Clarke, Volunteer (69th Regiment), at Waterloo, 187
Clausewitz, Karl von, 28
Clive, Lord, 175
Coglan, Sergeant (18th Hussars), at Waterloo, 137
Colborne, Sir John, at Waterloo, 130, 138, 173, 178, 180
'collective indiscipline', 276
colours, defence of, 186–8
combat, types of: at Agincourt, 144; at Waterloo, 144, 246; on Somme, 246–7
Combat Exhaustion, 335–6
Command Decisions (U.S. official history), 58
Commentaries (Caesar), 62–5, 66–7
communication system, Fourth Army, 264–8
compulsion, role of, 115–16, 184–5, 282, 330–31, 332
comrades in danger, rescue of, 300
'continuous operations' concept, 340–41
Cotter, Captain (69th Regiment), at Waterloo, 136
courage: in leadership, 190–93, 280; equated with morality, 280
Crane, Stephen, 287
Creasy, Sir Edward, 74, 120; his philosophy of war, 55–8, 60–61
critical distance, 166–7, 173
critical reaction to threat, 166–8, 173
crowdlike conduct, 100–101, 173–9
Crowds and Power (Canetti), 174

Crozier, Lieutenant-Colonel F. P., on Somme, 242–3, 282, 283
Crucifix Trench, 254
cruelty, deliberate, 47–9, 84–5, 108–12, 329–31
Cuirassiers, at Waterloo, 147, 148–9, 150, 152, 156–8, 204
Cutforth, René, 198

D-Day landings, 79
Darwin, Charles, 56
Davidson, Lieutenant-Colonel J. R., 30
De Lancey, Sir William, 133
De Lisle, General, on Somme, 261
Death of a Hero (Aldington), 286, 288
'Decisive Battle' idea, 57–61, 73, 342
Decisive Battles of India (Malleson), 57
Decisive Battles of Modern Times (Whitton), 58
Decisive Battles of the Second World War (German history), 58
Decisive Battles since Waterloo (Knox), 57–8
Decisive Battles of the World (Fuller), 58
Decisive Wars of History (Liddell Hart), 58
Delbrück, Hans, 32, 53, 55
Dickens, Colonel (Queen Victoria's Rifles), on Somme, 264, 266
Dickson, Corporal (Scots Greys), at Waterloo, 203–4
Dien Bien Phu, 14, 75, 317
Dirom (1st Guards), at Waterloo, 169, 170, 173
Divisions (on Somme): 4th, 219, 261, 267; 7th, 219, 245, 259; 8th, 219, 245, 254, 260; 9th, 220; 10th, 220; 11th, 220; 12th, 220; 13th, 220; 14th, 220; 15th, 220; 16th, 220; 17th, 220; 18th, 220, 241, 252, 259; 19th, 220; 20th, 221; 21st, 245, 253, 259; 29th, 219, 245, 250, 261; 30th, 222, 226, 252, 259; 31st, 248; 32nd, 226, 245, 249–50; 34th, 226, 245, 248–9, 259; 36th Ulster, 226–7, 245, 247, 250, 259, 282; 46th North Midland, 228, 245, 250, 255, 256, 260; 56th London, 227, 245, 247, 255–8, 267, 274, 283
Division, 6th Airborne, 300
Division, German, 19th Panzer, 293
Dnieper, river, 293
Douglas, Keith, 312 n.
Doyle, Sir A. Conan, 227
Drewe, Lieutenant (Inniskillings), at Waterloo, 131
drill, function of, 32–3, 195, 323
drink, and will to combat, 114, 115, 183–4, 245, 333
duelling, 322
Duffy, Dr Christopher, 31, 32–3
Dunkirk, 315
Duperier (18th Hussars), at Waterloo, 155, 184, 192
Durham Light Infantry, 15th Battalion on Somme, 253
Durnford-Slater, Brigadier, 332

East Yorkshire Regiment, 1st Battalion on Somme, 253–4
Edinburgh Pals, on Somme, 248

Eeles (95th Regiment), at Waterloo, 157
Eiger, North Face of, 306–7, 314
Eisenhower, General Dwight D., 337
Ellis, Colonel (23rd Regiment), at Waterloo, 200–201
Engall, Second-Lieutenant John, on Somme, 241
Engels, Friedrich, 28
Erlon, Jean-Baptiste Drouet, Comte d', 126–7, 132, 140, 154, 168, 170, 171–2, 175
Erpingham, Sir Thomas, 89
Ersatz Corps, 223
Evans, Sir George de Lacy, at Waterloo, 173
Evans (Union Brigade), at Waterloo, 154
Ewart, Sergeant (the Greys), at Waterloo, 146–7
exposure, 313–17
Eylau, 41, 45

Farewell to Arms, A (Hemingway), 286
Fatal Decisions, The (German history), 58
fatigue, 134–7
fear, dominance of, 70–72
Feilding, Rowland, on Somme, 263
Festubert, 210, 255
Fifteen Decisive Battles of the World (Creasy), 55–7, 60, 120
Finley, Dr Moses, on Homeric concept of honour, 193–4
First Day on the Somme, The (Middlebrook), 262 n.
Flers, 209
Flesh Wounds (Holbrook), 288
flight distance, 166–7, 173
Foch, Marshal, 302, 337
Ford, Ford Madox, 311
Franco-Prussian War, The (Howard), 41–2, 43
Fraser, Ensign Charles, 190
Frederick the Great, 314, 337
Free French Armoured Division, 2nd, 76
Freeman, Gillian, 29
French Revolution, evaporation of, 177–9
Fricourt, 245, 248, 250
Fuller, Major-General J. F. C., 58, 232, 245, 263–4, 266
Fusilier Brigade (7th Royal and 23rd Royal Welch Fusiliers), at Albuera, 35–9, 63

Gage, Cornet (Greys), at Waterloo, 146
Gallipoli (James), 31
Garcia Hernandez, 155
gas shells, 231, 232, 236
Gawler, Colonel (52nd Regiment), at Waterloo, 132, 145
Gettysburg, 308
Gibney, Assistant-Surgeon (15th Hussars), at Waterloo, 139, 143
Givenchy, 210, 255
Gommecourt, 227, 241, 243, 255–8, 260, 263–4
Goodbye to All That (Graves), 286
Goodwood offensive, 291
Gordon, Sir Alexander, 133
Gordon, General, 175
Gorlice-Tarnów, 276
Grant, Professor Michael, 68
Graves, Robert, 229, 286, 288, 289
Greek historiography, 65–9
Greenwell, Graham, 279–80

Grenfell, Julian, 279–80
Grimsby Chums, on Somme, 249
Gronow, Ensign Rees Howell, at Waterloo, 141, 142, 159, 322
group solidarity, and will to combat, 46–7, 50–52, 73, 185–8, 230
Guards: at Quatre Bras, 139; at Waterloo, 133, 141, 169, 171, 178–9, 189
Guards Armoured Division, at Arnhem, 300
Guderian, General Heinz, 336, 337
Guesde, Jules, 177
Guibert, Comte de, 69
Guillemont, 29–31
Gustavus Adolphus, 62, 67–8, 177

Haig, Earl, and Somme, 209, 215, 216, 246, 266, 267, 336
Halkett, General Sir Colin, 163, 202
Hall, Private Gilbert, on Somme, 241, 248
Hallam, Henry, 57
Hamilton, Lieutenant (Greys), on Waterloo, 146, 199, 202
Handbury-Sparrow, Lieutenant-Colonel A. A., 275 n., 280
hand weapons, 77; at Agincourt, 86–107
Hankey, Donald, 280–81
Harfleur, 80
Hawkings, Private Frank, on Somme, 241, 243
Hay, Colonel (16th Light Dragoons), at Waterloo, 195
Heavy Brigade: at Balaclava, 40, 42, 44, 63; at Waterloo, 126, 147
helicopter, 338–9
Hemingway, Ernest, 286, 288
Henry V: and Agincourt campaign, 78–85, 88, 89–90, 105, 106, 109–13, 145, 333; order to advance, 83, 89; and killing of prisoners, 84–5, 108–12; courting of risks, 336
Henry VI, reaction to stress, 337
Herodotus, 29, 65
Hindenburg, Field Marshal von, 54, 208, 209, 336
History of the Battle of Agincourt, The (Nicolas), 102 n.
History of the Irish Guards in the Second World War, 48
History of the Thirtieth Regiment, A (Bannantyne), 135 n.
Hitler, Adolf, 267–8, 300, 301, 315
Holbrook, David, 288
Homer, and concept of honour, 193–4
honour, officers' concept of, 191–4
Horne, Alistair, 61, 75
Hougoumont, 123, 126, 132, 133, 161, 162, 163, 164–5, 167, 168, 178, 179
Household Cavalry Regiment, 318
Houssaye, Henry, 137
Howard, Ensign (33rd Regiment), at Waterloo, 183, 200, 206
Howard, Major (10th Hussars), at Waterloo, 191–2
Howard, Michael, 28, 41–2, 43, 44, 61
Howarth, David, 119
Howett, 2/Lieutenant Stephen, 225 n.

howitzer shells, on Somme, 208, 231, 232, 234–5, 236, 238–9
Hugo, Victor, 29, 118
Human Aggression (Storr), 30 n.
hunger, 89, 135, 136–7
Hussars (at Waterloo): 7th, 129, 151; 10th, 136, 137, 155, 184, 191–2, 195, 196; 11th, 184, 196; 18th, 137, 141, 155, 192, 196, 204

If Germany Attacks (Wynne), 235–7
Imperial Guard, 178; at Waterloo, 127, 131, 132, 141, 163, 168–73, 178–9, 180
'improper violence', 47–50, 108–12, 202–3
In Stahlgewittern (Jünger), 311
infantry at Agincourt, 87–92; cavalry versus, 94–6; and cavalry, archers versus, 92–3; versus infantry, 97–107
infantry on Somme, 218–19, 229–30, 241–5, 246; artillery versus, 241–5; versus infantry, 251–8; versus machine-gunners, 247–51
infantry at Waterloo, 126–8, 144; artillery versus, 160–62; cavalry versus, 154–60; versus infantry, 162–94
Ingilby (gunner officer), at Waterloo, 140
Inherent Military Probability principle, 32
Inniskillings, at Waterloo, 128–9, 131, 133, 135, 161, 250
'internal desertion', 316–17

Jack, Major, on Somme, 243
James, Robert Rhodes, 31
Jaurès, Jean L., 177
Joffre, Marshal, 214, 336; and Somme, 215, 216
Jones, James, 288
Joynt, Lieutenant W. P., 47
Jünger, Ernst, 24, 311

Kampf als innere Erlebnis (Jünger), 24
Kee, Robert, 260
Kelly, Dawson, at Waterloo, 169, 172, 173–4, 202
Keowan, Lieutenant, at Waterloo, 198
Keppel, George (later 6th Earl of Albemarle), 139
'Kerensky Offensive', 276
Kharkov, 297
Khe San, 15, 310
Kiev, 291
killing: distaste for, 320–21, 323–4; officer-like activity, 190, 280, 321–2, 336
'killing zones', 104–5, 159, 310–11
King, Corporal 'Lofty', 332
King's German Legion, 155, 163; at Waterloo, 123, 126, 127, 132, 147, 196
King's (Liverpool) Regiment, 17th, 18th, 19th and 20th battalions, 222
King's Regiment, 1/7th and 1/10th (Liverpool Scottish) battalions, 29–30
King's Own Yorkshire Light Infantry, 9th and 10th battalions on Somme, 253–4
Kitchener, Lord, 175, 220, 221
Kitchener armies, formation and composition of, 219–29
Knox, Thomas, 57–8

Korea, 14
Kruuk, Hans, 283
Kursk, 290, 291, 297

La Belle Alliance, 132, 179, 206
La Boiselle, 245, 248
La Haye Sainte, 123, 126, 127, 128, 132, 133, 140, 148, 165, 167, 168, 186, 192
Lacedaemonians, at Mantinea, 65–7
Lancers, French, at Waterloo, 146, 150, 151, 202–3
Lawrence, Sergeant William, at Waterloo, 183, 186, 188
leaders: bond between followers and, in will to combat, 114; and personal risk, 336–7; and remote command, 336, 337–8; and resistance to stress, 337–8
leadership, 188–94, 277–81
Lee, Robert E., 54
Leeke, Ensign William, on Waterloo, 136, 137, 138, 141, 142, 147, 157, 160, 161, 179–80, 184, 191, 201
Legions, Roman, 62–5, 67–8
Leicester Regiment, 4th and 5th battalions, 228
Leipzig Salient, 232
Lejeune, Baron, 185
Liao-yang, 308
Liddell Hart, Basil H., 58, 254
Life Guards, at Waterloo, 147–50, 154, 179
Life of One's Own, A (Brenan), 274
Light Brigade, at Waterloo, 135
Light Dragoons: 16th, at Waterloo, 154, 181, 184, 195, 196; 23rd, at Waterloo, 191, 192, 195
Light Infantry, 52nd, at Waterloo, 127, 130, 132, 136, 137, 157, 158, 171, 178, 179, 180
Ligny, 122
Lincolnshire Regiment: 2nd Battalion on Somme, 254; 4th and 5th battalions, 228
Liverpool Pals, 221–2
Llewellyn (28th Regiment), at Quatre Bras, 130
London Regiment, 16th Battalion on Somme, 241
London Rifle Brigade, 227
London Scottish, on Somme, 255–8
London Territorial regiments, 227–8; on Somme, 255–8, 261
Loos, 210, 215, 255, 276
looting, 115, 182–3, 198, 281
Louis XVIII, 179
Loyal North Lancashire Regiment, 1/5th Battalion, 30
Ludendorff, General Erich von, 337
Luvaas, Jay, 27

machine-gun fire, 232–4; on Somme, 234–5, 247–51, 253–8
Macready, Ensign (30th Regiment), at Waterloo, 158, 189, 206
Maginot line, 75
Main de Massiges, 210, 215
Maisoncelles, 81, 85, 112, 113
Maitland, General Sir Peregrine, at Waterloo, 132, 162, 189
Malleson, G. B., 57
Mallet, Michael, 62

Mametz, 245
Man Could Stand Up, A (Ford), 311
Man on His Past (Butterfield), 302
Manstein, General Fritz von, 297
Mantinea, 65–7
Marder, Professor Arthur, 27
Marne battle, 74
Marshall, General S. L. A., 334 n.; historical method of, 71–3, 119–20
Marten (2nd Life Guards), at Waterloo, 154
Masters, John, 224–5
Maurice of Nassau, 62, 67, 68, 177
mechanized armies, encounter between, 338–41
medical service, 270–75
Mellenthin, General F. W. von, 293–4
Memoirs of an Infantry Officer (Sassoon), 286
Men Against Fire (Marshall), 71–2, 334 n.
Mercer, Cavalié, on Waterloo, 130, 136, 138–9, 141, 142, 143, 149–50, 152–3, 159, 180, 184, 189, 195–6, 317–18
Middlebrook, Martin, 262
military history: usefulness of, 20–25; deficiencies of, 25–35; sources, 31–4; 'battle pieces', 35–45, 61–7; 'outcome' approach, 45, 50; 'killing no murder?', 45–52; history of, 53–61; narrative tradition, 61–72; verdict or truth?, 72–7; accusatorial approach, 74; inquisitorial approach, 74–5
Military Legacy of the Civil War (Luvaas), 27
Minet, Captain E. C. T., 241
'minimum necessary force' doctrine, 49
Model, Marshal Walter, 337
Moltke, Helmuth, Count von, 20, 265
Monstrelet, Enguerrand de, 91
Mont St-Jean, 128
Montesquieu, Baron de, 314
Montgomery, Viscount, 300, 336
Montreuil, 264
Morris, Sergeant Tom (73rd Foot), on Waterloo, 149, 156, 183
Moscow, 291
motor riflemen, 293
Mottram, R. H., 298
mountaineering analogy, 304–8, 314, 319
Mountsteven (28th Regiment), on Waterloo, 169, 173, 189–90
movement, resistance to, 297–300
multiple-missile weapons, 77; on Somme, 208, 209, 216–19, 229–68
Murat, Marshal, 41
Murray (18th Hussars), at Waterloo, 141, 143, 155, 197, 204
Muter, Colonel (6th Dragoons), at Waterloo, 145

Napier, General Sir William, 35–9
Napoleon, 63, 177–8; and 'in column' formation, 33; and Waterloo, 120, 121–8, 131, 138, 139, 164, 167, 217
naval history, 26–7

Nervii, defeat on Sambre, 63–4
Neuve Chapelle, 210
Newcastle Commercials, on Somme, 249–50
Newfoundland Regiment, 219; 1st Battalion on Somme, 250
Ney, Marshal, at Waterloo, 127, 133, 170
Nicolas, Sir Harry, 102 n.
no-man's-land, on Somme, 259–68
noise, effects of, at Waterloo, 141–3
Normandy campaign, 76, 80, 335; length, 308, 310
North Staffordshire Regiment, 5th and 6th battalions, 228; and on Somme, 250, 260–61
Northumberland Fusiliers, 18th Battalion, 226

objective dangers, 310–13
officer-training, 15–20; military history in, 20–35
officers, and choice over killing, 321–2
officers, and leadership, 188–94; relationship with men, 188–9, 224–6; technical competence, 189–90; and courage, 190–93, concept of honour, 191–4
officers, of New Armies, 277–81
Ostermann-Tolstoi, General Alexsandr, 44

'Pals' Battalions, 221–6, 241
Panzer Mark IIIs, 291, 295
Panzergrenadiere, 292, 293
Papelotte, 123, 126, 133
parachute troops, 300
Passchendaele, 211, 275, 308; 'improper violence', 47
Patton, General George, 336, 338
Picq, Ardant du, 150, 314; on dominance of fear, 68–70
Picton, General Sir Thomas, at Waterloo, 126, 182, 192–3
Pleiku, 15
poison gas, 312
political parties, army structure of, 176–7
Polybius, 69
Ponsonby, Sir William, at Waterloo, 151, 183
Poperinghe, 211
Portarlington, Lord, at Waterloo, 191, 192, 193
Powell, Captain Harry, at Waterloo, 169
Pratt, Lieutenant (Inniskillings), and Waterloo, 129, 164
prayer, and will to combat, 115, 137–8, 333–4
prisoners, killing of: at Agincourt, 84–5, 108–12; at Passchendaele, 47; on Somme, 48
prizefighting, 320
Prussian Guard, at St-Privat, 41–2, 43–4, 45, 63
psychiatric casualties, 334–5, 341
'Public School Officer', 277–82
Public Schools Battalion, 219

Quatre Bras, 122, 130, 134, 135, 139, 155, 158, 187, 192
Queen Victoria's Rifles, 227, 266; on Somme, 241, 243, 247, 264, 282
Queen's Move encounters, 168–75
Querrieux, 264, 267

Ranke, Leopold von, 68

Rawlinson, Sir Henry, 266, 267
Red Badge of Courage, The (Crane), 287
Reed (71st Regiment), at Waterloo, 161
Regimental Aid Post, 271, 272
Regiments (*see also under individual names*):
4th, at Quatre Bras, 135
14th, at Waterloo, 130, 139, 153, 179, 181, 204
28th, at Waterloo, 130, 154, 189–90
30th: at Quatre Bras, 134–5, 158; at Waterloo, 135, 136, 145, 158, 164, 171, 180
32nd, at Waterloo, 136, 137, 162, 205
33rd: at Quatre Bras, 139; at Waterloo, 132, 171, 200
40th, at Waterloo, 135, 143, 158, 161, 188, 191
44th, at Waterloo, 182, 187
51st, at Waterloo, 130, 185, 190, 196
69th, at Quatre Bras, 139, 155; at Waterloo, 132, 136, 171, 187–8
71st Highlanders, at Waterloo, 135, 142, 156, 157, 161
73rd, at Waterloo, 169, 171, 172
79th Highlanders, at Waterloo, 142, 156, 158
92nd, at Waterloo, 136
95th Rifles, 163; at Waterloo, 132, 136, 157
Reichenau, Field Marshal Walther von, 337
religious observance, 114–15, 137–8, 241-2, 333-4
Remarque, Erich M., 286
Renaissance historiography, 61–2
Reserve Cavalry, French, at Eylau, 41, 42–3, 45
Reynell (71st Regiment), at Waterloo, 157
Ridgway, General Matthew, 336
'right to flight', 314–16
risk to individual, factors in: accident, 317–19; exposure, 313–17; length of battle, 308–10; objective dangers, 310–13; technical difficulty, 319–20
ritualized attack, 165–6
Robbins, Lord, 332
Robbins (7th Hussars), and Waterloo, 129
Robertson, Sir William, 220
Roman historiography, 61–5, 66–8
Romilly, Colonel, 322
Rommel, Field Marshal Erwin, 292, 300, 336, 337
Royal Inniskilling Fusiliers: 1st Battalion on Somme, 250; 9th Battalion on Somme, 247
Royal Irish Rifles, 9th Battalion, 242–3, 283
Royal Welch Fusiliers, 2nd Battalion, 229, 276
Rudyard, Gunner (Lloyd's battery at Waterloo), 132, 152
Rundstedt, Field Marshal Karl von, 337–8

St-Cyr, 17, 18
St-Privat, 41–2, 44, 45, 46, 63
Salford Pals, on Somme, 250

Saltoun, Colonel Lord, at Waterloo, 158, 179, 189
Sambre river, Nervii defeat on, 63–4
Sandhurst, 15–18
'Sandpit, The', 126, 132, 133
Saratoga, 331
Sassoon, Siegfried, 225, 229, 286, 288, 289
Saveuse, Guillaume de, 94, 96
Saxe, Marshal de, 69
Schlieffen, Alfred, Count von, 19, 63, 213, 315
Schwaben Redoubt, 232, 282
Scots Greys, at Waterloo, 136, 137, 142, 145, 146, 150–51, 179, 189
self-inflicted wounds, 275 n.
Sense of Proportion, A (Swinton), 265–6
Serre, 259
Seymour, Horace, 182
Shaw, Corporal (Life Guards), at Waterloo, 146, 183
Sheldon (28th Regiment), at Waterloo, 154, 172
shelling: on Somme, 208, 216–18, 231–40, 243–5, 251–8, 259; wounds caused by, 269–70
shellshock, 334–5
Sherman tanks, 295
Sherriff, R. C., 224 n., 278
Sherwood Foresters: 5th, 6th, 7th and 8th battalions, 228; on Somme, 263
Siborne, Captain William, on Waterloo, 118, 120, 140, 262
Simmons, Lieutenant George, 136, 205
single combat, 246: at Agincourt, 91–2, 100–101, 106, 144, 145; at Waterloo, 145–7; in knightly warfare, 322–3, 326
single-missile weapons, 77; at Waterloo, 140–94
'small groups' concept, 46–7, 50–51, 73, 230
Smiles, Samuel, 56
smoke, effects of, 323; at Waterloo, 131, 140–41
Somerset Light Infantry, 1st Battalion, 245
Somme, 29–30, 290, 302, 317; 'improper violence', 48, 49; the battlefield, 207–13; the plan, 213–16; preparations, 216–19; the army, 219–29; tactics, 229–30; bombardment, 231–40; final preliminaries, 241–5; the battle, 246–7; types of encounter, 246; infantry versus machine-gunners, 247–51; casualties, 249, 250–51, 258, 260, 261, 274, 285–6, 312; infantry versus infantry, 251–8; failure, 251–8; no-man's-land, 259–68; communication system, 264–8; wounded, 268–74; will to combat, 274–84; commemoration, 284–9; length, 308, 309; 'killing zone' and losses, 310–11
Soult, Marshal, 36, 37, 38
South Staffordshire Regiment, 5th and 6th battalions, 228; on Somme, 250–51, 260–61
Spanbroekmoelen, 211, 318
Sparks, Captain (London Scottish), on Somme, 258
square, importance of, 185–7

Stalingrad, 75, 290–91, 301, 307; length, 308
standing army, emergence of, 175–8
Stendhal, 29
Stilwell, General Joseph, 336
Storr, Dr Anthony, 30 n.
Strachan, Lieutenant (73rd Regiment), 195
Suffolk, Richard de la Pole, 2nd Earl of, 91, 113
Suffolk, Regiment, 11th Battalion, on Somme, 245
Sun Tsu, 166–7
surgical problems, 113, 268–74, 330
surrender, 328–9; at Agincourt, 105–6; at Waterloo, 196–7, 327; on Somme, 282–3
Swinton, General Sir Edward, 265
symbol, power of, 186–8, 321

Tahure, 210, 215
taking cover, attitude to, 179–80
Talavera, 308
tank warfare, 209, 290–97, 339–40
Tannenberg, 276
Taylor, F. L., 61–2
Taylor, Telford, 74
Taylor (10th Hussars), at Waterloo, 155
technical difficulty, 319–20
technology, congruence between civilian and military, 324–7
Territorial Force, 227–9
'territoriality', 126, 165–6
Thin Red Line, The (Jones), 288
Thornhill (A.D.C. to Earl of Uxbridge at Waterloo), 129
Thucydides, and narrative tradition, 65–7, 68, 69
Tiger tanks, 295
Tolstoy, Lev, 29, 75
Tomkinson, Lieutenant-Colonel William, 154, 196, 197
Tomlinson, Private (Sherwood Foresters), on Somme, 263
trench mortar bombs, 208, 231
trench warfare, 210–13, 229–74
Trotsky, Leon, 175
Turner, F. W. A., 275 n.
Tyneside Irish Brigade, 226; on Somme, 249
Tyneside Scottish Brigade, 226; on Somme, 249 n., 263
Tytler, Neil Fraser, 267

Ulster Volunteer Force, 227; on Somme, 242–3, 250
Under Fire (Barbusse), 311
Undertones of War (Blunden), 286
Union Brigade, at Waterloo, 126, 133, 154, 173, 185, 203
Uxbridge, 2nd Earl of, at Waterloo, 192

Vegetius, 61
Verdun, 75, 208, 210, 213, 214, 216, 230, 310; length, 308
Vietnam war, 54, 275, 309–10, 318, 333
Vimy Ridge, 208, 215
violence: improper, 47–50, 108–12, 202–3, 204; as medieval commonplace, 116
Vivian, Sir Hussey, at Waterloo, 131, 141, 147, 184, 192, 206

Volunteer (Territorial) Units, 227–9
Vormarsch (Bloem), 24, 287
Voyennes, 81

War and Peace (Tolstoy), 75
war literature, 75–6, 286–9, 311–12
War the Infantry Knew, The, 278 n.
Ward, S. P. G., 40
'wasting ammunition', as military virtue, 313
Waterloo, 29, 217, 238, 239, 302, 327, 328, 336; literary treatment of, 117–21; the campaign, 121–8; personal angle of vision, 128–34; casualties, 128, 161–2, 188, 204–5, 310, 311; physical circumstances, 134–44; types of combat, 144, 246; single combat, 145–7; cavalry versus cavalry, 147–51; cavalry versus artillery, 151–3; cavalry versus infantry, 154–60; artillery versus infantry, 160–62; infantry versus infantry, 162–94; disintegration, 195–7; accidents, 195–6, 317–18; aftermath, 197–200; wounded, 200–206, 268, 317–18; length, 308, 311; 'killing zone', 310; 'right to flight', 315
Waterloo: Day of Battle (Howarth), 119
Waymouth, Captain (2nd Life Guards), on Waterloo, 147, 148–9
weapons: types of, 77, 231–4; rising lethal power, 310–13, 329–30; officers and, 321–2, 336
Weller, Jac, 25, 119, 159
Wellington, Duke of, 322; and Waterloo, 117, 122–8, 131–4, 138–9, 157, 161, 163, 178, 179, 181, 194, 206; his view of events, 131–4; his comment on, 194; personal danger, 336; reaction to stress, 337
Wellington at Waterloo (Weller), 119
Wellington's Headquarters (Ward), 40
West Yorks Regiment, on Somme, 248, 250, 263
Whitton, Colonel, 58
Wilkinson, Brigadier-General L. F. Green, 29
will to combat, 69; at Agincourt, 114–16; at Waterloo, 139, 181–94; on Somme, 274–84
Williams, Brigadier-General H. B., 260–61
Wilson, Lieutenant (Sinclair's battery at Waterloo), 140
'win/lose' approach, 45, 46–7, 50, 73
Wingate, General Orde, 336
Wood, Captain (10th Hussars), at Waterloo, 136, 137, 147–8
wounded: at Agincourt, 112–14; Waterloo, 200–206, 268; on Somme, 268–74
Wray, Lieutenant Hugh, at Waterloo, 161, 191
Wyndham, Captain Henry, 190
Wyndham, Lieutenant (Scots Greys), 142
Wynne, Captain G. C., 236 n.

Yom Kippur war, 339, 341
York, Edward, 2nd Duke of, 88, 98, 113
York and Lancaster Regiment
12th Battalion on Somme, 265
Young, Brigadier Peter, 40, 42, 44
Ypres, 208, 209, 223, 255, 275, 312; 'improper violence' at Third Battle of, 47, 49
Yser, river, 208, 209, 213

Zhukov, Marshal Georgi, 337

More About Penguins and Pelicans

For further information about books available from Penguins please write to Dept EP, Penguin Books Ltd, Harmondsworth, Middlesex UB7 ODA.

In the U.S.A.: For a complete list of books available from Penguins in the United States write to Dept C S, Penguin Books, 625 Madison Avenue, New York, New York 10022.

In Canada: For a complete list of books available from Penguins in Canada write to Penguin Books Canada Ltd, 2801 John Street, Markham, Ontario L3R 1B4.

In Australia: For a complete list of books available from Penguins in Australia write to the Marketing Department, Penguin Books Australia Ltd, P.O. Box 257, Ringwood, Victoria 3134.

In New Zealand: For a complete list of books available from Penguins in New Zealand write to the Marketing Department, Penguin Books (N.Z.) Ltd, P.O. Box 4019, Auckland 10.